AF864515

MITA

Mobile Internet Technical Architecture

NOKIA

IT Press

IT Press

Published by Edita Plc, IT Press, PL 760, 00043 EDITA, FINLAND

Access the latest information about IT Press books from our World Wide Web site at **www.itpress.fi/english/**

Access the latest information about MITA at www.nokia.com/mita/

ISBN 951-826-499-6

Printed in Edita Plc, Finland

Contents

Acknowledgements

One might say, that putting this book together was a typical ITC project; almost all features are there, and almost in time. I would like to thank:

- the Editor's Board: Marko O. Keskinen, Sari Päivärinta, Jussi Ruutu and Marko Suoknuuti for their excellent work
- my superior, Toni Sormunen; especially not disturbing the creative process
- my family, Helena & the boys, for their understanding
- the good people at IT Press

Jyrki Kivimäki
Editor

Access the latest information about MITA at www.nokia.com/mita/

The authors/contributors per chapter:

Alakontiola Jukka
Alkula Matti
Asunmaa Petri
Bjorksten Margareta
Bouret Chris
Coulombe Stephane
Durand Julian
Eerola Esa
Gimenez Jose
Granberg Martti
Grassel Guido
Grundström Mika
Hakalin Petteri
Halonen Timo
Hamiti Shkumbin
Harriman John
Hasselman Thomas
Hjort Peter
Holma Harri
Horneman Hilppa
Hurst Leon
Hussmann Holger
Isomäki Markus
Kahanpää Helena
Kari Janne
Kasslin Mika
Keskinen Marko
Kilkki Kalevi
Kilpeläinen Janne M.
Kinnunen Jari
Kiss Krisztian
Kivimäki Jyrki
Kontio Markku
Lang Kari
Lassila Ora
Loughney John
Länsisalmi Atte
Melero Juan
Murto Juhani
Niemi Markku T.
Nikula Eero
Nykänen Petri
Oinonen Kari
Passoja-Martikainen Hanna
Perkins Charles
Pohja Seppo
Przybilski Michael
Päivärinta Sari
Raatikainen Kimmo
Raitanen Sanna
Rajahalme Jarno
Reilly James
Romero Javier
Ruutu Jussi
Sillanpää Kaarina
Sipponen Juha-Pekka
Stenvall Jyrki
Suoknuuti Marko
Systä Kari
Tallgren Tapio
Terho Mikko
Toroi Teemu
Toskala Antti
Veriö Inka
Wiklund Peter

Prolog

Prolog

The next few years will see the convergence of mobile communications and the Internet resulting in new services, new business models and new business opportunities.

The Internet and its content have become an everyday source of information, entertainment and other services for millions of people around the world. Today, however, there is one fundamental drawback to the Internet: the fact that it is restricted to a location, typically the office or the home - in the same way that making a phone call used to be only a few years ago.

In our vision, the Internet will become mobile, just as voice communication has done. The future is no longer about the Internet but about services on the Mobile Internet. Content will go wireless, just as voice has done.

This is a new environment, called the Mobile World, where the evolution paths of mobile communications and the Internet have converged resulting in new innovative services for consumers. The Mobile World can be perceived as a holistic, evolving environment of the future, but also as an environment on a more personal level that enables people to shape their own Mobile World through personalized communication technology. In the Mobile World delivering targeted, timely information and services is essential. The development of these services is driven by consumer behaviour.

The key to commercial success in the Mobile World depends on three main factors: First, understanding consumer needs, their lifestyles and attitudes, and consequently being able to provide a matching combination of service portfolios and products. Second, readiness of the business system to enable service consumption. And third, understanding that as the consumer needs and business systems evolve, the technical architectures need to match them accordingly. Those who deliver winning product categories and platforms with the right services and technologies will be the victors in the Mobile World.

In the quest to further develop and facilitate the next generation of products and services, Nokia is developing a technical architecture that enables seamless interoperability between key applications, network environments and the user identity/addressing. However, we will limit the complexity of the technical environment supporting services; consumers simply do not want to be concerned with the underlying technologies. In addition, Nokia supports open technologies, standards and relevant initiatives that support and facilitate the deployment of global technologies and services and stimulate market growth.

Bringing together all the key elements and managing the challenges they pose is at the core of this task. We will succeed by identifying the relevant communication modes, by defining the key technologies required to support them and by orchestrating the industry drive to develop a common wireless platform for services.

The Nokia solution to cover all these demands is the Mobile Internet Technical Architecture (MITA). For the players in the field, MITA is the essential framework when creating user-friendly Mobile World experiences. MITA supports network evolution to IP Convergence for voice and data, while underpinning key technologies such as IPv6 and Presence.

It is an architecture that comprehensively enables networks to be driven by services. In Summary, MITA enables seamless interoperation of the consumption of content in all networks and as a result competitiveness is high and access is easy.

The Mobile World is evolving as we speak. New consumer needs arise, and innovation on the Internet takes place at an accelerating pace. Thus, it is clear that technical architectures need to renew themselves accordingly. For this purpose input from readers would be greatly appreciated.

1.
Introduction

Introduction to Mobile World

The next few years will see the convergence of mobile communications and the Internet resulting in new technologies, new business models and business opportunities.

The Mobile Internet is not today's Internet accessed from a mobile device. We will not spend our time browsing Internet pages for content as we do today, although this will still be possible. Instead, we will use applications to access content, make transactions, do business, link up with friends and family, play games, watch videos, and listen to and download music. More important than anything else, we will use the Mobile Internet to control our lives and give us more time to do the things we choose to do.

We are moving towards a web-based business model where mobility and the Internet are unified. This will require new competencies from all parties involved in the industry. Understanding mobility and the unique characteristics of mobile business will be vital to build the networks and services of the future. Success in Mobile World will be about speed: speed of application and terminal development for refined consumer segments, speed of new service creation, and speed of cost-optimized network development and roll-out. It will be about the control of services and content residing in your own hands.

The most tangible element from an end user's perspective is the mobile device, which provides access to all of these services. To ensure the success of the created services in the Mobile World they must be highly user-friendly. Actually, the user should be able to ignore the underlying technologies and purely to enjoy the richness of the supplied services no matter what the access method is.

Along with the continually growing importance of mobile terminals as life management tools, will the Mobile World will become much more than a source for a quick weather information update. The Mobile Internet is not a fixed entity, yet it is starting to formulate its existence with the services and applications that will can be linked to it. Hence, as the business and technical environments develop, the available architectures will need to match them accordingly.

In the Mobile World, delivering targeted, timely information and services is essential. The development of these applications and services is driven by consumer behaviour. The key to commercial success lies in understanding consumers, their lifestyles and attitudes, and in creating the product-service combinations that match their wants and needs. Those who deliver winning product categories and platforms with the right technologies will become victors in the Mobile World.

The Nokia solution to cover these demands is the Mobile Internet Technical Architecture (MITA). For the players in the field, MITA is the essential framework for creating user-friendly Mobile World experiences. MITA supports network evolution to IP convergence for voice and data, while underpinning key technologies like IPv6 and Presence. It is an architecture that comprehensively enables networks to be driven by services. To summarize, MITA enables the seamless interoperation in the consumption of content in all networks and as a result competitiveness is provided with maximum access.

The Mobile World

Nokia has proceeded towards a vision of the Mobile World. In this vision a major part of all personal communication will be wireless - be it in the form of phone calls, data or images.

Perception of the Internet has been technology focused, i.e., people have been considered wishning to access the Internet and huge amounts of data while they are on the move. Yet, what people will do is something very different. To begin with, in the Mobile World the Internet will become practically invisible as people are not concerned about "accessing the Internet" but using services and their favourite applications. The special focus will be on consuming information and producing information. The consumption will apply to mobile brands, products, information and also to advertising.

Consumers want value for money, ease of use, and a diverse choice of personally relevant services. Without an open solution, they are forced into a confusing world of proprietary services and terminals. Therefore, the aim will be to limit the complexity of the technical environment so that the users really do not have to worry about the underlying technologies. As a result, a unified architecture will benefit all users in the Mobile World.

Mobile World Challenge

Interaction Modes, Identity and Network environments are the main entities that the industry needs to create an environment for where they can interact seamlessly and invisibly from the end-user's point of view. This environment can be described as the Mobile World challenge, which is illustrated in the following figure.

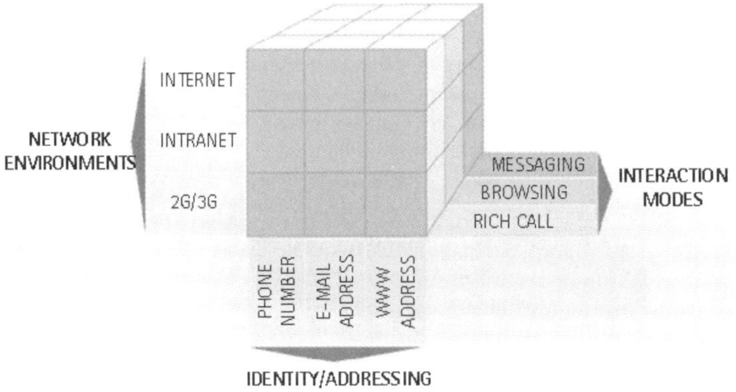

Mobile World Challenge

Also, from the MITA perspective, this setting presents those areas that require close consideration in order to achieve a fully functional technical architecture for the Mobile World.

Interaction Modes

MITA defines three types of interaction to simplify the definition of basic requirements for applications, terminals, and service provisioning technology. Interaction modes may be described as:

Messaging

Messaging is the interaction mode that is non real-time, i.e. it can be based on store and forward (SMS and MMS), store and retrieve (e-mail) or even store and push functionality in the network.

Browsing

Browsing is nearly real-time communication and it includes streaming audio and video. Browsing in 3G will be based on an evolution from today's WAP browsing to graphics and multimedia-enriched browsing. Nokia supports XML and the evolution towards XHTML from WML, the mark-up language used in WAP today.

Rich call

The Rich Call interaction mode maps to communication using a two-way real-time element. Basically, this means voice and video calls in their simplest form, and in more complicated cases, concurrent communication like file transfer is combined with the real-time communication element. Rich calls - that, is audio conversation supported by concurrent access to several media types - will greatly enhance personal communication.

Identity

An effective addressing system is vital for handling and simplifying the multitude of e-mail addresses, phone numbers and www-addresses for individuals and organizations. It will become even more so, with the ways of communicating exploding along with individuals using several terminals of different types.

The Mobile Internet Technical Architecture will provide solutions such as directory services that enable the interchangeability of address methods. The end user should be faced with as few differences as possible between the addressing methods.

Network Environments

The network environments: mobile networks (2G/3G), Internet and Intranet are the environments where most of the user interaction will take place in the Mobile World.

Furthermore, these different environments need to work seamlessly together and enable evolution towards unified service networks.

The interworking of networks, identity and the interaction mode are modeled by the Mobile World challenge as illustrated in the figure above, which is a simple diagnostic framework for highlighting the issues involved in interworking between layers.

MITA Concept

To make the Mobile World a reality, important technology choices are required if a technical architecture capable of supporting the myriad of services is to be built. For this reason, Nokia created MITA as a methodology with which to model the new environment and understand the technical issues inherent in building the optimal technical architecture. Nokia is using the approach to assist in making the right technology choices to deliver the optimal end user experience.

Central to the MITA approach is openness and interoperability. Situational analysis is not biased toward any network environment - fixed or mobile - and this maximizes inclusivity; Nokia recognises the importance of collaboration in this new business environment and the MITA initiative gives all the players equally open access to the emerging architecture so that the ultimate goal - a world in which subscribers can utilise the seamless functionality of all the Mobile Internet related applications and services - is achieved with the participation of the whole industry.

Because all the elements have been taken into consideration the outcome is highly beneficial to all parties including end users, mobile operators, application developers, service providers, infrastructure providers etc. Only through an open approach from all key players can the industry ensure a seamless user experience and rich service offering for consumers on a global scale. At the same time, this approach ensures open competition and rapid growth of the Mobile Internet market.

MITA Objectives

The primary objective is to provide a user-friendly Mobile Internet experience, with browsing, messaging and rich calls working seamlessly in any Network Environment and with any type of access. These essential aspects of a user-friendly mobile entity are identified in figure *Mobile World Challenge*.

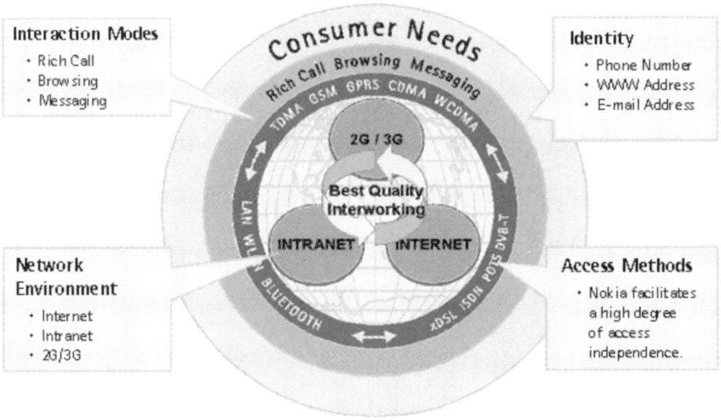

The scope of Mobile Internet Technical Architecture.

In order to lay a solid foundation for the evolution path towards future MITA compliant solutions, the development of current solutions must take future needs into account. This backward compatibility will provide valuable building blocks for the Mobile Internet, which will evolve into something much more versatile than it is today. Already, we see rapid development and can only estimate the future impact that a comprehensive technical architecture for the Mobile Internet will have.

To ensure far-reaching accessibility, Nokia supports open interfaces bringing new options for the Mobile Internet. Therefore, the Internet itself will not be changed but utilised in more versatile ways than before, while mobility has a principal role. For users to be able to access services independent of location and interface, will add to the value of the Mobile Internet. The Mobile Internet itself will not be a tightly set entity and, hence, the preferred business and technical architectures will need continual revision as the industry and technologies advance.

MITA Methodology

At the core of the MITA approach is a neutral methodology to help understand the critical technical issues in the Mobile Internet. This is achieved through abstract models that describe a conceptual framework for the Mobile Internet environment in progressively more detail and highlight the challenges of interworking between them. The MITA tools consist of:

- Architecture modelling principles
- Architecture concept models
- Architecture implementation models
- Architecture specifications

Collectively, the MITA tools are used to provide a consistent, unbiased situational analysis and a direction for making technology decisions. And as the approach is technology neutral, it is flexible and ensures that a consistent approach is applied as new technologies evolve.

Architecture Modelling Principles

The technical architecture modeling is based on principles, which should enable the creation of a future-proof end-to-end architecture with high possibilities for modularity of constituencies in the technical architecture.

The principles for modeling can be listed as follows:

- The architecture divides into independent sub-architectures while fulfilling business needs
- The architecture is open, modular, and hierarchical
- The architecture can tolerate varying rates of technological change in individual components. Change in one component should not render the whole architecture useless

The above principles dictate that layered models are applied to element design, network modeling, and identity structuring. A generic content delivery model is defined for a specification of Interaction Modes and related subsystems. In a similar way, access independent requirements are collected into and specified in a model of Access Independent Interface.

Architecture Concept Models

For the architecture there are two approaches worth closer examination. From a business architecture view point, there is the Mobile Internet Business Architecture (MIBA) and for a more technical approach MITA can be used as the concept model.

As a concept model, MIBA defines the interaction between different architecture elements. In this MIBA model, the terminal segment provides the end-user with access to services, the network segment provides terminals with connectivity to service networks and the server segment provides content and services for the end users.

Within each layer - content, connection, consumption - the MITA approach models each interacting entitiy as an element that can be desribed as consisting of three abstract layers: application, middleware, and platform.

Describing MITA elements in terms of constituent layers enable identification of the interfaces required for interworking between the layers in each element. Interworking is based on protocols implemented on the Mobile Internet layer and protocol-enabled content exchange between applications on the application layer.

By mapping the MITA element to the MIBA concept model, an end-to-end view of MITA is achieved. The MITA end-to-end view is a high-level reference model for more detailed technical architecture design. It identifies in principle all the elements of MITA that can be derived from the same baseline MITA element.

Architecture Implementation Models

MITA Subsystem Architecture further introduces an implementation view of the technical architecture. It is another viewpoint into the technical architecture, encapsulating a group of functionalities into independent, well-isolated Subsystems.

In general, a primary requirement is that an implementation of any Subsystem can be replaced with another implementation having zero or a minimum amount of modifications into other Subsystems or applications that utilize the services of the replaced Subsystem. In a similar way, the implementation technology of a Subsystem may be completely changed without influencing other related Subsystems. These modular implementation models enable the flexible evolution of the technical architecture implementation.

Architecture Specifications

Technical architecture work is divided into MITA specifications to achieve technical teams of manageable scope and to enable comparison to other models on the specification level. The MITA specifications are divided into three groups:

- Architecture specifications
- Key system specifications
- Architecture frameworks

The Architecture specifications are specifications that have a more permanent nature, whereas the Key system specifications represent issues where understanding end-to-end system aspects prior to technology selections is important. The third set of specifications includes frameworks for complicated issues such as security.

Compliance with MITA Principles

Applying MITA methodology results in three key criteria that must be adhered to for compliance with the MITA approach:

Access Independence

Services should not be tied to access methods. If this were the case new access methods would render the architecture useless. Therefore, a high degree of access independence is a key driver for the technical architecture. In order to achieve the global reachability of individuals and global access to content, access independent systems are required.

Securing Highest Quality Interworking Between Networks

There will also be separate network domains of wide area mobile networks, the Internet and Intranets. The interworking of these is yet another factor to ensure seamless end-to-end services. These network environments will be where most user interaction will take place in the mobile world. To ensure smooth communication between these environments for the end-user, fluent interworking on the technical level is required.

Securing the Best Interworking Between Multiple Addresses and Identities

The reality of access independent services will require the handling of multiple identities and addresses that have, historically, been closely tied to networks or applications. A MITA compliant solution recognizes this issue and proposes the best possible way to overcome the challenge.

The dissection of the applications should be done in such a way that network requirements are identified, and decoupled from application specific requirements. Requirements for similar applications are grouped under the three interaction modes. It is important to consider the differences between Network Environments, and how to enable applications to seamlessly work between any of the environments, i.e. to ensure service interoperability.

Applications and the Technical Architecture

Nokia classifes applications into four main categories: content, communication, productivity, and business solutions. Example services for each classification category are shown in the following table.

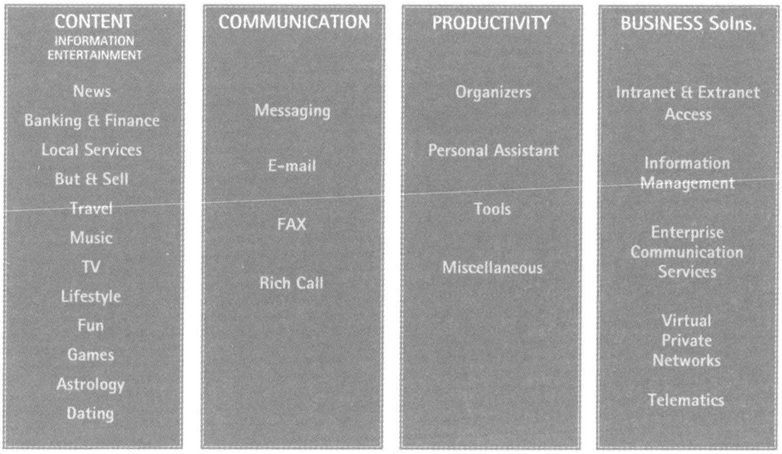

Mobile Internet Technical Architecture: Service Matrix

In applying the MITA approach to service creation, it becomes clear that rigidly integrated solutions are unacceptable.

For instance, a game can be downloaded to the terminal from the net and played independently. Here the interaction mode is browsing and the Network Environment is the Internet. If the game is played interactively between several people who send messages while playing and, for example, solving game related tasks, then the Interaction Mode is messaging and the Network Environment is the 3G network, possibly also interacting with the Internet. Finally, if the game is played in an on-line interactive environment, the situation is then a Rich Call Interaction Mode in the 3G network. These game situations highlight the fact that applications must be able to interact freely in all different interaction modes and over all networks.

Mobile Internet Business Architecture

To be able to create a winning solution for the mobile internet business architecture, Nokia has identified the drivers behind it as well as both the short term and long term ramifications that need to be focused on.

Clearly, the maintenance of profitability cannot be overlooked and solutions that facilitate increases in ARPU are therefore provided. In order to do this, the key constituents in the value chain need to be identified as well as their potential. Hence, end users, developers and content providers all have to be linked to service providers who in turn link them into the synergy of Consumption, Connection and Content, as illustrated in the following figure.

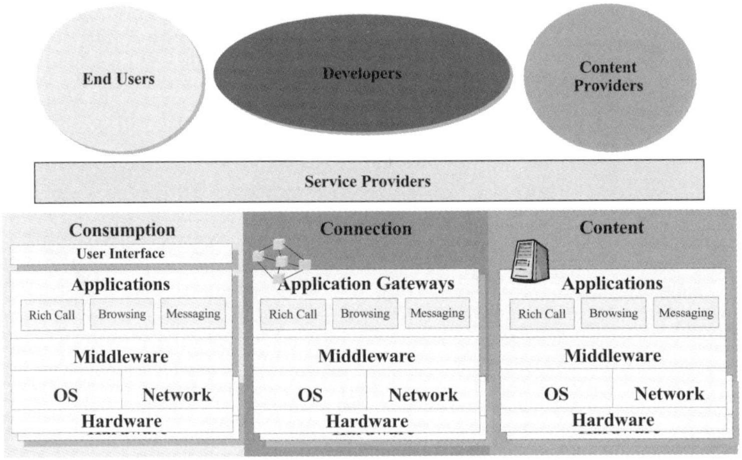

The Mobile Internet Business Architecture

The model contains the key constituencies on the top row, and the simplified architecture is presented on the bottom. The key constituencies in the analysis are end users, developers, content providers, and service providers. End user constituency covers both consumers and corporate users, and the service provider constituency covers all types of service providers, i.e., mobile service providers, Internet service providers, application service providers, or in some cases, corporates acting as service providers for their own corporate users.

The consumption aspect refers to end users who will be accessing the content provided by a variety of content providers. The connection, i.e. radio access between these two layers will be managed by the service providers.

Once all these elements have reached seamless interoperability, all the constituencies will also reach their objectives. End users are looking for cost effective and sophistiacted services within trusted and reliable coverage areas. Developers will be seeking for big volumes and continuity. For content providers, key issues will once again be cost efficiency and reachability on a massive scale. Furthermore, service providers are in this to maintain and maximize their user base and to collect subscription and transaction fees to generate profits.

Middleware Benefits

Middleware solutions provide services for consumers by bringing together the software solutions from applications vendors and operators. When the middleware architecture is based on MITA methodology, the following benefits are gained:

1) Support for different network access types from GSM to UMTS. This allows operators and service providers to build intelligence into their services today and enable their evolution to future technologies.
2) The interfaces are based on open industry standards. Operators can easily integrate the middleware to their current systems regardless their network configuration. They can also easily manage their application portfolio. Service providers can provide content using the same standard interface and ensure that applications can utilise their content.

3) Application developers can develop their application to utilise the intelligence of middleware enhancing the features with interoperability capability. This will enhance the end user experience and make users more willing to use the services while on the move, through the Mobile Internet.

Other benefits of middleware for end users include:

- Single login - user logs in once and all services become available
- Terminal identification through middleware. Users will always receive content optimised for their terminal.
- Preferences - the user can choose areas of interest and this way only relevant options are shown to the user, e.g. favorite restaurants, music etc.
- Navigation support - the user can use easy navigation

MITA Deliverables - End-to-End Solutions

It would be impossible to create winning solutions unless extensive end-to-end views were guiding the design of the technical architecture. Those architecture elements that require an end-to-end view are the interaction modes, Rich call, Browsing, Messaging, network interconnectivity, Key system elements and access methods. As we are in the middle of the development path towards a mobile world, coexistent multiple views are already a reality and some new views will be added for completeness.

These end-to-end views aim to cover all aspects of "Voice goes mobile", "Content goes mobile" and even "Servers become smart" views so that overall functionality will be guaranteed.

In order for Nokia to ensure the future compliance of this end-to-end view, we already have numerous solutions that fill the MITA model comprehensively and form the foundation for future applications and services. The following graph lists many of these solutions and quite clearly illustrates the overall view that has carefully been taken into consideration when aiming to provide optimised creation and delivery of services.

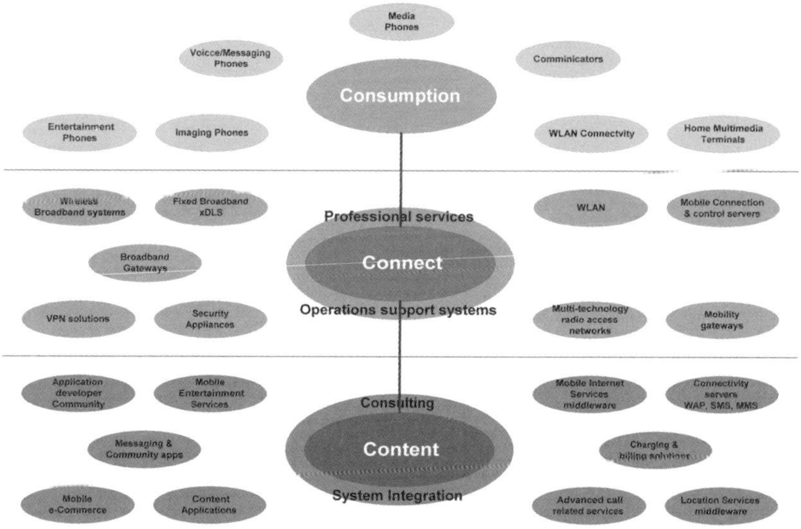

Nokia end-to-end solutions

Summary

Nokia has taken the global initiative to develop a comprehensive technical architecture for the Mobile Internet. The initiative, Nokia Mobile Internet Technical Architecture, aims to provide seamless interoperability between all interaction modes, any network environment and with any type of access. The ultimate objective of the initiative is to create a user-friendly Mobile Internet experience for everyone. This will be done by identifying the relevant communication modes, by defining the key technologies required to support them and by driving industry participation to develop a common wireless Internet platform.

In developing a comprehensive technical architecture for the Mobile Internet, Nokia aims to limit the complexity of the inherently technical environment; users do not want to worry about the underlying technologies. Nokia sees three key elements as fundamental to the Mobile Internet Technical Architecture: identity, interaction modes and network environments. Bringing these together and managing the challenges they pose is at the core of this initiative and will ensure high-quality and seamless interoperability in end-to-end services.

An open solution benefits all; profitable business scenarios call for interoperability, short development cycles, large volumes and, most of all, global reach. Unless there is a commonly accepted architectural solution, markets will become fragmented as well as requiring separate parameters, and the total volume will be much smaller than in a single global market.

Nokia invites all developers, service providers, operators and other industry players to participate in further clarifying and specifying the market requirements and technical solutions required to provide the Mobile Internet Technical Architecture needed for a superior end-to-end experience.

From this brief introduction we will proceed to a more detailed presentation of MITA. The following chapters will give detailed insight to the technological elements in MITA as well as present the current Nokia solutions and products in relation to MITA.

2.
Nokia Approach to Mobile World

Nokia Approach to Mobile World - Mobile Internet Technical Architecture

In order to master the end-to-end quality of future services, it is important to manage the three segments of the business landscape. However, the complexity of related technology must be hidden from the end user.

Nokia is defining the technical architecture for the Mobile World. The ultimate objective is to provide a user-friendly Mobile Internet experience, where browsing, messaging and rich calls work seamlessly in any network environment and with any type of access. The next figure highlights topics that are important factors in building a user-friendly Mobile World.

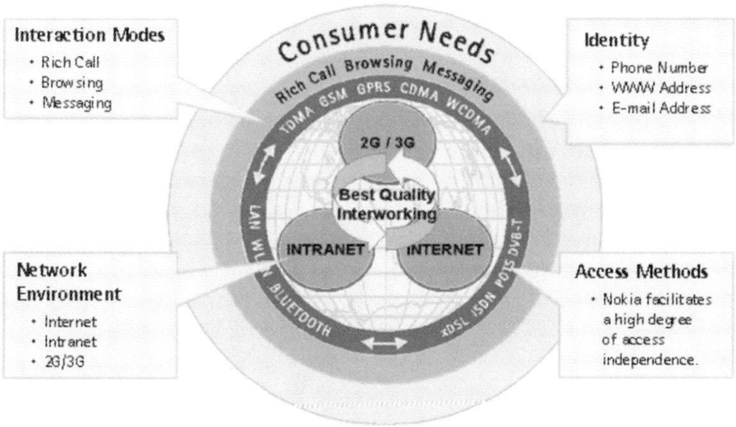

The scope of Mobile Internet Technical Architecture.

The purpose of the presented scope of the technical architecture is to:

- drive for consideration of multiple identification methods in all solutions
- drive for dissection of the applications in such a way that the network requirements are identified, and decoupled from application-specific requirements. The network requirements for similar applications are grouped under the Interaction Modes
- drive for consideration of access independence in creation of applications and solutions
- drive for consideration of differences between network environments, and how to enable applications to seamlessly work between any of the environments.

Identity

Identity is vital from the user's point of view. One characteristic of the Mobile World is the explosion of the volumes and ways of communicating. Individuals who now own one mobile handset or PC will use several devices of different type in the future. This means that an effective addressing system is required to handle and simplify the multitude of email addresses, phone numbers, and www-addresses for individuals and organizations. It also means that the well-served coupling between xSP's subscription, user identity, and an addressing method is facing a challenge of separation.

The address methods listed above are already widely used by end users. The technical architecture needs to provide solutions, such as, directory services that enable the interchangeability of address methods. The aim is to expose the end user to as few differences between the addressing/identity methods as possible: compare current mobile devices with the phone book where addressing is abstracted to "name". This means that clear and well-defined layers for naming, numbering, and addressing are needed for simplification over addressing methods and different network layers.

It is assumed that many layers of addressing grow on top of these addressing methods, establishing a new Human layer for addressing. For example, addressing for location based services using time, location, and customer relationship with the target service provider as elements of addressing: "All customers in the vicinity of McDonalds in Market Street, at lunchtime". Issues like end-user privacy and preferences, or security requirements will be taken into account in the implementation of these systems. In a similar way, the number of entities to be addressed is growing with new services and their needs (e.g., addressing of content, services, and information).

Network Environments

The separate network domains of wide-area mobile networks, Internet and Intranet remain. Although, from an aerial perspective, all networks seem to converge into a single IP network, the networks inherently have some characteristics, which differ from each other. Therefore, securing high quality interworking between the networks is needed for seamless end-to-end services. Each of the access networks has its own characteristics and requirements. For example, in the case of mobile networks, effective radio resource control management is needed and in corporate networks, security can be provided with Virtual Private Network (VPN) technology.

Mobile networks, the Internet, and Intranets are the environments where most of the user interaction will take place in the Mobile World. Different xSPs control these environments, and their design is driven by different requirements. Fussless communication between these environments by end users requires good interworking on the technical level.

From an overall perspective, it looks like there is only one unified network based on IP technology, but in practice there are many technical challenges to be solved before a seamless service experience can be provided. A few examples of the challenges are:

- Addressing and routing between access networks (e.g., network address translations and routing between IPv4 and IPv6 networks)
- User authentication, authentication, and charging
- User/service roaming
- Security issues (e.g., firewalls, virtual private networks, and policies)
- Quality of service and its linkage to other control issues (e.g., network policies, charging, and traffic control)

The Mobile Internet Technical Architecture divides the above complexity into four hierarchical network perspectives:

- Network domains
- Network environments
- Access networks

- Access technologies

At the strategic architecture level, there are two domains: Web domain and Mobile domain. At the technical architecture level there are three network environments: mobile networks, Internet, and Intranet and a number of Access networks:

- Mobile networks (e.g., Global System for Mobile Communication (GSM) network and Wideband Code Division Multiple Access (WCDMA) network)
- Fixed networks (e.g., Public Switched Telephone Network (PSTN) and any Digital Subscription Line (xDSL) network)
- Corporate networks (e.g., Local Area Network (LAN), Wireless Local Area Network (WLAN), and xDSL network)
- Home, ad-hoc and other private networks (e.g., Bluetooth and WLAN)

The relationship between the above network perspectives is illustrated in the next figure.

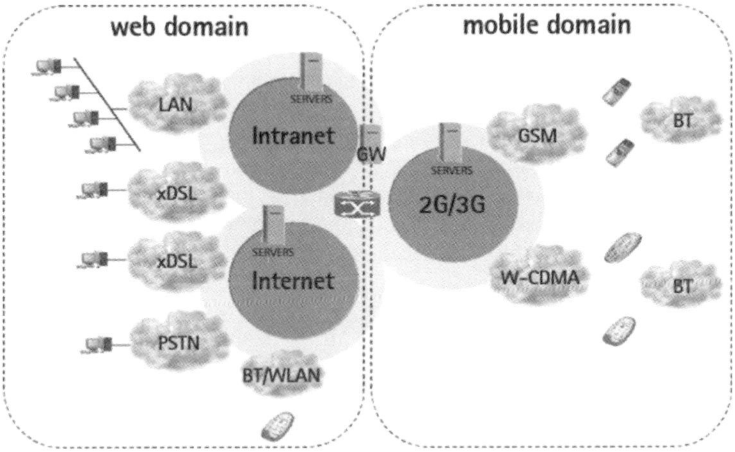

MITA Network perspectives.

While seamless interworking between different network environments is emerging, convergence between the same services regardless of their service providers will progress addressing harmonization requirements for service technologies. It is assumed that this will be a starting point for the creation of new unified service networks (e.g., web service, rich call, presence and reachability, distributed directories, and content delivery).

Access Independence

Services need to be access independent. Regardless of the efforts to create global standards for network access, there will always be alternative methods for accessing services. Differences, such as, wide and local area mobility vs. fixed access technologies remain. The concept of global reachability of individuals and global access to content requires access independent systems.

A high degree of access independence is a key driver for the technical architecture. In end-to-end solutions the emphasis in the technical architecture is to provide the best possible service based on the characteristics of the access technology utilized (e.g., available bandwidth, connection setup time, network delay, and symmetric/asymmetric performance). Services and their implementation should not be tied to any specific access technology. Otherwise, reusing the existing service base can not be accomplished with a new access technology.

Interaction Modes

Interaction modes are not very well utilized in the current networks and their communication applications. In most communication applications today, the User Interface, Communication Modes, and network technologies are vertically integrated providing one complete application for specific communication needs. Examples of such communication applications are:

- phone, SMS, and WAP clients in 2G networks
- email, web browsing, streaming, and Instant Messaging clients on the Internet
- file access and database transaction clients in theIntranet

This vertical integration means that implementations of the current communication applications cannot be freely and easily reused in new applications.

In the Mobile Internet Technical Architecture, Interaction Modes consist of characteristics and attributes related to requirements for end-to-end content transport and processing and of expectations for network service characteristics (e.g., QoS, security, privacy).

It is assumed that a small number of Interaction Modes cover the needs of most communication applications in the Mobile World. Based on this assumption, one primary objective in MITA is to find a minimum set of application and network independent Interaction Modes. So far, three Interaction Modes have been defined: Rich Call, Browsing, and Messaging. The primary characteristics and attributes of the Interaction Modes are listed in Table 1 and some examples of applications and their mapping to the Interaction modes are presented in Table 2.

Table 1. Characteristics and attributes of Interaction Modes

Interaction Mode	Characteristics and attributes
Rich Call	End-to-end delay minimization is critical attribute. This addresses requirements for both end-point internal processes and delivery network service expectations - a two-way interactive element is always present - it may use parallel multimode communication - synchronous delivery - connection-oriented session
Browsing	Near real-time one-way or two-way communication - the content creation process, delivery network, intermediary end-point(s), or processes in the receiving end-point cause more delays than the Rich Call Interaction Mode limits allow - one-way real-time communication - synchronous or asynchronous delivery · connection-oriented or connectionless session
Messaging	Non real-time communication - utilized intermediary end-points for store and forward, store and retrieve, and store and push communication - asynchronous delivery - connectionless session

Table 2. Application examples for Interaction Modes

Interaction Mode	Application examples
Rich Call	Voice and video calls - concurrent communication (e.g., file transfer is combined with the real-time communication element) - networked interactive games - real-time two-way chat or messaging - real-time streaming with backward control (e.g., video surveillance with remote camera control) - real-time multi-user game with multiparty call - point-to-point and multipoint conferencing
Browsing	Audio and video streaming, broadcasting, and multicasting - one-way push applications - games utilizing WAP technology - viewing and interacting with multimedia content
Messaging	Short Message Service (SMS) - email - Instant messaging - Play-by-mail games (e.g., email chess)

Classification of the Interaction Modes simplifies technology development as it helps to define the basic requirements of applications, terminals, and service provisioning technology.

Building solid support for the above Interaction Modes in the end-to-end architecture enables fast introduction of new applications with similar characteristics and requirements. One application can also utilize multiple Interaction Modes in communication at the same time.

There will also be commonalities and common services utilized by all Interaction Modes. Such services are, for example, profiles and settings, which are handled system-wide. From an end user point of view, these must be user friendly, requiring a minimum amount of setup, but still providing more elaborate settings for advanced users.

Services and Technical Architecture

In MITA, the definitions of end user service and application are:

- An application is a piece of software, which enables a service for an end user
- An application together with the content related to it create a service, which becomes an end user service once it has direct value to the end user

The main application categories are content, communication, productivity, and business solutions. These categories are classified as follows:

- Content - Information; news, banking and financial services, local services (city guide), buy and sell and travel. Entertainment; music, TV, lifestyle, fun, games, astrology, and dating
- Communication - Instant messaging, email, fax, Rich Call
- Productivity - Organizers, personal assistant, tools, miscellaneous
- Business solutions - Intranet and extranet access, information management, telematics, enterprise communication services and virtual private networks

In addition to the above main application categories there are special application groups. For example, applications like personal information management (PIM) with support functions (e.g., management, synchronization, or printing), group information services, and profile management can consist of a group of applications.

The technical architecture needs to support service creation for the applications listed above. However, rigidly integrated solutions should not be created so that the applications and services can be isolated from the networks and platforms underneath. This approach is illustrated in the next figure, where the main application categories do not directly map to category specific protocol stacks having support for only one identity, one access technology, and one network environment. Instead the main application categories share Interaction Modes and Key system functionalities.

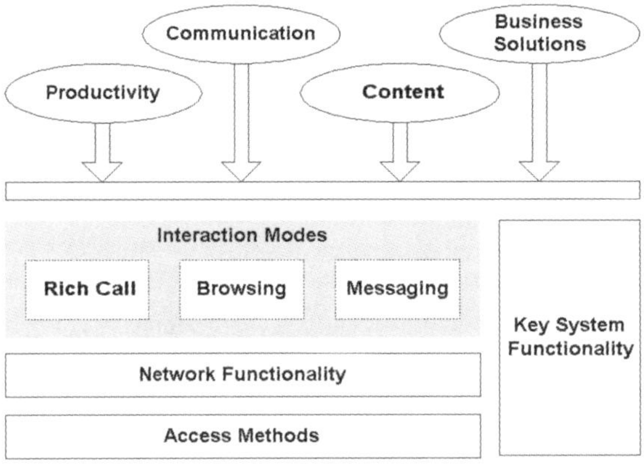

Mapping applications to the technical architecture

The previous figure shows how the applications do not directly mandate transport requirements to network functionality, but rather the application content processing and transport requirements are first mapped to Interaction Modes, and the most suitable one is selected. The applications are then implemented on interfaces provided by the selected Interaction Modes, independent to network functionality. Basically, this mapping decouples the network functionality and the implementation of applications. This approach can be highlighted with two examples:

1. Chat:

   Chat may be utilized as a Productivity or a Content/information application. In the Content/information category there may be different requirements concerning authorization or billing. In the Productivity category, security aspects may be of the highest importance. Regardless of the category, the underlying technical architecture should not need changes.

2. Games:

   Consider the transport requirements of the following game types: real-time multi-user game with multiparty call, game utilizing WAP technology, or play-by-mail type of game (utilizing SMS or email). Dissecting these games to their transport requirements and mapping the requirements to the Interaction Modes decouples the tight vertical integration.

The implementation of applications on top of Interaction Mode interfaces drives for best in class content processing and transport capabilities for all applications with similar requirements as Rich Call, Browsing and Messaging. The presented decoupling allows parallel development of applications and the enhancement and improvement of Interaction Mode implementation. In some cases, it may not be possible or even desirable to avoid vertical (and horizontal) binding. However, most such cases are handled in the Key system Specifications.

The Mobile Internet Technical Architecture aims at avoiding technical solutions, which translate to vertically tightly integrated stacks targeted too narrowly to specific end-user cases, or groupings thereof. If applications are mapped directly to network functionality, this may potentially create rigid silos where functionality designed specifically for one type of application will not be readily available for other types of applications, as illustrated in the next figure. In the worst case, these rigid silos cause multiple implementations of similar network functionality with slightly differing features.

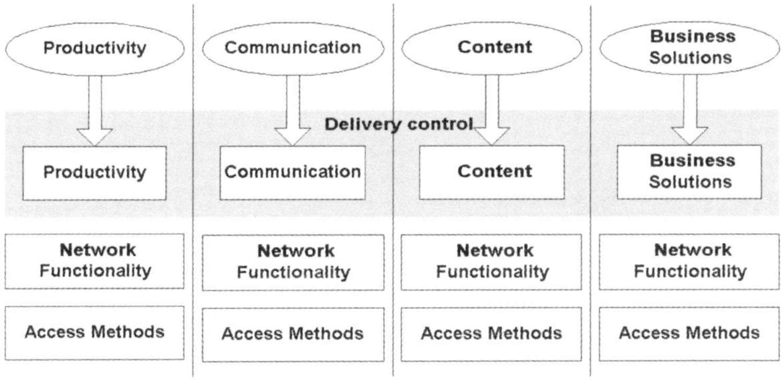

Strong application grouping.

The presented Interaction Modes enable a single implementation of network features for multiple applications and even for multiple platforms reducing the needed code space, improving the reliability of the implementation, and reducing the amount of maintained source codes.

End-to-end Views

Nokia's Strategic architecture addresses three primary End-to-end Views:

- Connecting people
- Content goes wireless
- Servers become smart

These End-to-end Views are starting points for MITA use cases and scenarios.

Connecting People View

The Connecting people view is the most familiar view, from a Mobile domain standpoint. In this view the end-points of the communication are both mobile devices, and the communication typically is voice in the Mobile domain. In the Web domain person-to-person, and specifically the peer-to-peer communication is already taking many other forms. A few scenarios for the communication alternatives in this view are illustrated in the next figure.

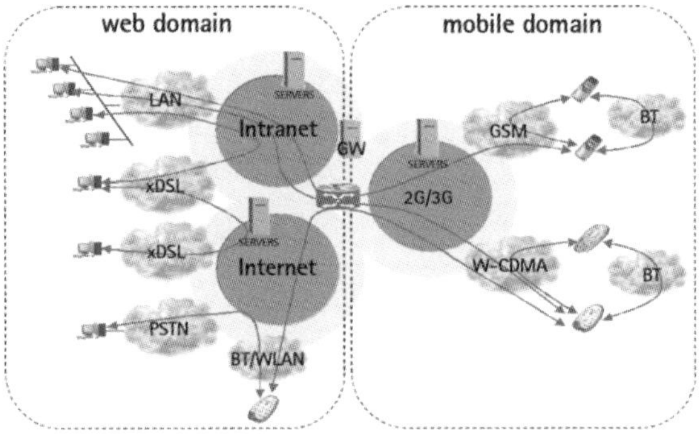

Some examples for the Connecting people view.

Content Goes Wireless View

The Content goes wireless view takes into account the current state of the Web domain, where content is located on a server, and combines this with the mobility of mobile networks in Mobile domain. Thus a client in a mobile device accesses the content on the servers. The end-points of the communication are a terminal and a server. A few scenarios for the communication alternatives in this view are illustrated in the next figure.

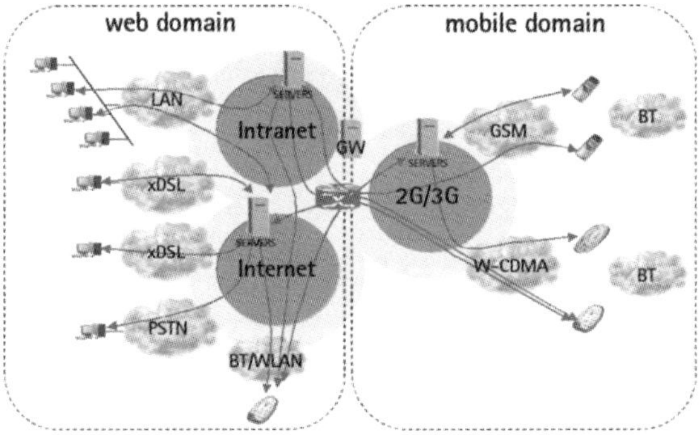

Some examples for the Content goes wireless view.

The Servers Become Smart View

The Servers become smart view looks forward to and anticipates the gradual evolution of the Web domain, and the technologies utilized in the servers. In this view the end-points of the communication are actually servers. The servers communicate with each other directly to accomplish tasks that do not necessarily involve terminals. A few scenarios for the communication alternatives in this view are illustrated in the next figure.

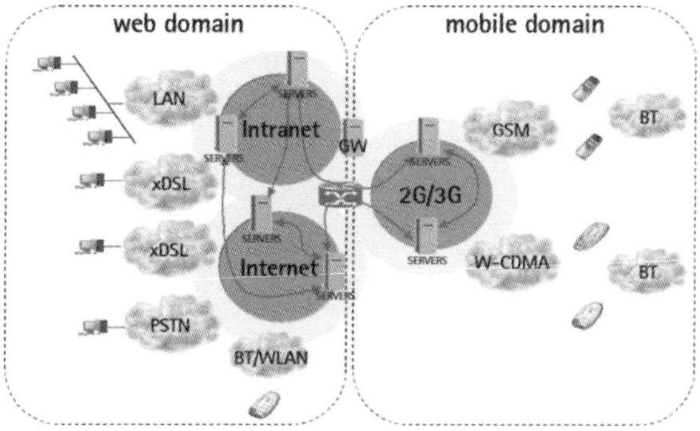

Some examples for the Servers become smart view.

Modelling the Technical Architecture

As the modelling of the technical architecture is done from a technical implementation perspective, fine granularity is required in the model. At the same time, the emphasis moves from business and general architectural principles to layering, to interfaces between the layers and to protocols that enable interaction between entities.

Principles for Technical Architecture Modelling

Technical architecture modelling is based on principles, which should enable the creation of a future-proof end-to-end architecture with high possibilities for modularity in the constituencies of the technical architecture.

The principles for modelling can be listed as follows:

- The architecture divides into independent sub-architectures while fulfilling business needs
- The architecture is open, modular, and hierarchical
- The architecture can tolerate varying rates of technological change in individual components. Change in one component should not render the whole architecture useless

The above principles address that layered models are applied to element design, the network model, and the identity structure. A generic content delivery model is defined for a specification of Interaction Modes and related subsystems. In a similar way, access independent requirements are collected into and specified in the Access Independent Interface (AII).

Layered Element Model

Both the business and the technical architecture require that a model of the technical architecture meets the presented principles and assumptions. In generic technical architecture modelling, one way to achieve these objectives is to divide an element of the technical architecture into layers: application, middleware, and platform. This is shown in the next figure. The presented element model is applicable to all instances of entities in all three segments of the technical architecture.

The application layer contains user interfaces (UIs), application logics, and related support functions. The middleware layer isolates the application and platform layers allowing the same applications to run on multiple platforms and provides tools for developers to implement new services on a solid execution environment. The platform layer consists of operating system specific software modules and hardware. Full separation of the presented layers requires that the interfaces between the application and middleware layers and the interfaces between the middleware and platform layers are well-defined. Typically, these interfaces are provided to developers in the form of Application Programming Interfaces (APIs).

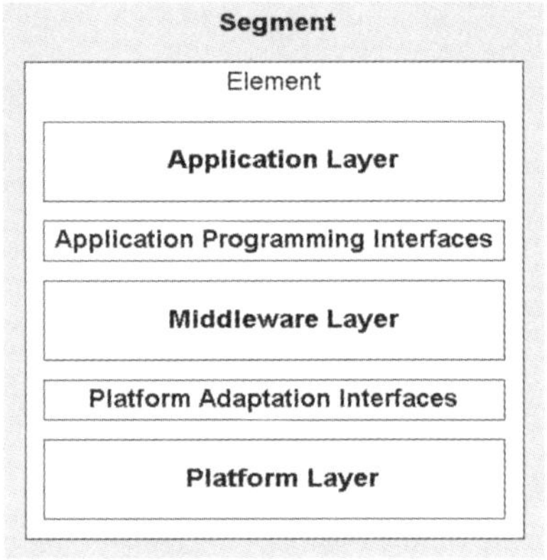

The layered element.

Interaction between two instances of the technical architecture elements is based on protocols implemented on the middleware layer and on exchanged application or service contents on the application layer, as shown in the next figure.

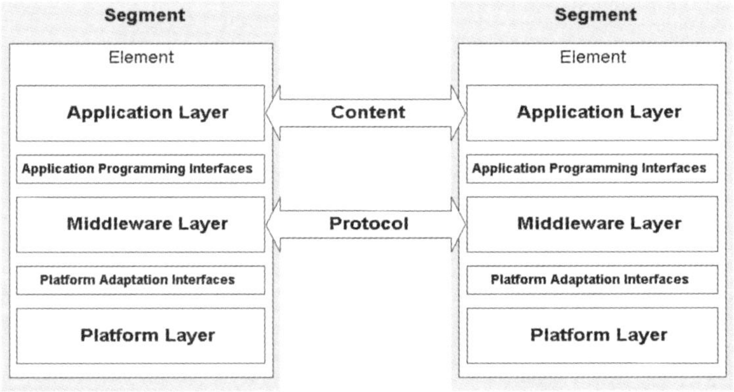

Interaction between Layered elements.

The division into layers and the interaction between elements are important for the Mobile Internet Technical Architecture. The division into layers enables identification of interfaces between the layers and between elements.

Layered Network Model

Typically, network protocols and network architectures are presented based on layered models. Two well-known layered models are the Open Systems Interconnection (OSI) model for generic network architectures and the Department of Defence (DoD) model for the Internet. The Mobile Internet Technical Architecture follows the principles of the DoD Model with one modification; Host-to-host and Internetwork Layers are combined to one Connectivity layer in the MITA model as illustrated in the next figure.

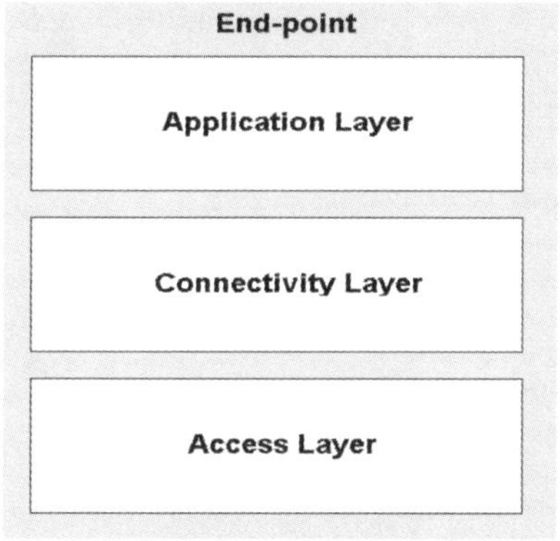

Network layers in a general MITA end-point.

In this model, the Access layer includes network hardware and link level device drivers in the operating system. The Connectivity layer includes routing and transport protocols, and the Application layer includes application protocols.

In a similar way, following the Internet End-to-end Architecture model, a general End-point-to-end-point interconnection model can be presented, as shown in the next figure.

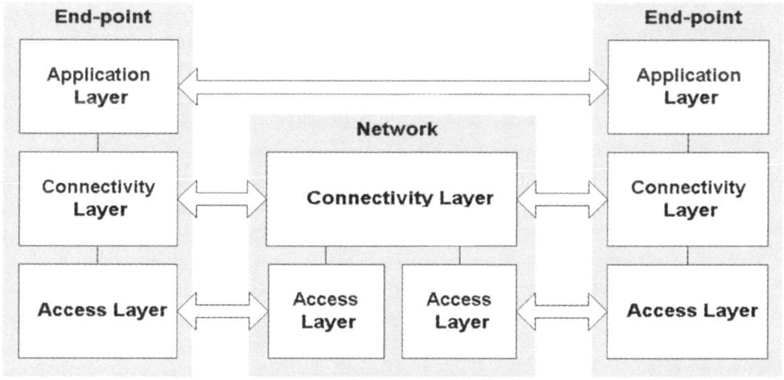

An end-point-to-end-point interconnection model for MITA.

The presented model simplifies mapping entities to network and end-point categories. Entities in the network category relay transport and application layer protocols transparently, whereas end-points terminate transport and application layer protocols. This means that entities in the terminal segment will belong to the end-point category. In a similar way, entities in the server segment will also belong to the end-point category. A major difference is that part of the entities in the network segment belong to the network category and part to the end-point category as servers. In Mobile Internet Technical Architecture, servers are classified as followis:

- Access network servers (e.g., Radio Access Network (RAN) servers)
- Connectivity servers (e.g., Home Location Register (HLR) or Dynamic Host Configuration Protocol (DHCP) servers)
- Application servers (e.g., Rich Call, Web, or email servers)
- Service servers (e.g., presence and reachability, content delivery, or directory servers)

Layered Identities Model

The third layered model for the Mobile Internet Technical Architecture categorizes identities and addressing methods. Each of the previous network layers has individual addressing methods and identities. It follows that three identity layers are required:

- Link layer addressing
- Addressing layer
- Naming layer

The Identity section introduced the new human layer above the naming layer. As a summary, all four Identity layers are illustrated in the next figure.

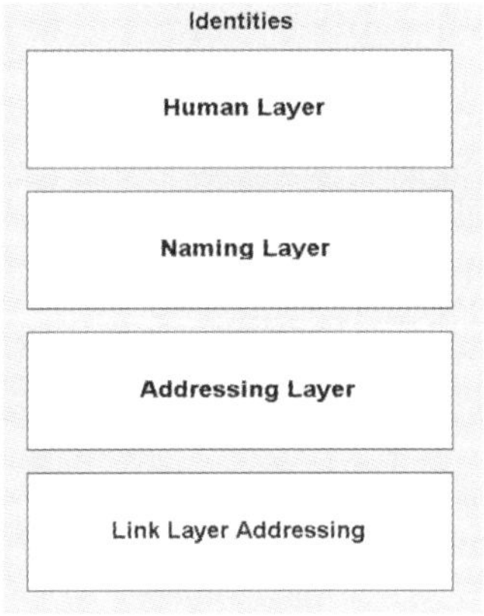

Identity layers in MITA.

The relationships between network layers and Identity layers are shown in the next figure.

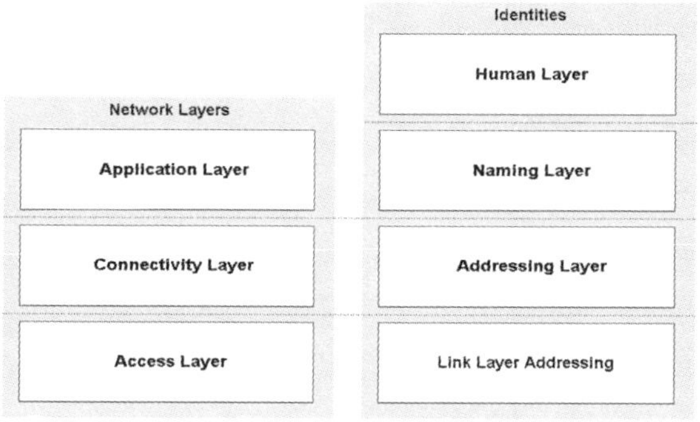

Relationship between network and identity layers.

The presented Identity layers address requirements for translation services between presentation formats on the different identity layers, as shown in the next figure. The following address translations are used in the technical architecture model:

- Translation from a Human layer name to a Naming layer name through search engines
- Translation from a Naming layer name to an Addressing layer address through directory and presence services
- Translation from an Addressing layer address to a Link layer addressing through address resolution and local directories

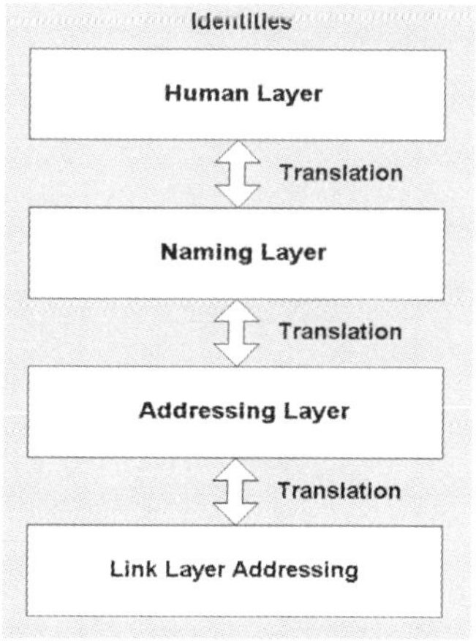

MITA Address translations.

The third technical viewpoint into the identity layer is the handling of end user subscriptions. There are two primary models:

- Unified subscription covering both Application/Naming and Connectivity/Addressing layers, as illustrated in the next figure.

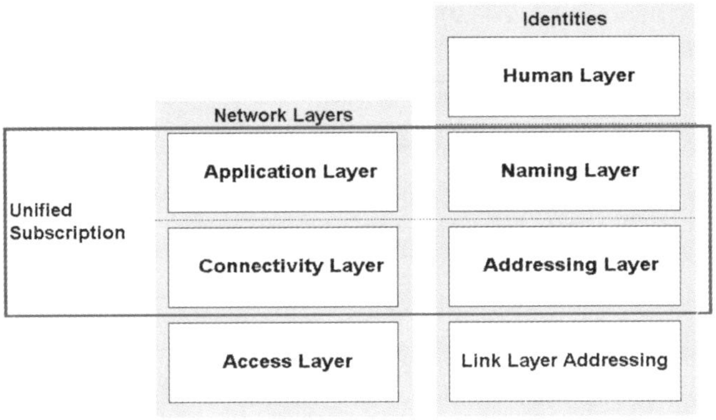

Unified Subscription.

- One Connectivity/Addressing layer subscription and multiple Application/Naming layer subscriptions, as illustrated in the next figure.

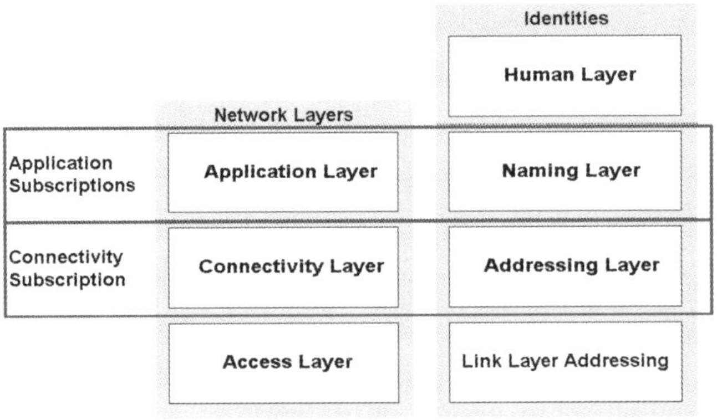

Separate Application and Connectivity subscriptions.

The management of identities is one of the key challenges in Mobile Internet Technical Architecture. For example, both of the above subscription models should be supported.

Content Delivery Model

A generic model for all content delivery scenarios is provided for technical architecture needs, as shown in the next figure. It consists of two network end-points: a source and a sink. The source end-point provides content for the sink end-point, which consumes the content. Between the source and the sink the content is delivered over delivery network(s). Both end-points have three processes. In the source end-point the processes are content creation, content delivery control, and network sending and in the sink end-point they are network receiving, content consumption control, and content consumption.

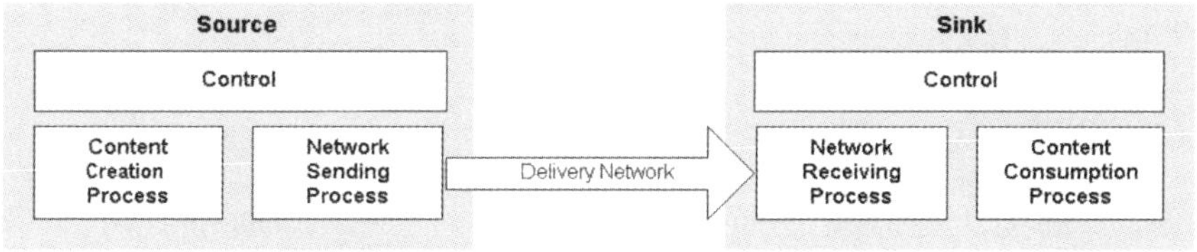

Generic model for content delivery.

In some scenarios, content may be delivered between the source and the sink via a third end-point or a group of end-points, called intermediary end-point(s), as shown in the next figure. In the simplest case, the intermediary end-point only receives the content from the source end-point and forwards it to the sink end-point. In a more complicated scenario, the intermediary end-point processes the content (e.g., reformatting or adapting the content) before sending it forward.

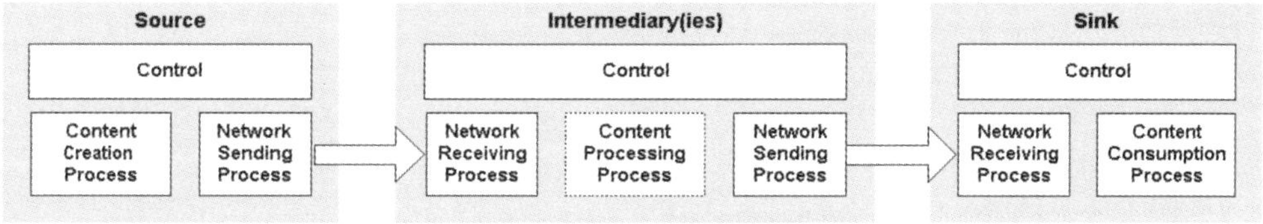

Intermediary end-point in the generic content delivery model.

Access Independent Interface

MITA architecture makes the user plane as independent as possible from the underlying access technology. MITA defines an Access Independent Interface (AII) to shield the network protocols from the actual access technology. The Interface prevents the end user from any interaction when using different access technologies. Three system entities are related to AII definition as illustrated in the next figure: the link layer driver, the protocols for transport, and a process for access independent mobility management.

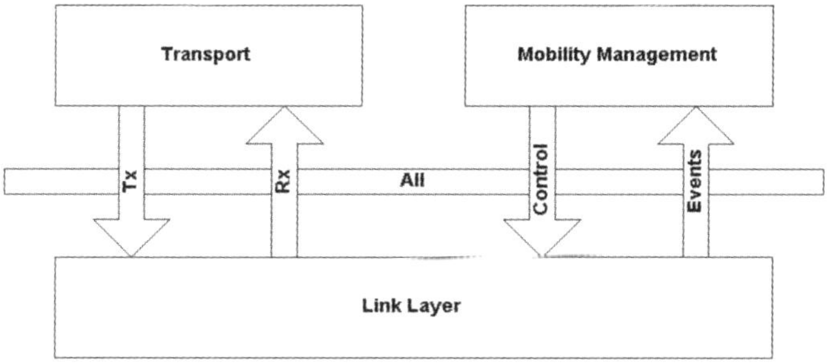

Access Independent Interface functions.

Four main groups of AII functions are addressed in the previous figure: sending functions (Tx), receiving functions (Rx), access control functions (Control) and access layer events (Events). The primary objective of AII definitions is to enable seamless usage of multiple wireless access networks simultaneously. It addresses many issues for deeper investigation (e.g., access independent mobility management, location of handover logic (e.g., link vs. network layer), and algorithms and methods to select the current wireless link from a set of available links).

Concept Models

The presented principles for the Technical Architecture Modelling are initially applied to the entities of Mobile Internet Business Architecture. Then, the fundamentals of Mobile Internet Technical Architecture are mapped to these principles and concept models for both MIBA and MITA.

MIBA Concept Model

In Nokia Mobile Internet Technical Architecture, the middleware layer has been named the Mobile Internet layer, as illustrated in the next figure.

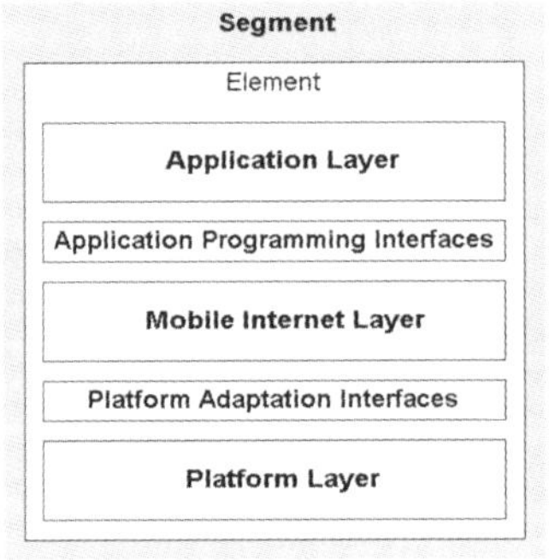

Mobile Internet Layer in the MITA Element.

The above three element layers are applied to each of the MIBA segments and their elements. An interaction between the elements is provided with protocols and content delivery within the payloads of the protocols. An outcome of this approach is shown in the next figure. In this MIBA model, the terminal segment provides for the end-user access to services, the network segment provides for terminal connectivity to service networks, and the server segment provides content and services for the end users.

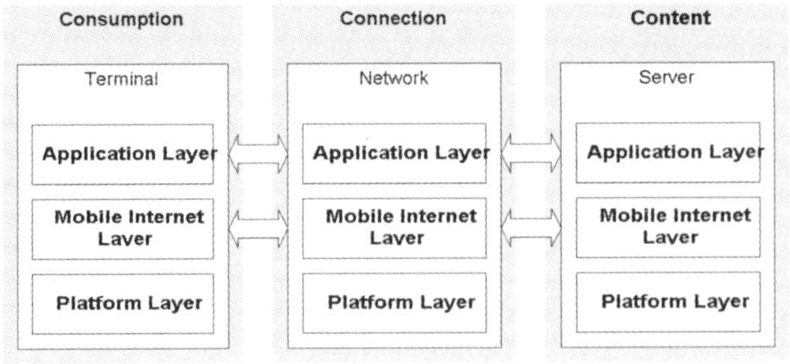

End-to-end View of Mobile Internet Business Architecture.

In some technical architecture models the application layer is further divided into application and content layers. In the Mobile Internet Technical Architecture, content has been seen as information, which may have multiple presentation formats and is handled in the content processing functions of the Mobile Internet Layer. Hence, content is not presented on its own layer.

MITA Concept Model

The basic model for Mobile Internet Technical Architecture is achieved through application of the principles and assumptions set forth in previous sections to the generic model of the technical architecture.

Interaction Modes have been supported dividing the application layer into two sublayers - one for the applications and their logic, and the other consisting of support functionality for the Interaction Modes, as illustrated in the next figure. There are groups of applications that support other applications on the application layer. For example, contact database functionality would be considered to be a supporting application for Phone and Email clients.

Applications access the supporting functions through Application Programming Interfaces. In the application development, lower layer interfaces of the MITA Element (e.g., Network interface functions or functions in SDK libraries) can also be accessed when necessary. The supporting functions on the application layer access the Mobile Internet Layer functionality below them through the Application Development Interfaces (ADIs).

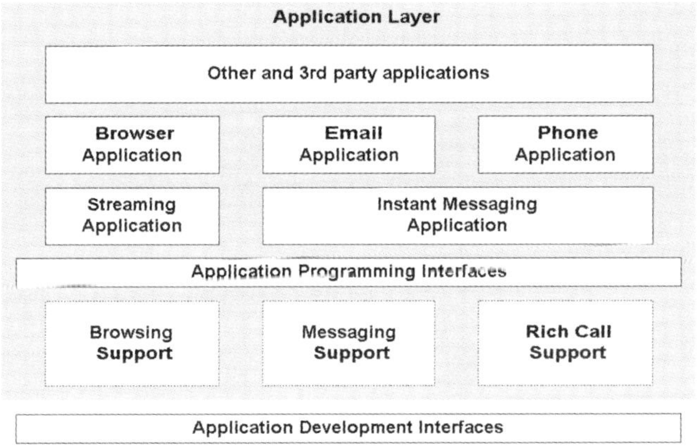

MITA application layer.

The Mobile Internet layer is divided into two functional stacks:

- Operating System (OS) and Platform support stack, providing local support for applications (e.g., application framework functions, execution environment, and UI support)
- Internet Protocols stack, providing Mobile Internet protocols above the link layer functionality for Interaction Mode support functionalities and applications

The architecture of the Mobile Internet layer is illustrated in the next figure.

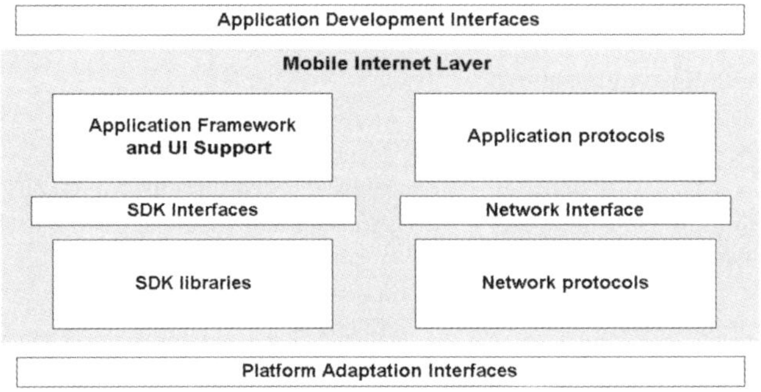

MITA Mobile Internet layer.

The Platform layer consists of operating system core modules, device drivers and platform hardware, as shown in the next figure. Platform Adaptation Interfaces (PAIs) are located between the Platform and the Mobile Internet layer, as illustrated in the previous figure and the next figure. In the Mobile Internet Technical Architecture the previously presented generic PAIs consists of the following primary interfaces:

- Platform Interfaces enabling access to local resources
- Access Independent Interface (AII) enabling access to network resources

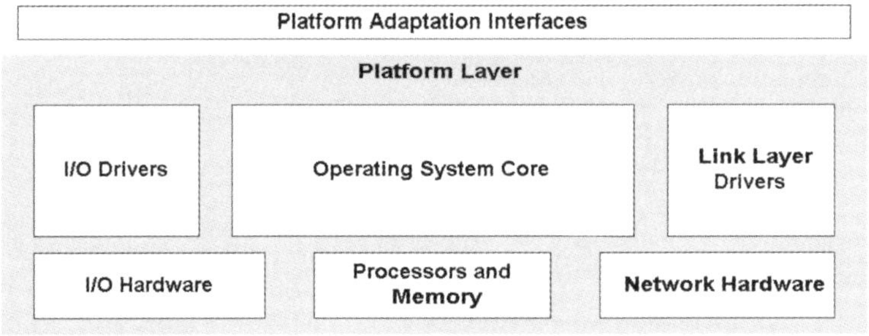

The Platform layer.

Network functionality maps primarily to the Internet Protocols stack of the Mobile Internet layer (e.g., TCP/IP and GPRS), but also partly to the device drivers (link layer) and to the network related hardware (physical layer) of the Platform layer. Operating System (OS) functionality maps partly to the platform support (including SDK libraries) of the Mobile Internet layer, and partly to the OS core and the hardware drivers of the Platform layer.

As the presented components are combined, a basic model for an element of the Mobile Internet Technical Architecture emerges, as illustrated in the next figure.

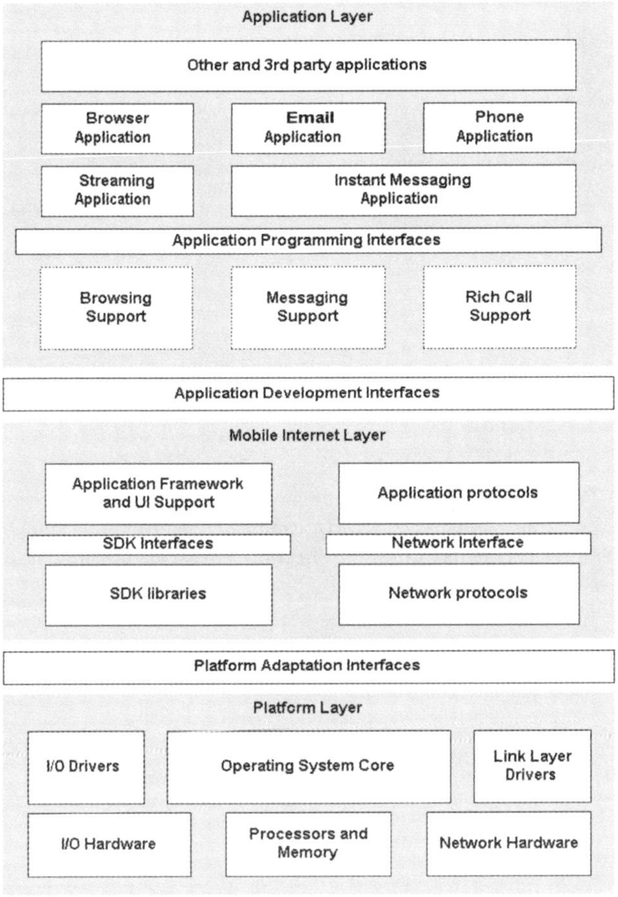

A MITA element.

The End-to-end view of MITA is achieved by mapping the MITA element to the presented End-to-end View of MIBA. An outcome of the mapping is illustrated in the next figure.

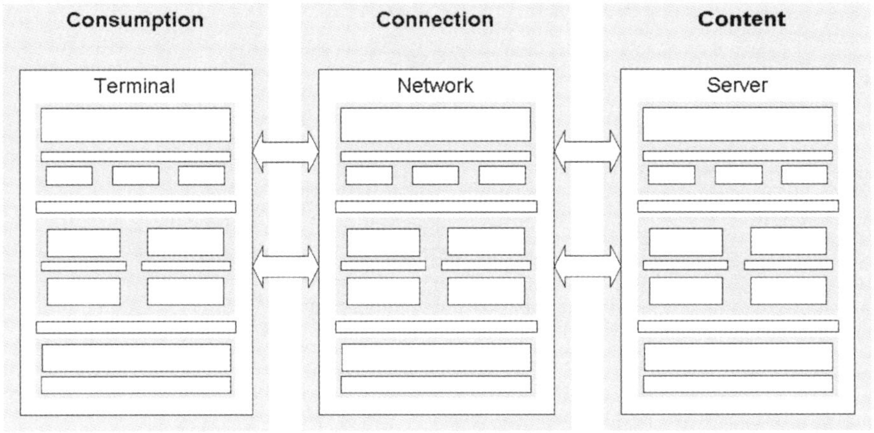

MITA End-to-end View.

The MITA End-to-end View is a high-level reference model for more detailed technical architecture design. It identifies that in principle all elements of MITA can be derived from the same baseline element. When the characteristics and requirements of different network environments are noted the above MITA End-to-end View extends to the MITA End-to-end matrix as illustrated in the next figure.

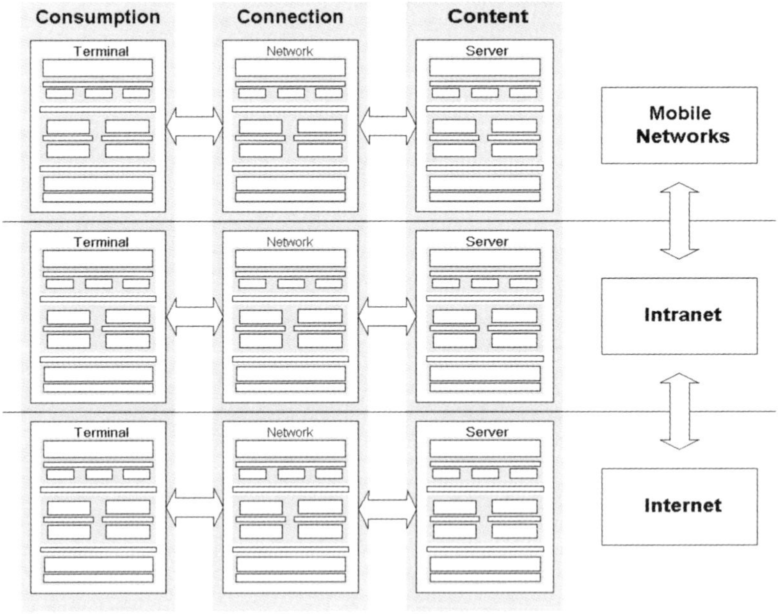

The MITA End-to-end matrix.

MITA Implementation Model

The concept model of Mobile Internet Technical Architecture consists of three functional layers presented in the previous sections. These layers were further divided into high-level functionalities: applications, Interaction Mode support, Internet protocols stack, OS and platform support stack and platform layer elements. The presented interfaces isolate these functionalities on a logical level.

The presented primary targets of Mobile Internet Technical Architecture require extending the presented modularity one step further by introducing an implementation view of the technical architecture, called MITA Subsystem Architecture (SSA).

Subsystems in MITA

The Subsystem Architecture is another viewpoint into the technical architecture, encapsulating a group of functionalities into independent, well-isolated Subsystems.

A primary requirement is that an implementation of any Subsystem can be replaced with another implementation having a zero or a minimum amount of modifications into other Subsystems or applications that utilize the services of the replaced Subsystem. In a similar way, the implementation technology of a Subsystem may be completely changed without influencing other related Subsystems. The next figure illustrates an outcome of the above requirements, which is that Subsystems have to have clear, well-defined interfaces:

- Support Interface for other Subsystems or applications that utilize the services of the present Subsystem
- The implementation of the Subsystem is based on Adaptation Interfaces provided by other Subsystems and/or the SDK libraries of the current platform
- Optional Protocol Interface, which allows interaction with other Subsystems over the network

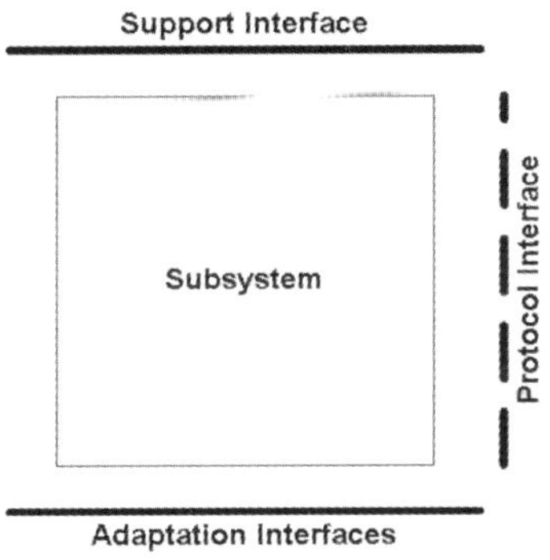

Subsystem Interfaces.

In MITA Subsystem Architecture two instances of MITA elements may be interconnected with an interoperable protocol interface on both elements, as illustrated in the next figure.

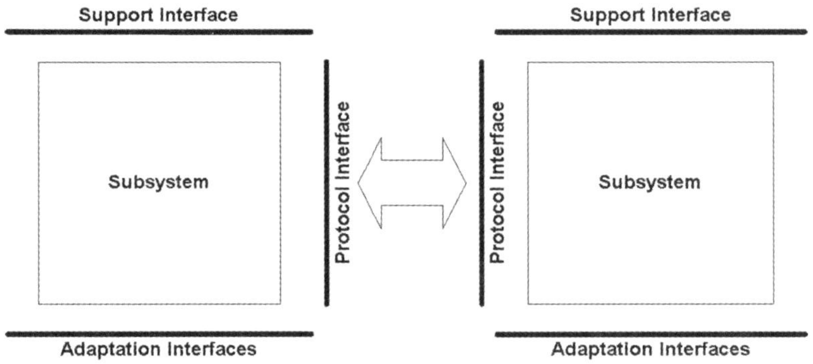

Interconnected Subsystems.

Categorization of MITA Subsystems

Mobile Internet Technical Architecture Subsystems are categorized into two groups:

- Intra-Element Subsystems encapsulate functionality within one MITA element
- Inter-Element Subsystems encapsulate interconnecting functionality over a network

The types of MITA Subsystems are illustrated in the next figure.

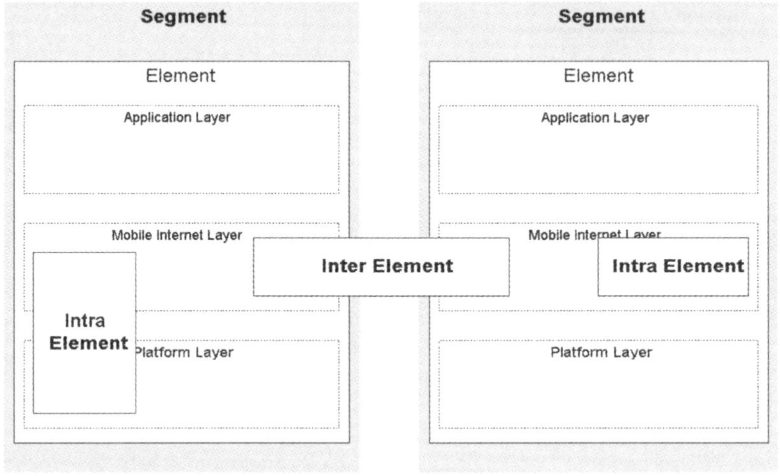

MITA Subsystem types.

These subsystem categories can be used in building more complicated subsystem topologies, for example, to client-to-server, server-to-server, or peer-to-peer communication scenarios.

Modular MITA Element

One essential requirement for the success of the MITA model is that an implementation of MITA Elements is based on the MITA Subsystem Architecture and the basic requirements for MITA Subsystems presented above are met.

A typical approach to build the Subsystem Architecture is a bottom-up model enabling the separation of application and middleware software from platforms first and then building common software functionalities on top of these baseline subsystems.

The final outcome of this Subsystem Architecture approach is a design of a MITA element, which has well-defined software interfaces and a modular software architecture. The next figure shows the basic architecture of a modular MITA element.

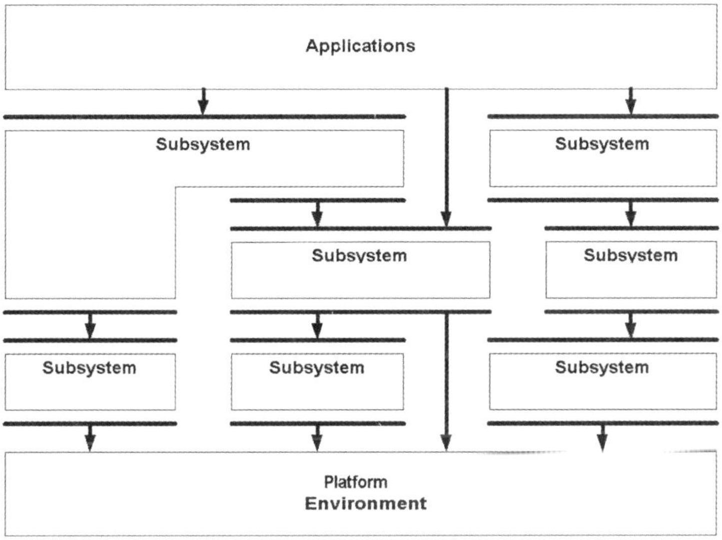

Modular MITA Element.

Mita Specifications

Specification Background

When the first analysis of the technical architecture was completed, one primary outcome was that following the typical design path starting from the existing technologies and R&D activities and searching for the optimal roadmap towards the Mobile World environment in gradual steps is not the route to the best technical choices. Other alternatives were studied. The outcome was that by following the development cycle in a reverse direction, new potentially revolutionary issues can be identified on time and the actions needed to tackle them can be launched as a part of normal evolutionary roadmaps.

Based on the above conclusion, MITA work follows three primary phases:

1. Set targets
2. Align with environment
3. Influence on roadmaps

The first phase studies current products and technologies, reviews Nokia roadmaps and the products and technologies that are emerging in a 3 year time frame and compares them to the Mobile World problem scope and MITA targets. An outcome of this phase is a technical vision, which gives direction for the technical architecture and gives it its first model.

In the second phase the technical architecture model is aligned with the existing R&D roadmaps in Nokia and the activities in standardization bodies. An outcome is a more detailed technical architecture and its first implementation view.

In the third phase, the implementation view is finalized and the required R&D activities are addressed for product development.

Every phase has a different relationship with the time frame, as illustrated in the next figure. The first phase addresses MITA on long term issues, the second phase shifts the focus to mid term aspects and the third phase completes MITA work with short term roadmaps, product development plans and reference implementations.

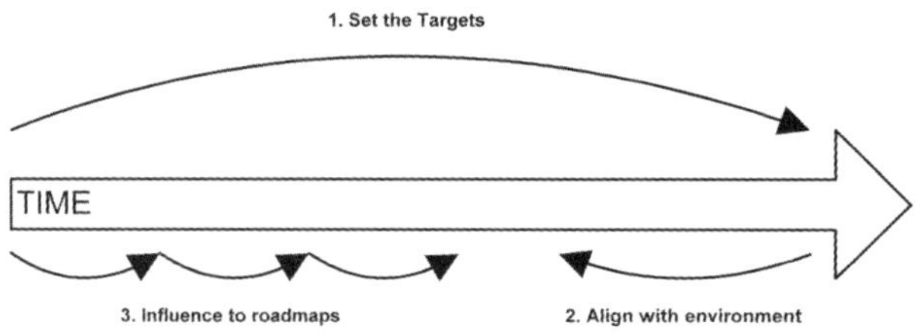

The MITA work phases.

Specification Groups

The technical architecture work is divided into MITA Specifications, to achieve technical teams of manageable scope and enable comparison to other models on a specification level. MITA specifications are divided into three groups:

- Architecture Specifications
- Key system Specifications
- Architecture Frameworks

The Architecture Specifications are specifications that have a more permanent nature, whereas the Key system Specifications represent issues where understanding end-to-end system aspects prior to technology selections is important. The third set of specifications includes frameworks for complicated issues.

Architecture Specifications

The Architecture Specifications are more permanent baseline specifications of MITA, covering the structure of the MITA Element as illustrated in the next figure.

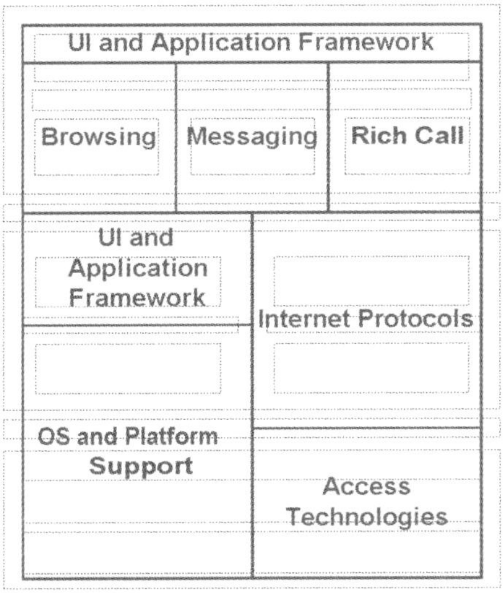

Architecture Specifications and the MITA Element.

UI and Application Framework in MITA

One of the most fundamental elements of Nokia products has been the user interface (UI). We expect that a perfect combination of revolutionary and evolutionary thinking is needed in order to create the UI guidelines for the MITA era. We believe that the targets are best met by focusing the effort on a carefully selected (limited) set of UI styles, and to the core applications selected according to the MIBA/MITA principles. With a focused approach, the accumulated knowledge of UI design inside Nokia can be brought into products in a timely fashion. The UI and Application Framework specification deals with topics that are generic to all Interaction Modes and the components of the MITA architecture.

Rich Call in MITA

Rich Call is one of the three Interaction Modes in MITA. Rich Call covers applications, which have as their cornerstone a real-time two-way communication component enriched with presence information, text, graphics, images and sounds before, during and after a call. Applications in this category are characterized by having demanding requirements for the underlying transport mechanisms and their integrity due to the real-time nature of the service. Typical applications in the Rich Call category are voice call and conference, video call and conference and real-time interactive games.

A Rich Call can be initiated from other Interaction Modes thus enabling services like click-to-talk and chat-to-talk. Therefore, from an architecture point of view, there is a need to communicate between the Interaction Modes. This also applies to Rich Call and other applications like contacts and presence management. Other Interaction Modes and applications may also be used simultaneously with Rich Call to provide for an even richer user experience.

Rich Call requires consistent behavior over various access networks and network environments. The access network may impose some constraints over the Interaction Mode, but it is desirable that the specific applications can scale their behavior to take this into account. The network environments and their interworking are also required to be able to connect Rich Call users residing in various types of networks (cellular, public Internet, corporate Intranet).

Browsing in MITA

Browsing is characterized by an interaction between an end-user and services in fixed or mobile servers. The Browsing Interaction Mode covers applications, which are interactive by nature, and in many cases, nearly real-time. There are two primary browsing schemes:

- Online browsing (e.g., Markup style browsing interaction, Interaction through executable programs, and Peer-to-peer browsing)
- Off-line browsing (e.g., Push of content to device and the same markup languages and executables are used, but the interaction is limited to local servers and services)

Processed content may be displayable or metadata:

- Displayable content includes text, graphics, animation, sound, etc.
- Metadata is not directly displayed, but can be used to expand the user experience by giving it a larger scope.

Browsing represents an interactive content consumption and manipulation experience. Initially the user consumes content in an interactive manner, but the experience will soon expand into manipulation of the content (true interactive applications).

The Browsing environment, and the browser in particular, will play a significant role as a platform, as the host of a browsing paradigm. The Browser is enabled with XML, and can thus offer XML services to other applications. The concepts of embedded links and metadata in markup documents translate very well into any XML content or content that can be translated into an XML representation, be it contacts, calendar or generic data.

One of the key roles of the browser, as the most important component of browsing, is to serve as a platform for enhanced functionality to be used by other applications. This functionality may include technology like animation services, audio services, video services, etc. In order to flexibly offer this kind of functionality, dynamic installations and the configuration of software (for example plug-ins) becomes important.

Messaging in MITA

The Messaging Interaction Mode covers applications, which have two-way, asynchronous delivery of content. For example, SMS, Multimedia Message Service (MMS) and email, involve a separation between submittal of content and the delivery to the recipient. The differentiation between Messaging and Rich Call is blurred by Instant Messaging (IM) applications that allow delayed delivery of chat conversions. The definition of Interaction Modes indicates that the Chat applications can be implemented either on the Messaging or on the Rich Call Interaction Mode.

Emerging technologies (e.g., SIP and XML) and applications (e.g., news, chat, and Internet Relay Chat (IRC)) address challenges to traditional messaging technologies (e.g., SMS and email) causing new interworking requirements. In a similar way, interworking between MITA applications (e.g., click-to-call, click-to-chat) and management and processing of the static and dynamic message content and attachments in various conditions address requirements for Messaging Subsystems.

Internet Protocols in MITA

The underlying fundamentals for Internet networking include architectural and technical simplicity, openness, ease-of-use and scalability. An open Internet End-to-end Architecture enables the efficient development of new services and applications and their fast deployment on a global scale. The Internet concept has to be enlarged to globally support user, terminal, and service mobility, which will enable new useful services and applications to all Internet users. As an outcome of the enlargement, existing Internet protocols may also face new requirements and needs for modification. This can be seen as an opportunity for the Mobile World industry. It is more a natural evolution than a disturbance phenomenon. These are the minimum technical and architectural requirements for the development of the Mobile World.

Internet Protocols Specification houses the Internet protocols and related interfaces in the End-to-end view of Mobile Internet Technical Architecture. It specifies and provides protocol interfaces between MITA elements.

OS and Platform Support in MITA

As an entity, MITA is not about operating systems. Yet, MITA is about how operating systems and applications can co-operate with each other in providing a rich end user experience in a consistent manner. MITA sets certain requirements for the underlying operating systems and support functionality provided by the operating system to the upper layers, such as, SDK libraries, application protocols, and network protocols.

In many cases, the underlying operating system libraries set the base for architectural paradigms. For example, the Linux operating system and Linux libraries and APIs are good examples of the procedural programming style. On the other hand, Symbian and Symbian libraries provide pure object oriented interfaces. It should be noted that programming languages (C, C++, Java) and programming paradigms are separate issues. However, in most cases procedural programming environments have been extended with high functionality object oriented designs.

MITA does not define a preferred operating system. Operating system selection is a business decision, not an architectural issue. However, the operating system should always be compatible with MITA requirements. These requirements are set by the upper layers and are collected into this specification for easy access. On one hand, this also means that MITA is independent of operating systems and programming paradigms. On the other hand, it means that the implementation of MITA in different platforms may vary according to platform capabilities.

Another part of the OS and Platform support Specification are issues related to the baseline development environments. The principles that were presented for operating systems also apply to the baseline development environments. This specification collects the generic requirements.

Access Technologies in MITA

A typical mobile device user is primarily interested in applications and services, while the actual access technology should be as transparent as possible and should not require any technical understanding, when using an application or subscribing to a service. Still, this transparency should not limit potential applications, as all of them might not work in an acceptable way over all access technologies. These technologies have different technical characteristics (e.g., available bandwidth, connection setup time, network delay, symmetric/asymmetric performance), which impact the user experience, i.e. the technologies differ in data rates, QoS, handover capabilities, latencies, etc.

The Access Technologies Specification defines the elements needed to allow the access from a mobile device to the IP network. Due to the different functionality of the various access technologies the platform layer varies between different forms of access. The Platform Adaptation Interface is supposed to shield the Mobile Internet Layer from the underlying technology, but access specific functionality is required within the network protocols in order to facilitate some access-specific functionality.

Key System Specifications

The Key systems are end-to-end and cross-organizational building blocks of the Mobile Internet Technical Architecture.

Naming, Numbering, and Addressing in MITA

Naming, Numbering, and Addressing (NNA) concentrates on the names and addresses of individual people or devices and how names and addresses appear on the end user interface. The end user favored nickname information is translated into globally unique names and finally to routable addresses. The translations between globally unique names and the related directory infrastructure belong to the Naming, Numbering, and Addressing Specification. The implications and requirements of the NNA environment (e.g., anonymous addressing and regulator aspects) are also studied in the Specification.

Naming, Numbering, and Addressing provides one unified translation infrastructure for the Rich Call, Messaging and Browsing services. An interaction with the Presence services enables networked and dynamic resolution of naming layer addresses. On the mobile device Naming, Numbering, and Addressing functionalities are positioned in the address book Subsystem. However it should be noted that the NNA services are not limited to address resolution requirements of the Rich Call service. Any application may utilize these services.

Presence in MITA

Presence deals with the issues of providing dynamic information about the status and availability of mobile users and mobile devices. Presence is a Key system for a number of specifications. The Messaging Specification utilizes presence services to enable instant messaging between mobile users. The Location Specification provides a set of location based services that leverage presence information. For Naming, Numbering, and Addressing and Reachability services the presence service provides dynamic and static information.

It should be noted that one end-user could have multiple application level identities. How to manage these and how to dynamically associate them to the current connectivity layer identities are issues which linking the Presence Specification to the Reachability Specification.

Reachability in MITA

The Reachability of terminals and people deals with the issues related to the question of how a terminal or a person is reached through the Web and Mobile domains. A Communication network is not a single network with seamless IP connectivity. A Communication network consists of networks such as IPv4 Internet, IPv4 Intranets, IPv6 Internet, carrier IPv6 based 2/3G networks and PSTN. There are also different types of gateways (e.g., network address translation (NAT), firewall, or SIP entities) between networks. Seamless person-to-person and terminal-to-terminal connectivity over these fragmented networks is the scope of the Reachability Specification.

Access Independent Connectivity in MITA

The Access Independent Connectivity (AIC) Specification deals with seamless access to personal services in all MITA Network Environments independently of their location and connectivity/access providers. The primary focus of the AIC Specification is to provide continuous connectivity for interconnected MITA end-points. The Access Independent Connectivity Specification provides access independent mobility management for the Connectivity layer.

Typically, mobility management is involved in connections between a terminal and a network. This statement is also valid for the AIC Specification. In MITA, mobility management has been divided between the AIC Specification and the Access Technologies Specification. Also, the Reachability, Service Discovery, and Presence Specifications deal with mobility issues.

Service Discovery in MITA

Service discovery consists of three types of discovery methods:

- Service discovery for ad-hoc networks
- Service discovery for finding services from the Internet
- The Presence and virtual community navigation layer is special type of service discovery

The Service Discovery Specification covers the Connectivity and Application layer service discovery technologies. The Access layer related service discovery technologies are covered in the Access Technologies Specification.

The Service Discovery Specification addresses requirements for the Access Independent Interface. When the AIC Subsystem identifies a need to change into a new access network domain, after proper authentication, the service discovery service for the Connectivity layer should be activated. In a similar way, the service discovery service provides control and event triggering interfaces for the Application layer Subsystems and applications.

Location in MITA

The Location specification defines software and protocol interfaces for location services. In practice, this refers to interfaces receiving location information for applications and to applications that utilize location based information. A goal in MITA are mobile location services architecture and interfaces that are able to deliver location information to applications regardless of the used positioning technology and of the division of functionality between terminals, networks, and servers. The same architecture must be flexible for several business models. In other words, a simple business model should not dictate it.

Location information is also an important element in many other MITA specifications. For example, location can be used to automatically update a user's presence and reachability information, and it can also be utilizied in service discovery for local services.

The Mobile location services architecture in MITA actually consists of several different architectures, where the positioning infrastructures are different but location interfaces to applications and middleware, in between, should be the same.

Terminal Management and Data Synchronization in MITA

Terminal Management, including both Bootstrap and Continuous Terminal Management, is very important. A good and correct configuration is the prerequisite for any application to actually be functional. Furthermore, it is to be expected that the complexity of the mobile devices increases with the addition of new and more powerful applications. At the same time, these increasingly complex devices move into mass-market environments. Thus, the need for Terminal Management will only increase.

It is highly likely that there will be multiple terminal management providers. Typically each provider has access to a certain subset of the parameter space. For example, there might be different management entities for the device's look-and-feel, for the browsing environment and for the messaging environment. The managing entities, for the above three, could be the device manufacturer, xSP and the corporation of the end user.

Terminal Management offers services to applications (e.g., provides them with configuration data). But similarly Terminal Management does have its own user interface, for example, to allow the user to control and interact with the management process.

In MITA, Data Synchronization will cover at least the following issues:

- Distributed applications and databases
- Linkage between presence and Contacts/Calendar
- PIM related issues (e.g., contacts, Calendar, todo lists)
- Controlled content download and management, using Smart Messages, or using a Terminal Management mechanism

Content Formats in MITA

Content format refers to a convention to package content. Agreements must be made on content formats in order to build the necessary interoperability between various machines, devices and applications. Given the limitations of the processing environment in mobile devices, we have to select a certain reasonably small set of content formats to be promoted and supported in Mobile Internet offerings. Content formats are to be agreed on in the areas of audio, still images and vector graphics, video and general-purpose documents.

Operation Support Systems in MITA

The Operation Support Systems (OSS) Specification deals with the Mobile Internet Management and Support architecture. The functionality of the OSS architecture is divided into management layers:

- Network management
- Service management
- Subscription management
- Subscriber accounting and identity management

By combining the management layers with the possible business players, two frameworks could be constructed:

- Management Framework
- Policy Framework

Privacy in MITA

In MITA, privacy is an additional layer above security. Security offers means for authentication and prevents bypassing the access control for personal information. The privacy layer is actually Privacy Rights Management and contains a rules database part and a generic access control part, which can be added to all personal information interfaces.

Interfaces are needed to inform services, applications or other users about privacy rules and to negotiate them, and also for delivering the identity of the other party to the access control function. Another set of interfaces is required for configuring the rule database. An interface with a UI entity is needed to deliver privacy notifications to user and to carry out queries to confirm the revealing of personal information.

Architecture Frameworks

Part of the MITA work deals with issues, which are complicated to solve without having common and unified models. These issues are structured under common frameworks, before more a detailed technical architecture is built. Current such frameworks are Directories, Security, and Mobile Internet Interfaces.

Directories in MITA

Directories are network repositories for information about people, places and things. There are numerous instances of directories in the Mobile domain. For example, in GSM technologies the Home Location Register (HLR) database is a directory of information about subscribers. Directories consist of three basic components:

- Schema
- Access Method
- Directory Store

Information in a directory, whether it represents data about a person or a device, is structured in a particular way. The characteristic defining the selection of information that is stored about people, places and things is called a schema. There is also a need to represent directory information when it is transferred around the Mobile domain. Exchange formats provide a standard exchange schema for such purposes. Significant MITA schemas include those that represent the mobile person object and the mobile device object. Other auxiliary schemas define managed mobile services for these basic objects.

Information in a directory is accessed using a directory access protocol. Directory access protocols provide methods for creating, modifying, searching and deleting entries. The administrators of a directory service can also utilize restricted administrator access methods to create new types of directory object types. There are a number of directory access methods that are needed for Mobile domain directories.

Information in a directory is maintained and managed similarly as information within a database. In fact, a directory is a specialized database application. Within the Mobile domain, there are numerous directory stores, each being used to capture information about a particular feature of the mobile person object or the mobile device object.

Additional components in the MITA directory specification include the use of directory synchronization to keep the numerous mobile directory stores updated with each other. Directories contain critical mobile assets including security information about mobile users.

Security in MITA

Security is about safety and trust. Safety is considerably easier to manage than trust, and without safety it is hard to create trust between unseen entities. It is important to understand that security addresses requirements to all entities of the Mobile Internet Technical Architecture. Security is required when calling someone or making an electronic monetary transaction with browser security functionality or features protecting customer information from being revealed or being altered during the transmission. Similar security requirements also apply to server-to-server transactions.

MITA security objectives are:

- Integrity
- Confidentiality
- Availability
- Authentication
- Authorization
- Intimacy

Integrity deals with the prevention of erroneous modification of information. Authorized users are probably the biggest cause of errors, omissions and the alteration of data. Storing incorrect data within the system can be as bad as losing data. Malicious attackers can also modify, delete, or corrupt information that is vital to the correct operation of business functions.

Confidentiality deals with the prevention of unauthorized disclosure of third party information. This can happen as a result of poor security measures or information leaks by personnel. An example of poor security measures would be to allow anonymous access to sensitive information.

Availability deals with the prevention of unauthorized withholding of information or resources. This does not apply just to personnel withholding information. Information should be as freely available as possible to authorized users.

Authentication deals with the process of verifying that users are who they claim to be when logging onto a system. Generally, the use of user names and passwords accomplishes this. More sophisticated is the use of smart cards and biometric authentication. The process of authentication does not grant the user access rights to resources - this is achieved through the authorization process.

Authorization deals with the process of allowing only authorized users access to sensitive information, protected services or resources and controlled actions. An authorization process uses the appropriate security authority to determine whether a user should have access to resources.

Intimacy deals with the protection of personal information by preventing disclosure to third parties intentionally or unintentionally. Putting the end user in a policy making position for deciding to whom and what to disclose.

In MITA the security functionality or a feature is divided into four categories:

- Fundamentals
- Protocols
- Applications
- Policies

Application programs use fundamentals and protocols to create security functionality that a customer can see. In most cases, applications make security visible to end users by providing feedback from the system security states. The Applications security category covers application layer and application framework security issues in the MITA Element.

The Protocol security category deals with protocols relating to data transmission (e.g., IPsec) and moving one's information from a device to another device. It covers Internet protocol security in the MITA Element.

Fundamentals are basic components of a device in the Mobile Internet without which no security can be implemented. Device and system security has to be developed from the ground up and the fundamentals provide the basis for creating security functionality in a system.

Security policies set the rules how technical security components can and will be used to create, and in some cases enforce, security in a system. In many instances in the past, there have been excellent technical solutions available for enforcing security, but unfortunately a political consensus among the players has not been reached. We must understand the importance of policies on the security landscape and accept that in many cases 80% policy and 20% technology make a functional security system.

Mobile Internet Interfaces in MITA

Mobile Internet Interfaces are fundamental elements of the Mobile Internet Technical Architecture. In previous sections both the layered model of a MITA element and the basic fundamentals of the MITA Subsystem Architecture were presented. These models identified many interfaces. Some of them are software interfaces within a MITA element and others are protocol interfaces between interconnected MITA elements.

Mobile Internet Interfaces are a group of open interfaces provided by Nokia for both software developers and system integrators, so that they can implement MITA compliant applications and systems. It should be noted that Mobile Internet Interfaces live longer than product generations and have similar features over product and server categories. Open software interfaces in the Mobile Internet Technical Architecture are called Mobile Internet Software Interfaces (MSIs) and open protocol interfaces are called Mobile Internet Protocol Interfaces (MPIs).

MSIs are divided into Application Programming Interfaces and System Software Interfaces (SSIs). In a similar way, MPIs are divided into two parts: protocol data units (PDUs), including protocol payload definitions, and content formats, describing presentation formats for the end-users' data.

In the Mobile Internet Technical Architecture all MITA elements have Mobile Internet Interfaces. In terminals, most of the open interfaces are software interfaces for application developers, whereas in servers, the open interfaces are both software and protocol interfaces enabling fast and rich service implementation by system integrators and developers. In network elements, Mobile Internet Interfaces are mainly protocol interfaces enabling different MITA End-to-end Views.

In this introductory overview, the most essential MITA elements and methodology were presented. In the chapters to follow we will describe a selected set of the technologies for the Mobile Internet and Nokia's end-to-end solutions for the Mobile World.

3.
Technologies for Mobile World

WEB DOMAIN

Basic Internet Protocols

Transport Protocols

As we have seen, the IP layer provides basic connectivity functions. Running on top of IP are a number of transport protocols such as TCP, UDP and SCTP. These protocols provide basic bearer services for applications and application protocols. We will look at these protocols in a little more detail.

TCP

Most of us are familiar with the Transmission Control Protocol (TCP) [793] whether we realize it or not. TCP provides the basic transport for protocols such as HTTP. TCP provides connection services, reliable delivery and data sequence preservation. Enhancements TCP has received over the years include:

- RFC 896 Congestion Control in IP/TCP Networks
- RFC 2001, 2883 Slow Start, Congestion Avoidance, Fast Retransmit
- RFC 2018 TCP Selective Acknowledgement Options
- RFC 2581 TCP Congestion Control

UDP

UDP provides unreliable, connectionless transport for protocols, such as RTP. UDP is a lightweight protocol - it does not provide congestion control mechanisms. User plane protocols such as RTP, GTP and FP and management protocols like SNMP, DNS, DHCP, LDAP and NTP run over UDP.

UDP provides 16-bit port numbers to let multiple processes use UDP services on the same host. A UDP address is the combination of a 32-bit IP address and the 16-bit port number.

The checksum in UDP over IPv4 either covers the entire datagram or is not used at all. In UDP over IPv6, the UDP checksum is mandatory and cannot be disabled. There is some interest in creating a UDP-lite for IPv6. The UDP-lite header replaces the "Length" field with a "Checksum Coverage" field. The length of the UDP-lite packet can be found in the length field of the IP header, so packet length information in UDP is not required. Checksum Coverage is the number of bytes counting from the first byte of the UDP-lite header that is covered by the checksum. Some payloads that are carried by UDP are tolerant of bit errors, so that discarding damaged packets can cause problems, especially in burst error situations.

SCTP

SCTP is a relatively new protocol. It was originally designed to be a transport bearer for PSTN and SS7 signaling over IP. However, it has been enhanced to serve as a general reliable transport protocol, along the lines of a next generation TCP. SCTP provides connection services, reliable delivery and congestion control like TCP does, but SCTP has some additional features. SCTP supports multihoming, which allows a single SCTP association to use multiple interfaces/IP addresses. This allows for additional protection against network failures.

SCTP supports multiple streams to avoid head of line blocking. SCTP is also message-oriented, which improves upon TCP's byte alignment. This means that protocols running on top of SCTP do not need to worry about message boundries - SCTP takes care of this. Addtionally, the SCTP-layer handles the fragmentation of the MTU (maximum transport unit). This should reduce the problem of message loss. SCTP provides for in-sequence delivery and as order of arrival delivery. In-sequence delivery provides the preservation of data order to the user. Order of arrival delivery passes the received data immediately to the user, which avoids re-sequencing delays.

SCTP also improves upon TCP connection setup security by using a secure four way handshaking procedure to prevent flooding attacks that have been prevalent in TCP.

Initial deployments of SCTP will be for transporting control plane signaling in Radio Access Networks as well as transporting SS7 over IP. Additionally, SCTP will transport AAA (Access, Authorization and Accounting) protocols such as Diameter. Other groups are looking at SCTP for applications such as IP Storage and Reliable Server Pooling.

The current status of SCTP is that a number of implementations already exist. Projects are underway to bring SCTP into the kernel of major operating systems such as Solaris, Linux and FreeBSD. These implementations of SCTP will include an enhanced socket API in order to take advantage of the new features provided by SCTP.

Related Protocols

DNS

If IP is the glue that holds the Internet together, DNS is the workhorse that makes everything happen. The primary job of DNS is to map Domain Names (URLs) to IP addresses and vise/versa via DNS records. DNS is, at its simplest, a distributed database.

New technologies such as ENUM are being deployed. ENUM is a technology that essentially maps telephone numbers (E.164 numbers) to DNS records. Under the domain name in DNS for the phone number, multiple records can exist, providing information such as a web page, email address or SIP address for the phone number. Thus the user performing the ENUM query can select the manner in which to contact the party in question (via email, SIP, etc.).

DHCP

The Dynamic Host Configuration Protocol is a configuration protocol running over UDP, which provides a mechanism for dynamically assigning IP addresses and configuration parameters to other hosts and clients. It is designed for large networks with complex software configurations. It allows servers to allocate re-usable IP addresses and configuration parameters for clients. Clients can obtain IP addresses for a fixed time period, from 1 minute to 99 years, or permanently. This saves users from directly having to configure IP addresses and other parameters on their machines when they attach to the network.

Common Internet Application Protocols

This section concentrates on the basic and most commonly used application protocols and their history on the public Internet. The Purpose is to give a short refresh of the most common Internet application protocols as the background for the most recent web service protocols.

Hypertext Transfer Protocol - The Access Protocol of the World Wide Web

The Hypertext Transfer Protocol (HTTP) is without doubt the most used, best known and widest spread protocol on the Web domain. HTTP is an application level protocol, which is light and fast enough for interactions between distributed hypermedia information systems.

HTTP has a long history in the World Wide Web global information initiative; it has been in use since 1990.

The first version of the HTTP protocol specification was HTTP 0.9. It was a very simple protocol implementation for data transfer across the Internet. The second improvement to the protocol specification was released in 1996, when HTTP version 1.0 was sent for comments to the IETF forum (Internet Engineering Task Force). Probably the most important improvement in version 1.0 was support for MIME-like formatted HTTP messages (Multipurpose Internet Mail Extensions). Other major improvements included the meta information about the transferred data in a message and several modifiers about request and response semantics. However, version 1.0 did not take into consideration the effects of hierarchical entities and functionality on the fast growing Internet.

There was still lack of support in the protocol for proxies, cache technology and the needs of persistent connections or virtual hosts. In other words, there was still need for improvement in the protocol, applications required the functionality to really determine the each other's true capabilities in the communication. Practical information systems require much more functionality than simple retrieval, search, front-end update and annotation. The current HTTP specification version 1.1 from 1999 tries to fulfil these requirements and solves the problematics of the blooming World Wide Web. Several additional specifications for HTTP concentrate on cases like authentication, state management and content negotiation and most of them are now included in specification version 1.1.

Like most of the Internet application level protocols, HTTP resides on top of the TCP layer in the TCP/IP stack, the default TCP port is 80. Of course, HTTP could also be used with some other reliable transport protocol in other networks. The difference from other application protocols is that in most cases HTTP is used to carry protocols inside its data segment. This is because HTTP is the most commonly used protocol on the Internet, so Internet and intranet infrastructures are built to support HTTP message pass trough (request / response) between the end points (server / client).

Secure HTTP (S-HTTP) provides secure communication mechanisms between an HTTP client and server in order to enable spontaneous commercial transactions for a wide range of applications. The intent is to provide a flexible protocol that supports multiple orthogona operation modes, key management mechanisms, trust models, cryptographic algorithms and encapsulation formats through option negotiation between parties for each transaction.

Other Common Internet Application Protocols

File Transfer Protocol

File Transfer Protocol (FTP) is the primary Internet standard for file transfer. FTP runs on top of the TCP/IP stack and uses separate simultaneous TCP connections for control and for data transfer. FTP provides a connection-oriented service and all data exchanged between the server and the client is guaranteed to not only be delivered, but also to be delivered intact as originally sent. FTP uses TCP port 21 by default. FTP was initially defined in RFC's for the ARPANET network in the beginning of the 70's. Since then, it has been the main protocol for transferring files trough TCP/IP networks.

Simple Mail Transfer Protocol

Simple Mail Transfer Protocol (SMTP) is the cornerstone of messaging interoperability across the Internet. SMTP runs on top of the TCP/IP stack. TCP port 25 is the default port for all SMTP operations.

In 1982, the IETF defined SMTP in RFC's 821 and 822. SMTP gets its name from the original protocol, which was quite simple and concentrated mainly on the task of sending plain text messages over an IP link between a client and a server.

Since 1982, SMTP has evolved to fulfil the requirements of modern messaging environments. The evolution of the SMTP protocol on the Internet has been very fast. Extended SMTP (ESMTP) and Multipurpose Internet Mail Extension (MIME) have enabled SMTP to deliver highly functional messaging systems.

Although a decade old, SMTP has proven remarkably resilient. Nevertheless, the need for a number of protocol extensions has become evident. ESMTP enables vendors to add extensions to the basic SMTP service so that they can handle different classes of messages or other functionality.

Sending plain text messages is not descriptive enough in many cases, modern messaging systems have to support graphics. Some early encoding technologies like Uuencode let users send attachments and formatted text, but these technologies were not very well integrated in many cases. It is important to precisely configure the tags that different attachments use before it is possible to confidently send an understandable attachment to a recipient.

MIME

MIME solves this configuration problem by defining the rules for the labeling and transmission of different data types within messages. MIME informs mail systems to process parts of the message so that recipients see exactly what the sender intended. MIME is also the basis for the transmission of streaming data such as audio and video messages.

Network News Transfer Protocol

Network News Transfer Protocol (NNTP) protocol was initially defined in IETF RFC 971 and released 1986. Since then, NNTP has become one of the most popular protocols on the Internet. Many implementations of the protocol have been created on several different platforms and operating systems. With the growth in use of the protocol, work began on a revision to NNTP in 1991, but that work did not result in a new specification. However, many ideas from that working group found their way into many implementations of NNTP. Additionally, many other extensions, often created by newsreader authors, are also in use. NNTP runs on top of the TCP/IP protocol like all the other application protocols in this chapter. NNTP uses TCP port 119 by default.

NNTP is a good example of a protocol which has become an Internet standard even without an unambiguous specification.

Terminal Emulation Protocol

Terminal Emulation Protocol (TELNET) was specified as a standard for the ARPA Internet community in 1983 (RFC 854).

The purpose of the TELNET Protocol is to provide a general, bi-directional communications facility. Its primary goal is to allow a standard method of interfacing terminal devices and terminal-oriented processes with each other. It is envisioned that the protocol may also be used for terminal to terminal and process to process communication.

Applications use the telnet protocol to connect over the network to a remote computer that is running telnet server software. The Telnet protocol provides two-way communication functionality that allows terminal devices and terminal-oriented processes to communicate with each other. Telnet uses TCP to transmit data and protocol specific control information. The default port for telnet is TCP port 23.

RFC 854 states: "The Telnet Protocol is built upon three main ideas: first, the concept of a 'Network Virtual Terminal;' second, the principle of negotiated options; and third, a symmetric view of terminals and processes."

Telnet is probably the most commonly used protocol for enabling process-to-process communication over TCP/IP networks.

References

[768] Postel, J. (ed.), "User Datagram Protocol", STD 6, RFC 768,

[793] Postel, J. (ed.), "Transmission Control Protocol", STD 7, RFC 793, September 1981.

[2916] RFC 2916, "E.164 number and DNS", P. Faltstrom, September 2000.

[2960] RFC 2960 "Stream Control Transport Protocol" R. Stewart, et al, November 2000.

Internet Protocol Version 6

The Internet Protocol (IP) has been the glue that has connected diverse networks together, and forms the basis of the Internet. It allows hosts to communicate over diverse networks. The current version of the Internet Protocol, IPv4, is well over 20 years old and is showing its age. In late 1990, projections showed that the address space of the Internet would become an increasingly limited resource, so the IETF undertook the task of specifying the next generation IP protocol, which was at that time called IPNG (Internet Protocol Next Generation). [RFC1752] A number of proposals for IPNG were submitted and much heated debate occured.

IPv5 was an experimental streaming protocol (Experimental Internet Stream Protocol, Version 2) first described in RFC 1190 and updated in RFC 1819. This work has been abandoned.

IPv4 offers a 32 bit address, which translates into approximately 4 billion possible addresses, however this number is greately limited by address allocation policies. As a short term fix, Network Address Translators (NAT) were proposed to provide private address space in order to gain additional addresses. Problems with NATs are well documented, however. IPv6 offers a 128 bit address, which translates to 340,282,366,920,938,463,463,374,607,431,768,211,456 possible addresses, which should be sufficient for the forseeable future.

IP Headers

The first figure shows the IPv4 header, while the second one presents the IPv6 header.

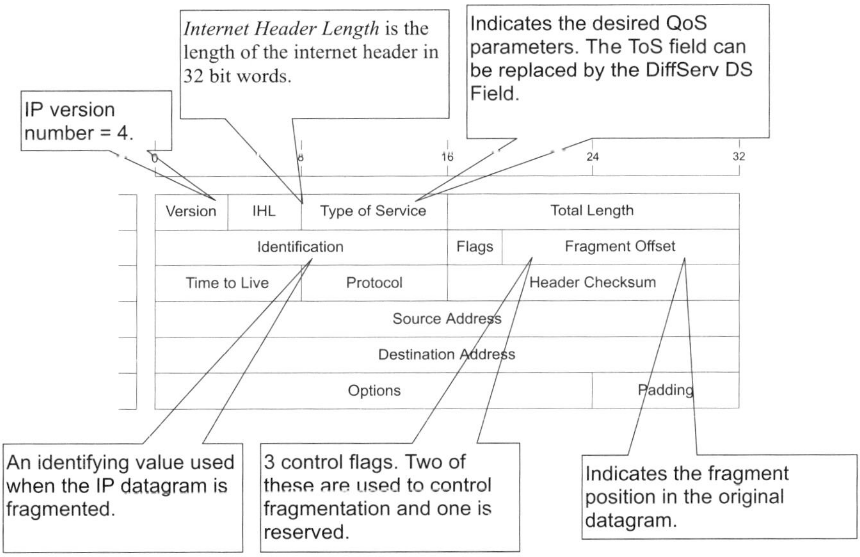

IPv4 Header.

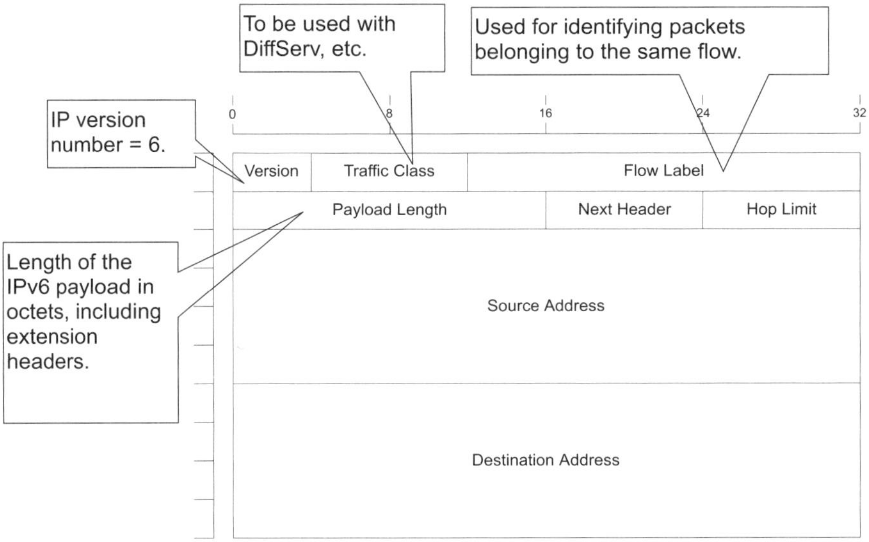

IPv6 Header.

As one can easily see, the IPv6 header is streamlined when compared to the IPv4 header. Unused or unnecessary fields in the IPv4 header were removed. The end of the IPv4 header contains extensions. These extensions are rarely used because they slow down routing considerably. With this in mind, IPv6 was designed to support extension headers. This allows the IPv6 header to remain a fixed length for simpler processing, while still allow extensibility. Currently, six types of of Extension Headers have been specified: Hop-by-hop Options header, Routing (Type 0) header, Fragment header, Destination Options header, Authentication header and Encapsulating Security Payload.

IPv6 Header Handling

In IPv6, optional information may be encoded in separate headers, which are placed between the IPv6 header and the upper-layer protocol header contained the IPv6 Packet. The "Next Header" field contained indicates the location of the next header, whether it is the head of the payload protocol or an extension header. Each Extension Header uses the same mechanism to indicate the location of the next header. Thus the Extension Headers form "a daisy chain" of of headers. A value of 59 in the Next Header field indicates that there is nothing following the header.

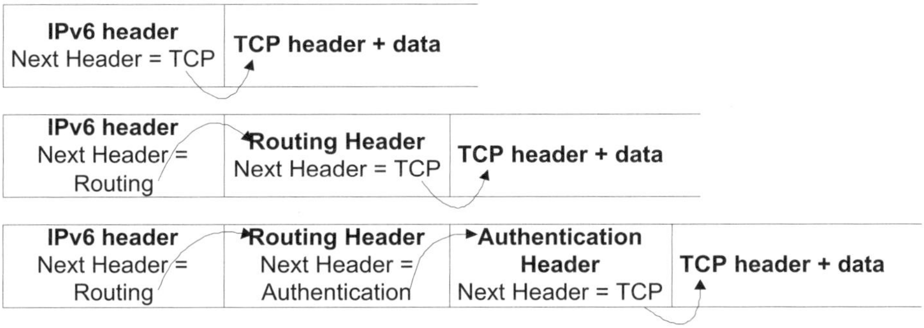

Extension Headers.

Other Enhancements

Other enhancements were also built into IPv6. Mobility support is a basic feature in IPv6, while Mobile IPv4 is purely an add-on to IPv4. A Basic IPv4 node does not support mobility features at all, even as a Correspondent Node. Every IPv6 node should support Mobile IPv6 Route Optimization (as a Correspondent Node) and a Home Address option. IPv6 has a Stateless Address Autoconfiguration feature, which allows nodes to be automatically configured, even without DHCP. IPv4 nodes need DHCP for automatic configuration. IPsec is a mandatory feature in IPv6. IPsec is an add-on to IPv4 and is not usually implemented. Additionally, IPv6 contains a flow label field which can be used to identify flows. Currently, the usage of the flow label is undergoing discussion in the IETF.

IPv6 Address Types

With 340,282,366,920,938,463,463,374,607,431,768,211,456 possible addresses, the astute reader would assume there would be different types of IPv6 addresses. The reader would be correct. "IP Version 6 Addressing Architecture" [RFC 2372] describes IPv6 addresses in detail. Please note that this speficiation is currently undergoing revision.

There are three types of IPv6 addresses: Unicast, Anycast and Multicast. Unicast addresses are the most common type. They form the bulk of addresses on the Internet, and generally define a single interface. IPv6 multicast addresses replace IPv4 broadcast messages, and define a set of interfaces. Packets sent to multicast addresses are sent to all of the interfaces defined by the address. Anycast addresses are new and define a set of interfaces. Packets sent to an anycast address are sent to one of the addresses defined by the anycast address.

There are several ways to represent IPv6 addresses as text strings. The preferred form is

```
x:x:x:x:x:x:x:x
```

The 'x's are hexadecimal values of the eight 16-bit pieces of the 128-bit address. It is not necessary to write any leading zeros. An example would be:

```
FEDC:BA98:7654:3210:FEDC:BA98:7654:3210
```

Additionally, a short-hand method exists for writing IPv6 addresses which contain long strings of zeros. The use of '::' indicates multiple groups of zeros. The '::' may only be used once in an address, but may be used to indicate leading or trailing zeros. Thus,

```
1080:0:0:0:8:800:200C:417A
```

may be represented as:

```
1080::8:800:200C:417A
```

IPv6 address prefixes are represented by the notation: ipv6-address/prefix-length, where the ipv6-address is an IPv6 address (as shown above) and the prefix-length, which is a decimal value specifying how many of the leftmost contiguous bits of the address comprise the prefix. An example, showing a 60 bit prefix, would be:

```
12AB:0:0:CD30::/60
```

Table 1: The Current Allocation of IPv6 Prefixes

Allocation	Prefix (binary)	Fraction of Address Space
Unassigned (see Note 1 below)	0000 0000	1/256
Unassigned	0000 0001	1/256
Reserved for NSAP Allocation	0000 001	1/128 [RFC1888]
Unassigned	0000 01	1/64
Unassigned	0000 1	1/32
Unassigned	0001	1/16
Global Unicast	001	1/8 [RFC2374]
Unassigned	010	1/8
Unassigned	011	1/8
Unassigned	100	1/8
Unassigned	101	1/8
Unassigned	110	1/8
Unassigned	1110	1/16
Unassigned	1111 0	1/32
Unassigned	1111 10	1/64
Unassigned	1111 110	1/128
Unassigned	1111 1110 0	1/512
Link-Local Unicast Addresses	1111 1110 10	1/1024
Site-Local Unicast Addresses	1111 1110 11	1/1024
Multicast Addresses	1111 1111	1/256

Additionally, IPv6 provides for scoping of addresses [IPv6 Scope]. The term 'scope' refers to the topological span addresses that may be used as unique identifiers for interfaces. For unicast and anycast addresses, there are three defined scopes: Link Local, Site Local and Global. Link local scope uniquely identifies an address on a single link only. Site local uniquely identifies an address within a site, such as an organization (e.g. an intranet). Global scope identifies an address uniquely on the global Internet. It is possible and likely that a single interface will have link local, site local and global addresses. It is important to remember that addresses are not visible (or usable) outside of their scope.

Conclusions

IPv4 offers approximately four billion possible global IP addresses, which is clearly insufficient. IPv6 offers more than enough global IP addresses for now and the forseeable future. Additionally, IPv6 builds upon the success of IPv4 and improves upon it. IPv6 is necessary and sufficient.

References

[RFC1752] S. Bradner & A. Mankin; "The Recommendation for the IP Next Generation Protocol," RFC 1752; January 1995.

[RFC2373] R. Hinden & S. Deering; "IP Version 6 Addressing Architecture,"RFC 2372, July 1998.

[RFC2460] S. Deering & R. Hinden, "Internet Protocol, Version 6 (IPv6) Specification", RFC2460, December 1998.

[RFC2993] T. Hain, "Architectural Implications of NAT", RFC2993, November 2000.

[IPv6SCOPE] S. Deering, et. Al; "IPv6 Scoped Address Architecture" A Work in Progress, <draft-ietf-ipngwg-scoping-arch-02.txt>, March 2001

Mobile IPv6

Mobile IP for IPv6 (Mobile IPv6) has been designed to handle the mobility management needs for mobile (typically wireless) nodes on the IPv6 Internet. The design of Mobile IPv6 has been motivated by a desire to remain as close as possible to the guidelines motivating the design of the underlying IPv6 protocol, and building on the experience gained from the design and implementation of Mobile IP for IP version 4 (RFC 2002). This chapter describes the relevant aspects of the design of IPv6, followed by the design of Mobile IPv6, and some of the most recent areas that are under discussion for further development and standardization.

The main motivation for producing a new version of IP has been to enable global addressibility for a much larger population of Internet devices than is possible using IPv4, which has been the mainstay of the Internet for almost two decades now. When IPv4 was first designed in the early 1980's, few people imagined that the Internet would grow beyond a hundred thousand nodes, much less the hundreds of millions of network devices that make up the Internet nowadays. One result was that the total size of the address space was set at 4 billion individual addresses. Worse, though, was the chaotic manner in which network were assigned. There were only three sizes of networks specified - class C networks with 256 nodes, class B networks with 65536, and class A networks with slightly over a million addressable nodes. Thus, mid-size organizations with a thousand computers would routinely consume 60 times the necessary address space.

By the mid 1990's, projections were made that this practice would consume the entire IPv4 address space within 10 years. This set into motion several complimentary activities which are still having dominant effects on the evolution of the global Internet.

Most immediately, the use of subnetting, originally offered as an optional feature of Berkeley Unix systems, was made mandatory. This feature, called Classless Inter Domain Routing (CIDR), allows networks to have sizes of any power of two, and thus allows much more efficient allocation of IPv4 addresses. As the use of CIDR spread across the Internet, the rate of allocation of the original classful network prefixes slowed substantially, effectively delaying the exhaustion of network prefixes for quite a few years. We might yet have another 10 years at current rates of consumption, except for the anticipated growth of wireless networking.

The consumption of network prefixes has slowed, but not the rate of actual address consumption. Thus, there is still great concern that not enough IPv4 addresses will be available to fulfil the needs of the Internet at its current rate of growth. The situation becomes drastically worse if we include the expected number of mobile terminals among the population of devices needing Internet addresses. While current mobile telephones do not use the Internet to any great extent, most people believe this will change, and 3G standards have been written with this expectation. Unfortunately, we cannot expect to achieve anywhere near 100% utilization of the IPv4 address space (nor the IPv6 address space, for that matter). Theories have been advanced suggesting that a natural limit to address utilization exists, and that the limit (the so-called "H-factor") appears to be about 25-30%. The H-factor arises because network sizes are constrained to be a factor of two, because of the natural effects of network aggregation, and because of the needs for network administrators to plan for growth. Consequently, unless something unexpected happens, the total number of computers assigned to have global IPv4 addresses is unlikely to ever exceed 1 billion. As the Internet continues its exponential rate of growth, that hard limit looks frightfully close.

IPv6, with its vast address space, offers a straightforward solution to this problem. Furthermore, care has been taken so that the IPv6 network prefixes can be assigned in a much more orderly fashion (initially, as well as ongoing). It seems realistic to assume that at some point in the future the Internet will be populated mainly by IPv6-addressable devices, with a smaller subset of devices that are addressable with both IPv6 and IPv4 addresses, and a smaller yet collection of legacy devices that for one reason or another can not be easily upgraded. However, just because we can design a future Internet with sufficient addressability does not at all mean that we will have a clear idea how to solve the new problems of scale that are almost guaranteed to arise. Examples abound to illustrate this unpredictability. For instance, the rise of the Web, the economic upheaval and transformation worked by the dot.coms, and the commercial exploitation of the Internet by ISPs are all having long-term effects that could not have been clearly foreseen ten years ago. Although we probably cannot predict the transformations that might be att ributable to an Internet with 4 billion nodes, we can pretty confidently state that IPv6 offers the best model for global addressability and end-to-end communications. This is especially important for voice applications, which need both of these important features.

Mobile IP

Mobility introduces a problem with today's Internet addressing model that can be stated very simply. An IP address is used for two different purposes: routing, and host identification. The routing prefix forms the initial bits of the IP address. The remaining bits indicate the specific identify of the host among the collection of network nodes collected together as part of the network which has been identified by the routing prefix. The routing infrastructure of the Internet operates to deliver packets to the specific network indicated by the routing prefix, and the router for that network delivers the packet to the host identified by the rest of the bits.

This model works fine for nodes that do not change their location. For such computers, the routing doesn't change. Since the identity of the host computer is also unlikely to change very often, if at all, the same IP address can serve both functions for an indefinite amount of time. When the network node can move, however, this simple relationship is no longer valid. The IP address should stay the same so that the host can keep its identity, but the IP address would have to change in order for packets to be routed to the new location of the host computer.

For network nodes that move around, Mobile IP solves this problem by associating a care-of address with the network node at each of the node's new points of attachment. The node can keep its permanent IP address (called a "home address"), but vary the care-of address to conform to the routing needs of the Internet. Thus, Mobile IP transforms (part of) the mobility management problem into a routing problem. Circumscribing the exact design space for network-layer mobility support has been a fascinating development over the last 13 years, and details are still emerging; some of these will be described at the end of this chapter.

A specialized router on the home network, called the "home agent", is given the responsibility of forwarding packets to the mobile node when it is away from its home network (i.e., when it has a care-of address at some new point of attachment). The home agent, then, has to receive information about every new care-of address acquired by the mobile node so that it can carry out this forwarding operation. The association between the mobile node's home address and its care-of address is called a "binding"; the association also includes a lifetime value, after which the care-of address information is considered expired and must be deleted. The message which contains the new care-of address (or the refreshed lifetime, in case the care-of address has not changed), and which the mobile node delivers to the home agent, is called a "Binding Update".

Using this terminology, we can say that the essential part of Mobile IP is the protocol by which a Binding Update is delivered. This turns out to be a very powerful tool. When a correspondent node (i.e., an arbitrary node within the Internet) needs to send a packet to the mobile node, it can just send it to the mobile node, which is identified to the rest of the Internet by its home address. The mobile node's care-of address is not typically used to identify it, but only to forward packets to its current point of attachment to the Internet. In this way, communications to the mobile node can continue without disruption even while the mobile node changes from one care-of address to the next. As far as the application on the correspondent node is concerned, the mobile node is still addressable at its home address, and that does not change, so communications can proceed transparently. In fact, even the transport protocol (e.g., TCP) does not detect the change of a care-of address.

From the above description, it is possible to formulate a pretty good plan for designing Mobile IP.

1. The mobile node must detect that it has moved.
2. The mobile node must discover or configure its new care-of address
3. The mobile node must inform the home agent about its new care-of address
4. The home agent must forward packets from the home network to the mobile node's care-of address.

IPv6 offers native features that enable a mobile node to accomplish some of these needs directly, and the other requirements, although not already provided, are still greatly simplified in IPv6.

This powerful method of associating a topologically correct, but changeable, care-of address to the mobile node's care-of address goes a very long way towards solving the mobility management problems for wireless Internet nodes. However, there are new classes of problems, which are uncovered after the routing problems are solved, including resource discovery, profile management, context and environmental controls, billing, and adaptive programming. None of these problems are realistically solvable using Mobile IP, and they can all be of great importance for improving the user experience. Even given that they are important problems, they must be considered beyond the scope of the discussion in this chapter.

IPv6 Design Points

As has been described, the main reason for using IPv6 addresses for the future Internet is that the available address space becomes huge. The (128 bit) IPv6 address space is 2^96 times as large as the IPv4 address space. Given that the IP header required modifications to accomodate address fields of this size, a certain amount of effort was expended to repair problems that had been identified over the 10 years of experience with IPv4. Some of these efforts were made to correct what were considered to be design flaws, and some of the others to enable improved performance.

Router Advertisement

Although Router Advertisement messages are specified for both IPv4 and IPv6, in the former case, advertisements are not extensively used on today's Internet. Furthermore, advertisements play a much more important role in the address autoconfiguration features for IPv6, as will be described shortly. Since home agents are typically considered to be routers, and since mobile nodes need to find out the identity of the home agents on the home network, the Router Advertisements are specified to be used for this purpose, and extended with additional information that is needed for Mobile IPv6. For instance, there is a new bit which indicates that the router is also willing to serve as a home agent for (mobile) nodes on the network. There is also another extension that supplies the global IP address of the home agent, for situations where that information is not otherwise available.

Router advertisements also supply information about how long the IPv6 addresses may be used. Every IPv6 address has a lifetime; if the lifetime expires, use of the address must be discontinued. Typically, each new Route Advertisement is considered to renew the lifetime, so that a node, which is attached to a network with typical lifetime parameters in the advertisements may continue to use the same IPv6 address indefinitely. This use of address lifetimes helps with renumbering.

Address Autoconfiguration

One of the important improvements in IPv6 is the development of Stateless Address Autoconfiguration. This relies on first acquiring a "link-local" address. Link-local addresses have the special routing prefix "FE80::/64", which means a bit-string 64 bits long, with 0xFE80 in the first two bytes, followed by 48 bits of zero. A packet with link-local destination IP address cannot be forwarded by any router. Once the node has exclusive use of a link-local address, it can form a globally valid IPv6 address by replacing the FE80::

prefix by the routing prefix; the routing prefix is obtained from the advertisements issued by the local (default) router(s).

This method for getting IPv6 addresses works as follows:

1. The node first attempts to acquire a "link-local" address. It forms a 64-bit interface identifier, and probes on the link to find out if anyone is using the IPv6 address formed by prefixing FE80:: to that interface identifier. This process is called "Duplicate Address Detection (DAD)".

   It is very unlikely that the link-local address selected by the mobile node would already be in use, since there are 16 quintillion such link-local addresses available. However, if DAD fails for this reason, the mobile node can easily retry the process with another choice for its interface identifier on the new link.

   Typically, using the nodes MAC address on every new link will assure a unique link-local address.
2. The node monitors the link for a Router Advertisement. When the advertisement is detected, the mobile node extracts the routing prefix, along with the lifetime information for the addresses on the link, the MTU for the link, the layer-2 address for the default router, and any other pertinent information. The advertisement also specifies whether Stateless Address Autoconfiguration can be used. If so, the mobile node just replaces the upper 64-bits of its routing prefix with the advertised routing prefix.
3. If the advertisement prohibits use of Stateless Address Autoconfiguration, the mobile node must use some other means of acquiring an address which is topologically correct for the current point of attachment. The exact means is beyond the scope of the discussion in this section, but one possibility would be to use DHCPv6 (DHCP for IPv6).

The mobile node can use this method for obtaining addresses each time it establishes a new point of attachment to the IPv6 Internet. Note that, using these rules, the mobile node has to wait for the completion of Duplicate Address Detection (DAD) at every new point of attachment. For some applications, this delay is unacceptable. A recent effort has been undertaken to eliminate this delay, as part of an overall specification for Fast Handovers. That work will be described in a later part of this section.

Security

One advantage of using IPv6 is that every network node is guaranteed to be able to compute authentication data as well as perform encryption and decryption. These are powerful features for enabling e-commerce and keeping the Internet safe from the threats of many kinds of malicious attacks. This is in line with recent efforts by the IETF leadership to ensure that all newly standardized protocols are able to be operated securely.

The mobile node must be able to deliver new care-of addresses to all intended recipients securely. In addition, for the particular case of delivering care-of addresses to the home agent, the protocol must also be reliable in addtion to being secure, because the home agent must always have up-to-date information about the mobile node's care-of address.

In order for the recipient to be assured that the care-of address it has received is truly the same as was intended to be reported by the mobile node, Mobile IPv6 requires that all such messages be accompanied by some data that can only have been produced by the mobile node - and, thus, is unforgeable. For this reason, Mobile IPv6 in its original design mandated the inclusion of an IPv6 Authentication Header (AH) in every packet reporting a new care-of address. Recent activities surrounding the exact formulation for inclusion of authentication data will be described in a later part of this section. Since IPv6 network-layer information is organized as a series of extension headers, it is quite feasible to include additional header information in those packets. These additional headers are organized independently and located after the main IPv6 header. The Mobile IPv6 care-of address is inserted into one of these optional headers, and then AH follows the previous headers with authentication data which assures that all of the header information (not just the Mobile IPv6 data) is authentic.

The other part of security, which is required in any operation setting is key exchange and distribution. This is a much harder problem; in contrast to the mandated features for authentication and encryption, no specific protocol has been mandated for key distribution for IPv6 network nodes. Probably, the most widely implemented such protocol for IPv6 network nodes has been Internet Key Exchange (IKE), but that protocol is considered too heavyweight for the general needs of Mobile IPv6. For most of the last 5 years, the mobile-ip working group was not considered to have the responsibility for developing such key distribution protocols, and we basically assumed that they would be developed elsewhere. Now, as part of the ongoing re-evaluation of what is acceptable for standardizing major new protocols, it has been required to identify a key distribution protocol. Further details about key distribution for Mobile IPv6 are found later in this section.

Destination Options

As mentioned previously, some of the features of IPv4 were redesigned during the development of IPv6. One such redesign involved the layout of network-layer options. In IPv4, the options were directly included as part of the main header. Parsing the options is quite time-consuming, compared to parsing the other header fields. This caused most IPv4 routers to be processed along the "slow path" of the intermediate routing points. Consequently, very few IPv4 packets carry any optional fields at all, and in the end one could almost say that IPv4 effectively does not offer any useful optional protocol features.

In contrast, IPv6 options are more easily parsed, since they are not located inside the main IPv6 header. Moreover, specific measures have been taken to eliminate all processing requirements on intermediate nodes, in many cases. This has been done by introducing a new class of network-layer options, called "destination options". Other options, called "hop-by-hop" options, retain the IPv4-style option semantics for features that need to supply information to all intermediate routers. Actually, other IPv6 options (e.g., IPsec options and the IPv6 fragment header) could just as well have been classified as "destination options", but they have their header number and so do not have to be organized as sub-parts of the general Destination Option header structure.

ICMPv6 Improvements to Encapsulation

Once the home agent has the care-of address information for the mobile node, it is able to reroute packets as they arrive on the home network so that they arrive instead at the care-of address. This is done by the use of IPv6 encapsulation, which is very similar to IPv4 encapsulation for most purposes. Tunneling works very well in the usual case, as long as some care is taken to avoid exceeding the MTU for all links along the path from the home network to the mobile node's current point of attachment (i.e., from the home address to the care-of address). Sometimes, however, things do go wrong. In this case, the home agent will likely receive an ICMP error (perhaps, Host Unreachable) indicating that the care-of address may no longer be a suitable locator for the mobile node. Such events should be reported to the node that originally sent the packet to the mobile node at its home address - i.e., the correspondent node.

This tunneling operation is done, using encapsulation, in practically the same way for both IPv4 and IPv6 mobility support by the home agent. However, for IPv4 there is an important difference, because of the way ICMP works for IPv4. For either version of IP, ICMP is specified to return the entire IP header of the undeliverable packet, along with some part of the payload. For IPv4, this is specified to be at least 8 bytes of payload. For IPv6, it is at least 512. Unfortunately, 8 bytes of payload for an encapsulated packet is not enough for the home agent to determine the IP address of the sender of the encapsulated packet - i.e., the IP address of the correspondent node. This happens because the encapsulated IP header appears as payload at the intermediate node (i.e., router) that generates the ICMP message back to the home agent. If the home agent discards the encapsulating (outer) header, the subsequent 8 bytes of payload is not enough to contain the source IP address of the encapsulating (inner) header.

In contrast, IPv6 processing by the home agent for such ICMP messages is practically guaranteed to supply the address of the correspondent node. Thus, the home agent can, with very little overhead, relay the ICMP message to the correspondent node, and hope that the correspondent node will cease producing traffic for the now unreachable mobile node.

In order to have the same effect for IPv4, the home agent has to store quite a bit of state information for each tunneled packet (called "soft-state"), for at least the duration indicated by the round-trip time (RTT) between home agent and care-of address. The stored state information would typically include most of the header information for each tunneled packet. Additional processing overhead would also be required to match up the incoming ICMP messages with the collection of stored state for the recently tunneled packets to the care-of address indicated in the ICMP message. If, on the other hand, the IPv4 home agent does not implement the collection of soft-state as has just been described, correspondent nodes (especially with UDP applications like streaming audio) are likely to continue to pump out volumes of data that can not be delivered, thus adding unnecessarily to congestion at the home network.

Mobile IPv6 Protocol Overview

With the above IPv6 features in mind, it is now very easy to describe the Mobile IPv6 protocol. Recall that the basic operation of Mobile IP requires that the mobile node acquire a care-of address at each new point of attachment, and that this care-of address has to be supplied to a router (called the "home agent") on the mobile node's home network. With this general description, one can outline the general requirements for a Mobile IPv6 protocol as four general subrequirements:

Discovery	The mobile node has to find out that it has moved.
Address Configuration	The mobile node has to configure a care-of address
Binding Update	The mobile node has to inform its home agent about the new care-of address, and
Tunneling	The home agent has to reroute packets from the home network for delivery at the care-of address.

In this section, we will describe the protocol operations designed to provide a solution for the above four sub-requirements.

Discovery

Mobile IPv6 allows a mobile node to make use of the basic IPv6 Router Solicitation and Advertisement messages defined in RFC2461. When a mobile node can no longer detect any periodic broadcast Router Advertisements at a particular point of attachment to the Internet, it can either configure a new care-of address from another Advertisement that it has already received, or it can issue a Router Solicitation in order to trigger the transmission of an advertisement for its own use.

Mobile nodes should typically make use of mechanisms from RFC2461 for movement detection only when they do not have better mechanisms from lower-layer protocols. With many well-designed physical media, the mobile node can obtain information about the noise level and signal strength for packets received from various access points. When one access point begins to drift out of range, the mobile node can attempt the process of establishing a link with some other access point that has a better signal. The link establishment could trigger the transmission from the mobile node of a Router Solicitation. Then the Router Advertisement from the new router would have all the information needed by the mobile node to discover the necessary details about its new point of attachment.

Address Configuration

Once the mobile node establishes its link, it can configure a care-of address as described previously. No changes for Mobile IPv6 are necessary.

Binding Update

Once the mobile node determines its new care-of address, it has to report this information to its home agent. The message in which the mobile node sends this information is called a Binding Update; the terminology literally means to update the home agent's "binding", which it keeps for the mobile node. A "binding" is, naturally, the association between a mobile node's home address, which identifies the mobile node to all the rest of the IPv6 Internet, and the mobile node's care-of address, which locates the mobile node at its current point of attachment. The home agent maintains a Binding Cache with one entry as needed for each of the mobile nodes that it serves but which are currently attached to some other network other than the home network. As mentioned earlier, the Binding Update has to be verified (using cryptographic techniques) as authentically transmitted by the mobile node. Once the home agent receives a Binding Update from one of its mobile nodes, the home agent takes steps to ensure that it can redirect packets destined to the mobile node's home address, as described in the following subsection about Tunneling.

The mechanism by which mobile nodes can inform their home agents about the care-of address is a very powerful tool, forming the basis for all mobile networking support on the network layer. This tool can also be used to enable correspondent nodes to deliver packets directly to the mobile node without requiring the services of the home agent.

Tunneling

In order to redirect packets from the mobile node's home address to its care-of address, the home agent has a bit of work to do. Its work items include the following:

1. Capturing packets that are addressed to the mobile node's home address.
2. Participating in DAD to protect the mobile node's addresses on the home network against reallocation to another network node.
3. Redirecting the packets that it captures when they are forwarded to the home address, by using encapsulation.
4. Other tasks (not described here) to inform the mobile node about renumbering events on the home network.

The home agent performs the first task (1) by following the protocols specified in RFC 2461. In particular, the home agent has to listen for Neighbor Solicitations and respond on behalf of the mobile node with a Neighbor Advertisement, with the 'O' bit set. That bit setting requires the mobile node's erstwhile neighbors to update their Neighbor Cache with the information provided by the home agent. Naturally, the result is that the mobile node's neighbors on its home network are instructed to use the home agent's layer-2 address as the appropriate resolution of the mobile node's home address.

In order to preserve the functionality of Duplicate Address Detection, the home agent must also respond to address probes for the mobile node's link-local address. Otherwise, another new node on the home network could conceivably autoconfigure the same link-local address as the mobile node had already done when it was attached to its home network. This would violate a basic presupposition of the IPv6 protocol, with undoubtedly horrible effects. Experience dictates that infrequent, unpredictable address-level conflicts must be prevented by taking all due measures during the protocol design stages. Mobile IPv6 has done so by a natural use of the already existing Duplicate Address Detection and Neighbor Advertisement features.

The last piece of the tunneling problem, after all the preconditions have been met on the home network, is determining the encapsulation protocol. Here the natural choice is IPv6-within-IPv6 encapsulation, RFC 2473. This choice enables the home agent to avoid making any changes whatsoever to the network-layer packet as it was transmitted by the correspondent node. This is important, because changes to the original packet would likely make it impossible to maintain the correctness of authentication data that may have been supplied by the correspondent node.

Route Optimization

If we wish to design for the future mobile Internet, we have to make scalability a major goal. There are many dimensions in which scalability is important, but the first major dimension is to ensure that all network entities involved with the transaction will be able to handle the processing load. Otherwise longdelays will be unavoidable at the overloaded network element. Another more subtle way to improve scalability is to involve as few network entities as possible in the protocol operation, at least in all of the typical cases of interest. Rarely performed operations may reasonably involve additional processing or more network entities.

Applying this general design principle to Mobile IPv6, we can see that the home agent represents a single point of failure, unless additional steps are taken, which are currently outside the scope of the Working Group document. The home agent could very well become a point of congestion, if large numbers of correspondent nodes wish to communicate with many of the mobile nodes on the associated home network at the same time. For the applications envisioned for the future, such as streaming video and multicast radio to hundreds of thousands of wireless users per home network, this would amount to snatching defeat from the jaws of victory. Fortunately, we can utilize the Binding Update to avoid this major scalability problem. If the mobile node supplies a Binding Update to its communication partners (correspondent nodes), then all subsequent traffic from the correspondent node can travel directly to the mobile node's care-of address without the intervention of the home agent.

The procedure is conceptually almost trivial and requires practically no additional protocol, because the mobile node can use the same Binding Update message format with any node on the Internet as it does with its home agent. Just as the home agent does, the correspondent node maintains a Binding Cache with entries for mobile nodes that have supplied care-of addresses. The actions required from the correspondent node are much like that of a home agent, except that the correspondent node does not have to tunnel packets to the mobile node from other correspondent nodes, whereas the home agent does have to do this. The other difference is that usually the correspondent node does not have to acknowledge the Binding Update (i.e., send a "Binding Acknowledgement"), whereas the home agent always does.

Even though the protocol for receiving Binding Updates is essentially the same, the operational details are quite different because of the needs for key distribution. Since the correspondent node has to have a security association with the mobile node before it can verify the authenticity of the Binding Update, it has a prior need to exchange key information with the mobile node; the key exchange is the basis upon which the security association is built. The original supposition within the [mobileip] working group was, as suggested earlier, that such key exchange algorithms would be designed as needed and probably in other working groups. Recent events have precipitated a pioneering effort to create a lightweight key exchange protocol which creates a minimal security association between mobile node and correspondent node - a security association which is only intended for the simple task of verifying authentication data sent along with Binding Updates.

Note that, as a side-effect, these route optimization techniques help to avoid any need for the home agent to relay ICMP messages back to the correspondent node. This results from the fact that when correspondent nodes send packets to the mobile node's care-of address without the assistance of the home agent, then ICMP messages are directly sent to the correspondent node instead of the home network.

This use of the Binding Update has the following additional advantages:

- Reduces network utilization, since fewer links are traversed. This could mean much less congestion in the routing fabric of the Internet.
- Reduces the load on the home agent, and the potential for complete isolation of the mobile node due to loss of the home agent.
- Reduced round-trip times for application endpoints
- Better predictability of routing paths, which could be crucial for QoS management.

- More symmetric delay characteristics for incoming and outgoing traffic between mobile node and correspondent node.

Work continues on achieving the necessary lightweight key exchange protocol.

Ingress Filtering

On today's Internet, there are many security concerns. Packets received from arbitrary correspondent nodes should not be trusted without some further assurance about the origin and integrity of the data within the packets. This applies equally strongly to the fields in the IP headers of packets; IP source addresses can often be easily forged by mischievous or malicious users to give misleading information to applications running on the computer receiving the forged IP header. Since IP addresses typically identify end nodes, as well as containing routing information about the end nodes, a forged header could allow masquerading nodes to cause a wide range of damaging effects. Ingress filtering is one of the techniques that has been developed to reduce the number of opportunities for mischief and malicious damage. While engineers debate its effectiveness, no one debates the fact that ingress filtering is being deployed at many sites in today's Internet.

Stated simply, an ingress filtering border router prevents packets from entering the global Internet if the source IP address does not match any of the expected routing prefixes for the network domain originating the packet. This allows the network administrator for the domain to inhibit malicious nodes from utilizing that domain as a source of masquerading attacks against network nodes in other domains. Note that ingress filtering does not offer any additional protection to the nodes in the administrator's domain; it may offer some legal protection, if it can be proved that masquerading attacks could not have emanated from the administrator's domain.

Ingress filtering is problematic for Mobile IP, because the whole point of the protocol is to enable communications with correspondent nodes to proceed undisturbed even as the mobile node acquires new care-of addresses. Since applications typically use the Source IP address to identify the mobile node, and the mobile node uses its home address to identify itself for persistent connections, the easiest procedure would be for the mobile node to insert its home address as the Source IP address of its transmitted packets. Unfortunately, Ingress Filtering causes this not to work.

In order to avoid this problem without introducing a need for further negotiation, Mobile IPv6 specifies the Home Address destination option. This new option contains the home address, to be supplied to the application at the correspondent node, while at the same time allowing the care-of address to be used as the topologically correct Source IP address for all transmitted packets. This is perfect, because the care-of address is, by definition, an address that matches the admissibility criterion at the ingress filtering border router.

The Home Address destination option introduces additional complexity at the network layer for several reasons. The recipient of a packet containing this option has to supply the home address to the higher-level protocol layers (e.g., the transport layer) instead of the address in the Source IP field of the IPv6 header. This necessitates a change in the traditional way of programming IP-level processing. More importantly, it also necessitates changes in the way that security data is verified. For instance, IPsec headers must be processed "as if" the home address were the address originally contained in the Source IP address field of the IPv6 header, instead of the care-of address, which was typically in that header field.

Since it is important for all nodes to supply the correct IP address to higher-level protocols, the Mobile IPv6 specification mandates that the Home Address destination has to be implementated by all IPv6 nodes. This unequivocal mandate will have to be seriously studied by all IPv6-related IETF working groups when the Mobile IPv6 specification is promoted to Proposed Standard.

Home Agent Discovery

If a mobile node is away from home, and its home agent crashes or discontinues operation, the movement of the mobile node would cause it to become unavailable to all those Internet nodes except the ones that receive explicit signals from the mobile node. Thus, in order to remain available for answering incoming application connectivity, the mobile node should have a way to discover another home agent on its home network, which could then take over the duties for the mobile node's previous home agent.

There is a reserved range of well-known anycast addresses on every IPv6 network. This enables more useful implementation for IPv6 features compared to IPv4, and is another example of how IPv6 makes it easier to design the features needed for use with mobile networking. Members of an "anycast group" agree to carry out (by whatever means) the following semantics:

- Each packet sent to an anycast address is to be processed as it were received by exactly one of the members of the anycast group.
- Packets (currently) may not be transmitted with an anycast address as the Source IP address of the IPv6 header.

To enable Home Agent Discovery, a well-known anycast address has been allocated for use by all the home agents on any network. The availability of the well-known "All Home Agent's" anycast address makes the procedure very straightforward.

Mobile IPv6 defines two new ICMP messages for the purpose. The Home Agent Address Discovery Request message is sent by the mobile node to the All Home Agent's anycast address. One of the home agents on the home network receives this message, and responds to it with an ICMP Home Agent Address Discovery Reply message. The remaining details about the operation of this discovery mechanism are involved with how the home agents maintain lists of all the available home agents on the home network. In this way, any home agent in the anycast group can supply the greatest amount of useful information to the mobile node. Constructing the list is simple and does not require an additional protocol, because home agents are already required to advertise their presence on the home network.

Renumbering

IPv6 renumbering is intended to simplify the process by which a network domain can change one or more of its network prefixes. This is a huge problem in today's Internet; renumbering can take many days to accomplish, and the results are usually quite chaotic as unpredictable bugs turn up long after the scheduled end of the renumbering process. Consequently, renumbering rarely occurs. This has the further consequence that the routing tables for the Internet become more and more difficult to aggregate properly. It would be much better if the addressing and aggregation properties of the local administrative domain could follow the business processes of acquisition and merging, as well as retargeting for new Internet Access Providers, without the fear of disruption. The processes of address autoconfiguration in IPv6 (e.g., RFC 2461 as explained earlier) are designed to expedite this renumbering task. When a router does not renew the lifetime of a previous routing prefix, and it begins to advertise a new routing prefix, the nodes on the network take that as a signal that they should reconfigure their IPv6 addresses so that they will properly match the newly advertised IPv6 routing prefix.

If the home network is renumbered while the mobile node is attached to some other network, all mobile nodes home addresses on that network may become invalid. This is important information that has to be disseminated in some efficient manner. The exact algorithms are fairly complicated, because of the need to avoid sudden bursts of control information at those times when renumbering events are initiated. The general idea is to allow the home agent to supply to the mobile node all the information which, if it were on the home network, it would receive in local Router Advertisements. The home agent does this by sending a ICMP Mobile Prefix Advertisement message to the mobile node. If the mobile node detects that the lifetime of its home address is almost due to expire, it can trigger the transmission of the advertisement by sending a ICMP Mobile Prefix Solicitation message to the home agent.

Recent Directions

The basic Mobile IPv6 specification, while incomplete, has features relevant and needed in a very broad realm of applicability. But it is known that there are quite a few situations where additional features are needed that cannot be easily included as part of the base specification. These new areas include:

- Interactions with QoS
- Interactions with AAA (espcially third-part authorization)
- Firewall management
- Localized Mobility Management
- Seamless Mobility
- Context Transfer
- Paging
- Mobile Routers

In this section, brief sketches of these actitives will be presented; for further details, please consult the mailing list archives for the [mobile-IP] and [seamoby] working groups. Many aspects of these new features are geared in some way to improve system performance for handovers. There are several general ways to classify handovers, and it is important to consider specific solutions in light of each of these different classification schemes:

Intra-domain vs. Inter-domain

Handover signaling can be restricted to occur only between network entities in the same administrative domain, or alternatively it might be allowed between entities in independent administrative domains. In the former case, security needs can much more easily be met, and the manual configuration of security associations between network elements is feasible. Indeed, it might even be possible to rely on aspects of the physical security of the domain to simplify or eliminate the need for security altogether. Between domains, no such simplifications are likely, and manual configuration is decidedly unscalable.

Mobile-controlled vs. Network-controlled

Deciding when to change points of attachment from one access router to another, and which new access router to use for connectivity, are important parts of the handover process. These decisions can either be taken by the mobile node itself ("mobile-controlled"), or by network elements within the access network on behalf of the mobile node ("network controlled"). Usually, the decision occurs based in information provided by either it or the network, so (to give one example) we might describe a particular handover protocol as being "mobile-controlled, network-assisted" if the mobile node makes the decision based in information provided to the mobile node for that purpose by a network element within the access network.

Predictive vs. Reactive

The handover can be attempted in anticipation of certain network events (e.g., loss of signal strength or excessive load). This is called a "predictive handover", and it always involves an attempt to identify, with a high degree of certainty, the identity of the new access router. Alternatively, any necessary handover operations can be attempted after the mobile node establishes a physical link at a new access router.

In any discussion about the various handover schemes, it is usually very instructive to compare the features of each scheme according to the mentioned classifications. If it is possible to have a single scheme work equally well according to all eight potential sub-classifications defined by the above three major classifications, system complexity in support of handovers could be reduced in very important ways. Otherwise, the mobile node and network elements are likely to be deployed with similar but different solutions to nearly alike problems, adding to development time, expense, maintenance problems, and increasing the likelihood for errors.

Seamless Mobility

Of all the recent technology areas listed above, Seamless Mobility is perhaps the most crucial. There are two fundamental aspects of importance. A handover should be fast, meaning that a mobile node should experience very little delay between the time that a link becomes unavailable at one access router, and the time at which a link becomes established at a new access router. The first objective for fast handovers is to meet the delay criterion, which will enable handovers for voice connections, i.e., Voice over IP (VoIP). The second objective for seamless mobility is to make the handovers smooth. This has been interpreted to mean that the number of dropped packets should be zero, or if that is not possible, absolutely minimized. Seamless mobility is defined to require handovers that are both fast and smooth. Note that the two ideas are not independent, because when a handover is very, very fast there is less opportunity for packets to be lost or damaged. Nevertheless it has proved beneficial to consider them separately.

The current discussions about fast handovers for Mobile IPv6 center around the Internet Draft produced by a design team in early 2001. The design team document focuses only on predictive handovers since reactive fast handovers are expected to be handled reasonably well by features already existing in the Mobile IPv6 base specification.

Voice over IP with SIP

It is assumed that in the future almost all fixed and mobile telecommunications networks will be based on Internet technology. The main benefits of running voice over the Internet Protocol (IP) are reduced costs and new services. In the beginning of the Voice over IP (VoIP) hype the emphasis was mostly on cost savings. While this is still considered to be true especially in the long run, the focus has recently been shifted to new services. After all, services are what the customers and the network providers are interested in. Especially, services combining several communication types and modes will lead the way in future networks. Voice itself will be just one, although important, piece in the whole communication architecture.

Session Initiation Protocol (SIP) is an emerging Internet Engineering Task Force (IETF) standard for setting up multimedia sessions in the Internet. Its basic capabilities are setup, modification and teardown of any communication sessions, so it is a signaling protocol. SIP also provides user mobility, meaning that a user is reachable via a single address regardless of his current point of attachment to the network. SIP suits combined services, as it borrows many features from HTTP and SMTP, which are currently widely used on the Internet for web browsing and e-mail. SIP is also modular and extensible, so it can be used for other things than basic session establishment, as is discussed in the Presence and Instant Messaging with SIP chapter.

Besides session setup signaling, VoIP requires several other protocols and mechanisms. The actual voice is carried by the Real Time Transport Protocol (RTP), which can be used to carry video and other real-time media. In order to transport the RTP stream over the network the Quality of Service (QoS) has to be guaranteed in some manner. Techniques such as Differentiated Services (DiffServ) and Multiprotocol Label Switching (MPLS) can be used for this purpose.

This section presents an overview of the protocols used for voice and multimedia communications over the Internet. IETF's Internet multimedia architecture is then introduced and the role of each protocol is briefly described. Finally, a summary provides some conlusions.

Internet Multimedia Architecture

The IETF has worked on the Internet Multimedia Architecture already for several years. The overall idea is that there should be a specific tool for each task and that a large system can be built by combining relatively simple building blocks. Layering is another guiding principle, as with all Internet protocol designs.

A following figure shows a simplified overview of the architecture as a layered protocol stack. At the bottom is the IP layer which runs over every network techology and provides the basic connectionless packet delivery service for anything above it. This is the famous "IP over everything and everything over IP" paradigm of the IETF. In order to support multimedia services, seamless mobility and conferencing the IP layer needs to be enhanced compared to this simplistic view. Mobile IP allows terminals to move freely between different wireless networks. Differentiated Services brings different levels of service to the network, which allow for example voice packets to be prioritized over e-mail traffic. IP Multicasting is an interesting technology that allows a sender to deliver a packet simultaneously to multiple receivers easing the scalability of large conferences or media streaming. Even security can be implemented on the IP layer by applying IP Security (IPSec), which provides confidentiality and integrity protection for all traffic.

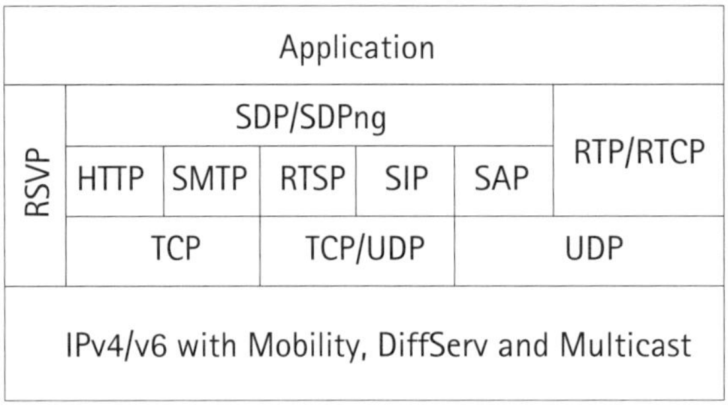

Internet Multimedia Architecture

Above the IP layer reside the transport protocols, which operate end-to-end between hosts or terminals. The Transmission Control Protocol (TCP) offers connection oriented reliable delivery with congestion control, while the User Datagram Protocol (UDP) provides a simple connectionless datagram service. Recently, a new protocol, Stream Control Transport Protocol (SCTP) has been designed to enhance TCP functionality. The Resource Reservation Setup Protocol (RSVP) can be used to signal for QoS reservations on the IP layer through all the routers on the traffic path. RSVP runs directly on top of IP, and can be seen as a part of the IP layer functionality. RSVP suffers from scalability problems as it operates on a per packet flow level and its use for end-to-end reservations over the Internet is quite limited. At least in the core network, DiffServ and MPLS are enough to achieve guaranteed QoS.

The actual application layer protocols are situated on top of the transport and utilize it based on their needs. Voice and other multimedia content such as video or animation are transported by the Real-Time Transport Protocol (RTP), which runs on top of UDP's connectionless service. RTP offers synchronization of the media streams it carries by including a sequence number and timestamp headers. With voice data the actual speech frame is inserted as payload for RTP and sent to the recipient. A specific payload format has been defined for each codec to be used within RTP. RTP is accompanied by the Real-Time Transport Control Protocol (RTCP) which provides e.g. statistics about each sender and recipient in an RTP session. An RTP session can be either unicast (point-to-point) or multicast. With multicast, the IP layer takes care of delivering the packets to each recipient so RTP does not need to take care of it.

It is important to understand that in IP networks signaling and media do not always follow the same path. Media, such as voice, is sent over RTP directly between the communicating parties, while signaling often goes via servers. For media QoS is especially important, signaling does not usually have strict real-time requirements.

A group of signaling protocols exist for various purposes. The figure shows the transport that each of them utilizes, in some cases there is more than one option. These protocols have the same look and feel:

- Common addressing, user@domain.
- Similar text-based headers and encoding according to the RFC 822 message format originally developed for SMTP Internet mail. Protocols parser implementations can utilize the same code base.
- Same request-response model and response codes.
- Utilization of Multipurpose Internet Mail Extension (MIME) for flexible payload formats.
- URIs and URLs for addressing.
- DNS for address mapping and inter-domain request routing.
- Common security and authentication frameworks mechanisms can be used.

This similarity allows for easy combining of these protocols to do services, which need to use more than one of them. For example, an SIP message can carry a Hypertext Transfer Protocol (HTTP) Universal Resource Locator (URL) or even a Hypertext Markup Language (HTML) document or it is easy to send e-mail from the network server if a callee was not reached to inform him about the missed call by using the common addressing format.

SIP is used to establish, modify and terminate sessions. It provides personal mobility by allowing a user to dynamically register to the network with his communication address (SIP URL). A session is usually a number of RTP streams to be exchanged. In a basic VoIP scenario a session is simply a combination two unidirectional RTP streams between the calling and called party. SIP and session establishment is covered in more detail in the next part.

Session Announcement Protocol (SAP) is used to advertise multicast sessions to larger audiences. It has been used for a long time in Mbone, the experimental multicast overlay network. The Real-Time Streaming Protocol (RTSP) provides the possibilty to control streamed media by sending commands such as PLAY, STOP and PAUSE to the server where the media is located. In a VoIP environment RTSP can be used to control a voicemail server in order to record or listen to messages.

HTTP and SMTP are used for web browsing and e-mail transport, respectively. As they are quite generic and flexible, especially HTTP is being used for various other purposes, even as a transport for remote procedure calls.

In order to describe the parameters and characteristics of the media sessions, each of the signaling protocols uses a common description language, SDP (Session Description Protocol). SDP is carried e.g. in SIP message payload or within an e-mail. SDP describes the codecs and addresses to be used to send audio and video. Recently, an effort to redesign SDP to include even more capabilities has begun in the IETF. It is referred to as SDPng (SDP next generation).

An application software can use all of these protocols to provide a rich, comprehensive and understandable service for a user. The application uses SIP for establishing voice and video sessions, RTSP for streaming video or voicemail announcements, SMTP for fetching and sending e-mail and HTTP for browsing web content. It is up to the application designer, both on the terminal and server side, to best combine these protocols and to design an interface for the user. Certainly, the possibilities are much greater than with traditional telephony or even with second generation wireless systems.

An example of a simple user interface combining voice, messaging, web and streaming is shown in a following figure.

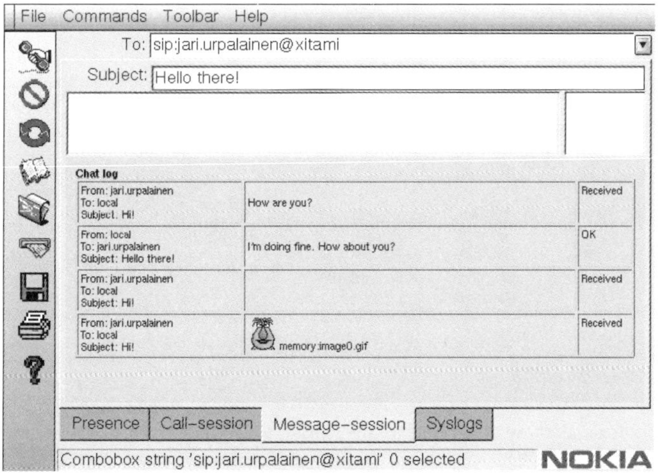

An example of a "Rich Call" user interface.

Session Initiation Protocol

As explained in the previous chapters, SIP is used to setup, modify and tear down multimedia sessions. It is agnostic to the type of session to be established, and actually uses SDP to describe the characteristics of a session. Usually the session is a combination of speech, audio and video streams, but it can also contain shared applications such as a whiteboard or text messages. Even network gaming sessions could be established with SIP as long as all the applications understand the required parameters for the game.

A basic "SIP network" is composed of four types of elements: User Agents (UA), Proxy servers, Redirect servers and Registrar servers. User Agents typically reside in end-points such as IP phones, personal computers or wireless handsets. They initiate requests and provide responses. Usually UAs also have an interface to media handling and to the actual application software providing the user interface. For example, VoIP calls are carried out between UAs.

Proxy servers are intermediatories which receive and forward requests providing them with e.g. routing or other services. Redirect servers simply respond to a request by asking its originator to redirect it to a new address. Registrar servers keep track on the actual points of contact of the users by accepting registrations from the UAs. Registrar servers and the SIP registration procedure in general provide user mobility as the user is able to be reachable from any location via a single address. In this sense, they resemble HLR (Home Location Register) functionality in GSM networks. Each user is part of a domain and each domain runs at least one registrar server which knows the location of its users (if they are registered). In some sense, SIP registrations also cover Mobile IP functionality, but it must be understood that SIP is not capable of providing real-time handovers and not all communication is based on SIP sessions.

SIP uses an address format common to Internet mail, i.e. user@domain. An example is sip:john.doe@operator.com. The domain part is used to find the correct domain for the user: DNS can be queried for the address via which the SIP registrar within a domain is reachable. This is analogous to e-mail routing, which has been up and running for years already. The user part is to distinguish between individual users within a domain. In addition, traditional E.164 telephone numbers can be used within SIP as tel: URLs. An example is tel:+3581234567. A mechanism called ENUM can be used to map the E.164 number to a corresponding SIP URL. This can be done by a UA or a proxy.

A sample SIP network is shown in a following figure. It consists of User Agents (in this case mobile terminals) and a number of domains operating SIP proxy, redirect and registrar servers. In addition, there can be servers dedicated to certain applications, such as presence. In order to connect to the traditional telephony network a gateway is required.

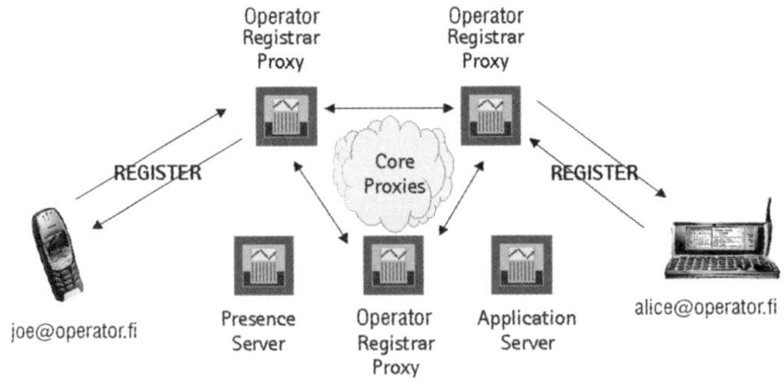

A sample SIP network with three domains.

Basic SIP includes six requests or methods. These are INVITE, ACK, OPTIONS, CANCEL, BYE and REGISTER. Each request can be responded with a number of response codes in a similar manner as in HTTP or SMTP. The basic response classes are: 100 - Informational, 200 - OK, 300 - Redirection, 400 - Request failure,

500 - Server failure and 600 - Global failure. Each class has a number of specific instances, such as 180 Ringing. An example of an INVITE request is presented in the Presence and Instant Messaging with SIP chapter. Responses look very similar to requests.

The most important headers are "Request-URI": which defines where the request is to be sent next, "To": which contains the recipient (callee) address and "From" which contains the sender (caller) address. A SIP message may contain any payload conforming to MIME specification.

INVITE is used to setup and modify sessions. It is a special request that is followed by an ACK after receiving the final response, while other request transactions end at the first final response. An example of basic session establishment is shown below.

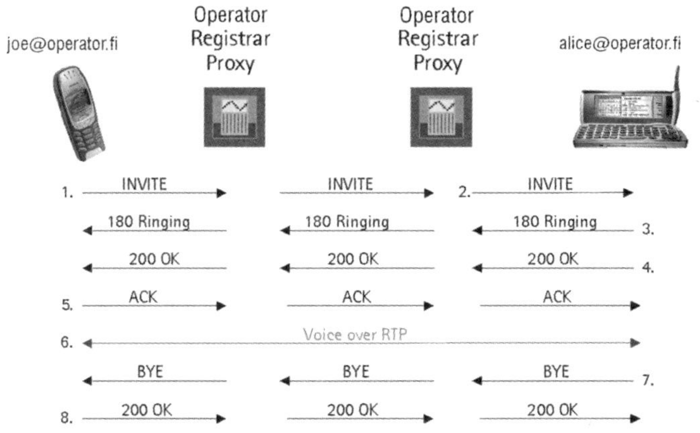

Basic SIP session establishment and termination.

1. Joe sends an INVITE to Alice using his outbound SIP proxy server. Within INVITE an SDP payload describes that Joe wants to establish a two-way voice call with an AMR codec.
2. SIP proxies route the request to an address where Alice is currently registered. The second proxy is co-located with Alice's registrar server.
3. Alice is willing to accept the call and her terminal supports AMR, so the phone starts to ring and 180 Ringing is sent to Joe.
4. When Alice picks up the phone a 200 OK response with the appropriate SDP is sent to Joe.
5. Joe responds with ACK.
6. The RTP sessions carrying voice coded with AMR are established. The packets flow directly between the two terminals.
7. Alice hangs up and sends BYE to Bob to terminate the session.
8. Bob replies with 200 OK.

Note that the RTP streams do not follow the same path as the SIP messages did, but flow directly between the terminals. It is possible to send the subsequent SIP requests directly between UAs. On the other hand, proxy servers in the middle may ask to remain on the signaling path for the duration of the call. This might be useful if the proxy offers some services to the call. The actual services can be implemented in various manners, such as with Java servlets or using the Call Processing Language (CPL).

It is possible to impose so called Rich Call services in the call setup. SIP can carry small images, or icons in the INVITE payload and these can be shown to the callee. It is also possible to send the callee an URL which points to further information, such as the caller's web page. The same can be applied in the reverse direction as well. For example, a shop can respond with a web page to calls that come after opening hours to provide caller's with some information about their products and services.

OPTIONS is used to query the capabilities of a UA. CANCEL can be sent to cancel a pending request. BYE is used to terminate an on-going session. REGISTER is sent by a subscriber to his registrar server to update his contact information.

SIP supports both parallel and sequential forking. A user can be registered from several contact points at the same time. In this case, the proxy service logic decides what to do with an incoming request. In case of parallel forking the proxy forks the request simultaneously to each destination.

SIP can support the same basic and digest authentication mechanisms as HTTP. Proposals exist to embed EAP (Extensible Authentication Protocol) in SIP to offer a flexible way to introduce new authentication mechanisms and to use backend protocols, such as Diameter. Any SIP request can be authenticated.

It is possible to extend SIP in several ways. New headers and methods can be invented. The basic principle is that such additions should only require support in UAs, so proxies can remain untouched. A number of SIP extensions have been defined including REFER method for call transfers and other refererrals, SIP Events framework for subscriptions and notification of events using SIP and QoS interworking to allow QoS resource reservation within a call setup (this is very important as in the simple example presented above the phone starts to ring even if the resources are not availble). Finally, even presence and instant messaging extensions for SIP are under definition in IETF, as explained in more detail in Presence and Instant Messaging with SIP.

Conclusions

The Internet provides an excellent platform to integrate all communication services. IETF has defined a multimedia architecture, which allows an easy combination of different services including real-time/streaming voice, audio/video, text/multimedia messaging, shared applications, web and e-mail. There are plenty of protocols to carry out the tasks of establishing and negotiating the required media sessions as well as transporting the actual media.

SIP is one of the key protocols in this future communication puzzle, as it provides the means to establish generic sessions, and it is already being extended to include various other capabilities. SIP also works well with the other Internet protocols.

References

Session Initiation Protocol, RFC 2543
Real Time Transport Protocol, RFC 1889
Session Announcement Protocol, RFC 2974
Session Description Protocol, RFC 2327
Real Time Streaming Protocol, RFC 2326
Resource Reservation Setup Protocol, RFC 2205
E.164 Numbers and DNS, RFC 2916
3GPP Release 5 IP Multimedia Subsystem, 3rd Generation Partnership Project; Technical Specification Group Services and System Aspects; IP Multimedia (IM) Subsystem - Stage 2; (3G TS 23.228 version 5.0.0)

Quality of Service

The origin of IP is in computer communication. The basic idea of the Internet was to connect existing computer networks rather than to build a new one. The result was a best effort forwarding service for packets, rather than a connection-oriented service with well defined quantity and quality characteristics. There were three main reasons behind this focal selection. First, the networking environment was so diverse that it was not reasonable to assume that all networks could meet any complicated quality or quantity requirements. Second, the very nature of the service was a co-operation between organisations rather than a commercial service for customers with readiness to pay for specific services. Third, the main need for communication was data communication between computers without any strict quality or quantity requirements.

Now, 30 years after the original considerations, it is evident that the decisions related to the fundamental characteristics of the Internet protocol were very successful. Further, it is amazing how well the simple principles of IP have met the changing needs of communication services – or at least no other networking technology has been more successful. The diversity of networking technologies is even more prominent nowadays due to the increased mobility and new wireless access technologies. In contrast, the co-operative service model is not generally valid anymore, and various new applications with different requirements try to exploit the transmission capacity of IP networks. In particular, from the traditional telecommunication perspective it is difficult to believe and accept the fact that best effort technology can be used as a basis for serious commercial services. Moreover, the adaptivity of many applications to varying conditions has proved to be better than what the designers of traditional telecommunication services have assumed.

Still, there is a pressure to change the characteristics of IP to the direction of a more controlled service, both because of the changes in service models and application requirements. In addition, wireless access appears to bring about the need to control the sharing of resources more thoroughly than in fixed networks. Integrated Services, Differentiated Services, and Multiprotocol Label Switching (MPLS) have tried to provide tools that make IP technology more suitable for different business models and for different applications. So far, all these efforts have had only a limited effect despite the vast specification work. One reason for the lack of success is that any significant addition to the best effort model endangers the fluency of traffic between different network domains. This is a very serious risk if we remember that interoperation is the origin of the IP technology and that the main strength of the Internet is truly global coverage. Therefore, we must be very careful when designing new features even if some needs, such as improved quality, seem to be of great importance.

This chapter addresses the question of how the network operator should approach the Quality of Service (QoS) issue. The starting point here is the set of profound changes in the use of IP technology itself. The roles of network operators, service providers and end-users are considerably different than 20 years ago. Furthermore, these changes are and should be reflected to the level of applications, flows, and packets in order to be really effective. The vital question is what will the relationship be between, on the one hand, the changing roles of service providers and end-users, and on the other hand, the treatment of flows and packets inside the network.

We may begin considering this issue by defining the basic objectives for a QoS system on four levels:

1. Support the mission of the service provider or network operator
2. Serve the principal needs of end-users
3. Meet the quantity and quality requirements of applications
4. Transmit as many packets as possible

The fundamental assumption is that in the final criterion that defines the actions of the network, a hierarchical structure is formed where the first objective is the most important and the last objective is least important. Each service provider may define their mission in whatever terms and all other objectives will be considered in the light of that mission. In communication services, the mission most probably includes an objective to serve the needs of end-users. It should be stressed that when the needs of different end-users are conflicting, we have to look at the upper level in the hierarchy (i.e., the mission of service provider) instead of a lower level (i.e., the applications requirements) in order to solve the conflict.

Correspondingly if, after solving the conflicts between end-users, there still are conflicts between different flows serving different applications, we have to look primarily at the needs of each end-user rather than trying to maximise the use of network resources. In this sense, the hierarchy appears to be imperative and universally applicable. If and only if the service provider explicitly states that the importance ordering of the objectives is different than the one described above, the structure of the hierarchy could be different. That may also mean that one (or two) of the requirements is ignored completely, for instance, because individual end-users cannot be distinguished.

From this perspective, QoS is more about the mission of the service provider, that is, typically business, than the fulfilment of the special requirements of applications.

In order to meet these objectives the network operator has to proceed in phases:

1. Definition of the fundamental mission of the network operation as clearly as possible
2. Design of an appropriate service model including billing (if adopted)
3. Building of an appropriate network with enough capacity and a suitable management system
4. Use of suitable mechanisms for differentiation between the packets, flows and end-users of the network operation
5. Offering additional information to end-users for adapting to varying network conditions

These phases also form a hierarchical structure. Without an explicit mission statement, all other phases of the project are vulnerable. Without a service model, it is very difficult to determine the needs for forwarding capacity and a management system. Most importantly, no mechanism can rescue the service provider if the network does not have enough capacity. Differentiation mechanisms are reasonable only within an appropriate service model, and serve the fundamental objectives set for the service model, but their capability to improve the situation in the case of insufficient capacity is limited. Finally, the service provider may decide to offer additional information about network conditions for end-users. However, this information is meaningful only if all preceding phases have been successfully implemented and any mechanisms related to the information do not disrupt the function of the overall service model.

In general, it is extremely important to be careful when adding new features. A widely adopted approach in IP is that the service model is based purely on the best effort principle and on sufficient capacity inside the network to transmit almost all packets. In this case, the last two phases can mostly be ignored. This alternative will always be seriously considered when more complicated systems are developed.

Let us assume that a network operator wants to design a QoS system by using the results of the previous discussion. First, the operator defines its mission to be the maximisation of revenue. The service model consists of a high quality voice service over an IP network with time-dependent billing, e.g., 0.1 €/min, and web browsing and e-mail services for residential customers with two flat rates, e.g., 10 €/month and 30 €/month (phase 2). This implies that the service provider needs to dimension the network according to the expected number of simultaneous voice users plus some extra bandwidth for data services (phase 3). Further, the operator decides not to provide any specific information about the state of network (phase 5).

Consequently, the main issue to be addressed is the type of service differentiation rules and mechanisms to be implemented in order to realise the service model and to fulfil the mission (phase 4). In order to ensure that the customers are kept satisfied with the voice service, and that there is an appropriate incentive for customers to pay extra for improved data services, the following two rules seem practical:

1. To give preference for voice flows over data flows, but only on the condition that the amount of resources used by each voice connection is strictly limited.
2. To give preference for those customers that have paid more when dividing resources among data connections.

The exact way to accomplish these tasks by means of technical mechanisms is an important issue but yet secondary when compared to the service provider's mission and overall service model. Before going into the technical details of the implementation, we have to identify the most probable missions and service models in order to have better knowledge about the requirements set for the differentiation mechanisms.

For simplicity let us assume that the business part of a service provider's mission is maximisation of revenue. It should be noted that the target of maximisation can be something other than money.

The main characteristics of a typical service provision system are

1. Each customer has a (typically business) value for the service provider. The value depends on the price paid by the customer. In addition to a flat rate, the price may include a time-dependent or volume-based component related to some specific service elements.
2. Each customer "owns" a number of flows, each of them consuming a certain amount of network resources.
3. The total value of the customer is divided, according to some principle, among all the flows or packets owned by the customer.
4. The value of each flow or packet defines how the network shall treat the flows or packets. Particularly, every flow or packet with high value shall be favoured over other flows or packets with lower value.

If these principles are implemented in an appropriate way, the most essential part of a service providers' objectives are likely to be satisfied. In order to accomplish this, there must be a clear link between customer value and flow handling and/or packet marking in a way that all network elements are able to make reasonable decisions. Thus, the most important characteristics of differentiation mechanisms are:

1. Necessary information about customer relationship, most notably billing and on-going traffic process shall be gathered, and further processed in a way that the resultant data can be used by QoS mechanisms. Particularly, the data must have clear indication of the business value or importance of all flows and packets.
2. Means to transmit the gathered and processed information through the network to all network elements making QoS decisions. Flow-related information, this means signalling messages, and for packet-related information, specific bits are needed in the packet header.
3. Network elements must have tools to make efficient discarding decisions. The business value of a flow defines its relative value when the network elements make a decision to either accept or discard entire flows. Correspondingly, the value of each packet determines whether or not it should be discarded during a congestion situation.

These are the main needs that IP technology encounters in a changing business environment and a changing relationship between end-users and network operators. The other part of the change pertains to applications. Certain applications, most notably voice, have requirements that cannot always be fulfilled in a best effort network with a high utilisation factor. The problems are related to transmission delay and packet loss in the service. To understand the relationship between these characteristics and the best effort principle, we have to consider the protocol that is used to control the traffic load in IP networks, that is, TCP (Transmission Control Protocol). What TCP essentially does, is to rapidly decrease the bit rate when it notices that one or several packets are lost, and gradually increases the rate when no packet loss is observed. Thus, the load level on each link does not permanently exceed the capacity of the link and an excessive amount of re-sent packets do not overload the network. However, on a bottleneck link several percent of the packets can be lost even during normal operating conditions. In addition, viable functioning of the system requires that the network nodes are able to buffer a sufficient number of packets, which may result in a large delay variation. Consequently, both delay and packet loss characteristics of a best effort IP network are not good enough for some critical applications.

Despite this apparent defect in IP, we have to be careful with the practical consequences. First, even though the functioning of applications is an important issue, dividing of resources is primarily a business issue. Second, almost any application can be designed to adapt to changing capacity without requiring a strictly defined bit rate. Third, in many situations the quality of service can be kept high enough even for the most demanding applications, particularly when the capacity of the link is very high. Fourth, any differentiation mechanism tends to consume resources that could be used to increase the packet forwarding capacity in a pure best effort network.

In conclusion, a special delay requirement of a packet may be of importance inside a network, whereas bit rate and packet loss requirements related to individual flows or packets are usually of minor consequence in IP networks. Note, that the business importance of a packet does not have a direct relationship with the quality requirements of applications. The fact that certain applications, like video applications, have strict packet loss requirements does not directly imply business importance. It is imperative to make a clear distinction between the business importance (of end-users, flows and packets) and quality requirements of applications.

Regarding the needs of applications, the main inference is that in addition to (business) importance, each packet may contain information about its delay requirement.

We have defined the basic approach used to satisfy the essential needs of the business and of applications. Methods exist to use the network resources even more efficiently. First, the service provider can offer additional tools for the end-user to inform the network of their own preferences. The user of a mobile terminal may, for instance, wish that the transmission of an email should gain preference over the update of antivirus files. Note that this favouring can be effective only within the capacity available for that particular end-user, and does not usually have any significant effect on how resources are divided among other end-users (though this issue ultimately depends on the mission statement of the service provider). Another possibility is to give end-users or applications additional information about the probable capacity available for a new or existing flow. That information could also be conditional in the sense that the available capacity is a function of the price and/or special quality requirements.

However, according to the fundamental principles presented earlier in this chapter, this kind of information has to be considered as extra information either for the end-user or for the network without any significant effect on the service model. In addition, it seems that without a certain level of differentiation already available inside the network, it is not possible to efficiently use such methods. It does not make sense to support end-user preferences without controlling the sharing of resources between different end-users. Also, it is somewhat questionable to offer information about the state of the network to end-users unless the network can give incentive for the end-users to adapt to the situation, because an incentive usually implies differentiation between adaptive and non-adaptive flows.

These considerations are basically well known, and moreover, there are various mechanisms to accomplish the tasks. What has been missing is a harmonious system covering all the necessary parts but excluding all the unnecessary ingredients. The discrimination of necessary and unnecessary is the key task, and a particularly hard one within the framework of the Mobile Internet. Mobile Internet will consist of several different networking technologies, network environments, and services. In some parts of the Mobile Internet the Quality of Service mechanisms and principles may be based on an approach that is not suitable for other parts of the network. For example, radio technologies often support a natural flow or connection-oriented QoS while fixed Internet has been designed for packet-based approaches. This implies that the only reasonable way to provide real end-to-end QoS must be based on a common measure, and fulfilling operators' business goals is exactly this measure.

The necessary traffic control actions can be implemented in various ways, including the specifications made by IETF, like Differentiated Services, RSVP (Resource Reservation Protocol) and MPLS (Multiprotocol Label Switching). The key point is that the fundamental objectives of the system have to be preserved independently of the technological choices.

Let us finally outline the main characteristics of service provisioning in a network environment with three main layers: the optical core, the metropolitan region, and radio access. Using the example in this chapter the main technical characteristics can be as follows:

1. Basic principles

   The flat rate paid by a customer defines share of the resources available for that customer. The customer can increase the share by temporarily paying more, for instance 0.1 €/min to guarantee that sufficient capacity is available for voice connections.

   In addition, flows may request a real-time service. This can be done automatically for all voice connections paid by the customers.

2. Wireless links and the radio access network

   The incoming traffic of every customer is measured In every access node. The measurement result and the share paid by the customer are compared and the outcome defines the importance of the incoming packet - the higher the share or lower the bit rate, the higher the importance of the packet will be.

   The quality and reliability of voice connections over the radio link are guaranteed by means of appropriate resource reservation. On the other hand, the data connections share the remaining radio capacity in a dynamic manner depending on the packet importance marking.

3. The Metropolitan area with medium-size links and routers

   There are no resource reservations for individual flows.

   During a congestion situation the packets with lowest importance are discarded first. In addition, among the accepted packets, the packets that have required real-time services are transmitted before other packets. If and when voice packets obtain the highest importance marking, they are protected from the negative effects of data connections.

4. The Core network based on optical transmission and high-capacity routers

   The Optical core network can be based on pure best-effort services with an appropriate traffic measurement and management system that guarantees the sufficiency of network resources to transmit virtually all packets.

This kind of harmonious QoS system efficiently serves the service provider even though the technical approaches in different parts of the network are dissimilar. It should be stressed that in this framework QoS is more a business than technological issue. Every technological choice serves the mission of the service provider, every technological choice works consistently with all other choices, and the total outcome will be cost-efficient. Thus, we do not need one universal QoS technology; instead we need a consistent and comprehensible QoS framework that is able to combine business objectives and differentiation mechanisms in an efficient way.

Linux

Linux® is an operating system that is being widely used in the Web domain for different functions. It is known to be a robust, versatile and efficient system that can be used for running a wide range of different services. These include Web servers, e-mail servers, and other Internet services. Linux is also widely used in client side applications, especially in research and academic uses and in organisations that are cost conscious; Linux is free of charge.

Linux is not only free; it is also Open Source. This means that the operating system source code is available to anyone to view and modify it according to his or her needs. This is clearly an asset to the developer community as there are no limitations to optimisation or modifications. A variety of are tools available for this and it is easy to develop new tools if needed.

The Linux operating system can be used in a wide variety of system categories. There are, at least, three categories where Linux has a role: servers, embedded systems, and desktop terminals. In the server category, which is clearly the focal point in the Web domain, Linux has a significant role. There are plenty of products that can use Linux as their operating system platform such as the Apache™ Web server. Currently, Linux is being used in 27% of new server installations according to IDC and this figure is estimated to grow in the future [1].

Linux is also gaining popularity in embedded systems. An embedded system is a device category that requires some tailoring in the actual operating systems distribution. As an example, Nokia Media Terminal™ is a product that falls into the embedded system category. It is an excellent example of an open platform product designed for viewing and recording Digital television. It uses Linux as a platform, including kernel functions (device drivers, memory management, networking and file system support) and some basic user services (Windowing system XFree86 and Mozilla™ browser). Developers are free to modify and add functionalities and applications according to their needs for this open platform. [2]

In the desktop environment, there are several commercial companies that package the Linux with a wide selection of applications and a up-to-date windowing system with installation support. The end user should not be required to be competent with Unix to be able to use and install Linux and applications. These companies are in the business of making appropriate distributions of Linux using the standard kernel and packaging it with suitable add-ons. Users can get the distribution on a CD-ROM at a competitive price or free-of-charge from the Internet.

A number of technical justifications make Linux an excellent choice for servers and the services platform. From a cost point of view, the initial investment in a server system based on Linux is modest compared to some other systems since there is no license fee for the operating system itself. In addition, cost of ownership is also typically less or at same level as with proprietary operating systems. Also, as Linux is not tied to any single company one can choose the service provider onto business principles.

The work to develop Linux was initiated by Linus Torvalds. It is a robust operating system based on something that was considered to be a hobby. It is quite a remarkable effort from a community of individuals – developers – loosely referred to as the Linux Community. The way Linux has been developed and is evolving is something that industry has not seen so far. It is entirely a product of teamwork by volunteers with a passion and enthusiasm to develope the best operating system in the world.

Origins of Linux

Linux originated from the desire of Linus Torvalds to have a variant of Unix for the Intel 80386 microprocessor. A good starting point for this would have been Minix, a Unix variant written by Professor Andrew Tanenbaum for educational purposes. The Minix source code was available, but it was not open in the sense that users could modify it. The source code was copyrighted by Addison-Wesley. There was a port of Minix for the 386, but it was not officially sanctioned by Tanenbaum. He was reluctant to accept additions to Minix because he used it as a classroom example of an operating system.

As Linus Torvalds could not get the additions he wanted to be included in Minix, he started developing his own operating system. Linus and Andrew Tanenbaum had interesting discussion on Internet newsgroups that reveal some of their assumptions at the time [3]:

- It had a monolithic kernel instead of a microkernel like in Chorus™, Mach, or Minix.
- Its design was too close to the Intel® architecture.

Today, most operating systems are (still) monolithic, so history has shown that Linus was right on the first point. The benefits of microkernels have not yet outweighed their performance penalties.

As for the second point, the first version of Linux really was tied to the Intel architecture – after all, it started from the desire to utilise the 386 features to the fullest. Linus argued that writing the platform specific parts of the kernel was no big deal, and he was later able to prove his point by porting Linux to other architectures. This may either prove the superiority of the original Linux design or the programming capability of Linus Torvalds.

Linus Torvalds foresaw the importance of the Intel platform. Since Linux was originally designed for it, it had very good performance.

Other developers were soon attracted to Linux. Unix was a real operating system, but there were no free variants of it available at the time. Intel 386 was becoming very popular, but the operating system it usually came with, Microsoft's MS-DOS™, was rather rudimentary, and definitely not open. Tanenbaum had an operating system with source code available, but he was not eager to accept additional features to it. BSD was troubled by lawsuits. GNU Hurd was far in the future.

One of the first targets for Linux was to be able to use the GNU tools. Many of these had been created by Richard Stallman or had been supported by his Free Software Foundation. Stallman had originally set out to create a free, Unix-like operating system. For this, he needed several programming tools. His first project had been Emacs, a free text editor that had become unexpectedly popular on Unix systems. Stallman followed by creating the GNU C Compiler (gcc), which has been one of the best C compilers on any platform. The Free Software Foundation also sponsored bash, a Unix shell. Although Stallman has created many successful tools, his initial dream, the operating system, is still not complete.

Therefore, Linus Torvalds had the basic tools available – he only needed to port them to Linux and to make sure that Linux was able to run them. Getting more and more tools to run on Linux was an early goal for the operating system project and also directed later work. This differentiated Linux from other free Unix-like operating systems.

Jon Hall orchestrated a Linux port to Digital Alpha. Working for Digital at the time, he invited Linus to speak at the Digital Equipment Corporation User Society [4]. During the meeting Linus explained that he had been thinking about porting Linux to Alpha, but he did not have the equipment available. Jon Hall arranged for a Digital Alpha machine to be sent to the University of Helsinki, and Linus used the opportunity to port Linux to it.

By 2000, Linux has reached a 27% market share of all server operating system shipments [5]. It is second only to Windows that has a 41% share. Linus himself has indicated that there are limits to the scalability of the kernel to larger machines with more CPUs [6]. There are also doubts about how well Linux will fit to the average desktop computer.

Background

The work has continued with several key persons looking after specific parts of the operating system, namely, networking support, process management, file systems and device control. The kernel itself is only visible to the hardware and the Linux distributions cover the rest of the ground, namely, the Graphical User Interface (GUI), windowing system, applications such as Emacs, ftp; packaging the collection of entities in the required way. Naturally, the environment, be it desktop or router, will mandate the use of different applications and protocols.

Furthermore, one can also modify the kernel according to specific needs as Linux is distributed with source code. Thus is possible to have a build that contains only the needed parts of the kernel modules as well as adding completely new or modify ones.

The Linux kernel has been ported to several different microprocessor architectures including PowerPC™, ARM™, and Sparc®. There are no limitations in the kernel design that restrict it to be used only in Intel based processor architectures.

Features

Most of the features (e.g. multitasking, virtual memory and networking) that can be found in modern Unix operating systems can also be found in Linux. It supports kernel threads through the system call clone. Threads in Linux are just like processes, except that all threads in a process share the same memory space. Threads have individual priorities and are scheduled as separate entities.

Monolithic

Linux has a monolithic operating system kernel: all subsystems of the kernel belong to the same memory space and use function calls to communicate with each other. The alternative is a microkernel architecture where message passing and process abstractions are implemented at the lowest level of the kernel. All other parts of the kernel are then in separate memory spaces and use the message passing system.

Hence, monolithic refers to the choice of the basic communication mechanism inside the kernel, and is not intended to imply that the kernel would resemble a bowl of spaghetti. In a monolithic kernel, errant pointers can write garbage to unintended places, but there are many other ways to crash a kernel with badly written code. Developers seldom need to write new code into the kernel.

Modules

Linux also provides kernel modules, a dynamic mechanism to add functionality to the kernel. This is necessary for supporting low-memory configurations or when two different devices would otherwise clash. It is also convenient for device driver development, as a driver can be created as a kernel module and then loaded without recompiling the kernel and restarting the machine. A kernel module can later be statically linked to the kernel to speed startup and execution.

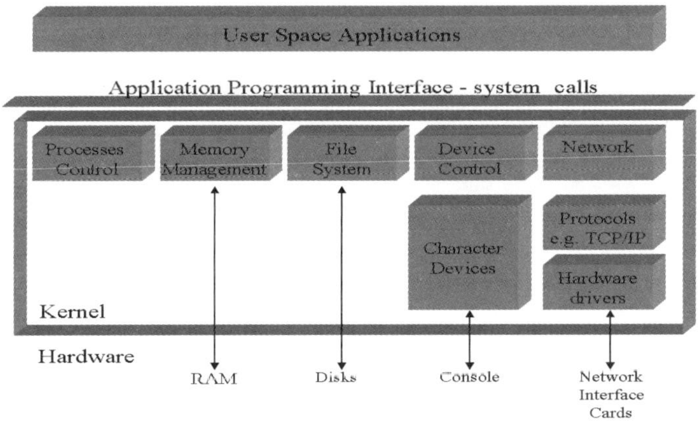

Kernel parts illustrated [7]

Kernel modules can also be used for real-time programming. Since they run in the kernel space, they have direct access to all hardware devices. They cannot be pre-empted by other processes unless they explicitly yield control. There is a patch to the Linux kernel (RTLinux™) taking care the execution of real-time task is quaranteed even if the kernel disables interrupts. Thus the interrupts are only disabled from the kernel point of view and in reality at the hardware.

However, there is a downside: the kernel module mechanism is not as flexible as a microkernel. A microkernel allows one to replace the scheduler and the virtual memory policies. Linux has three different hard-coded scheduling policies in the scheduler, and if these are not enough, the scheduler code must be modified. Still, kernel modules can be used for creating device drivers and implementing new file systems.

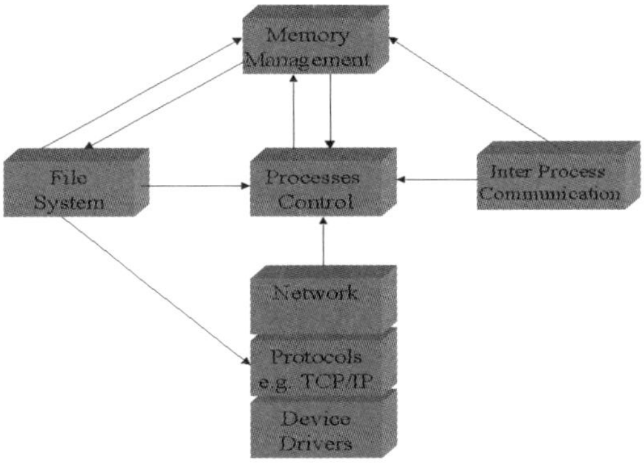

Conceptual presentation of a kernel [8]

Kernel httpd accelerator

The kernel httpd accelerator is an example of modules. As Linux is often used as a Web server, the kernel has some support for this. The kernel httpd accelerator is a kernel module that interrupts Web page requests before they are passed to a web server. For static Web pages, it is able to read the page from the file system and send it back to the requestor. Dynamic Web page requests are passed on to a 'real' Web server.

This speeds up the responses to static Web page requests and reduces the load on the Web server. Even Web pages that display information customized for the reader usually consist of a number of large images and some dynamically generated text. These images can then be served by khttpd.

Distributions

Distributions are important because it would be too difficult to collect all the individual pieces of software separately and to make sure that they will run together. In a sense, distributions take the individual pieces of software and make them a whole system. Important functionality that a distribution must provide are the following:

Installation Help

The installation program must be able to detect the available hardware and tune each application to it. The installation program must also support other operating systems that may exist on a computer.

Upgrades

The most popular distributions, Red Hat™ and Debian™, have good package management systems. They can find newer versions of installed components from the Internet and manage the dependencies that a new piece of software has. Package management also helps in removing programs and their dependencies. The latest business idea that Red Hat and Ximia™ are implementing is automated, remote upgrades.

Automatic Settings

The installation programs in distributions must also be able to set default values for all applications. This is important especially for firewalls and other security-related parts.

A distribution also provides a known reference point for other programs. A program may for example state that it runs on Red Hat 7.1.

Most companies that provide Linux-based products have selected a third-party distribution as a basis. Only in the case of some embedded devices that require heavy modification to the kernel need to have a proprietary distribution in place.

Real-time Support

Linux is not designed to run real-time tasks. This is because providing guaranteed execution times is often incompatible with making the system run normal tasks as efficiently as possible. One example with this is how often kernel tasks call the scheduler: it is more efficient for a kernel task to finish whatever it is doing than to try to the same job in as small pieces as possible. However, as Linux is freely available and since it is becoming increasingly popular in embedded use – and to watch video clips - there have been efforts to make Linux more real-time capable.

The first approach to improve the responsiveness of the kernel is to insert scheduling points and to reduce the time the kernel blocks interrupts. This is known as a low-latency patch. In practice, it works well for tasks that do not require absolute control of the hardware. The benefit is that the applications can still use the normal Linux kernel, albeit a patched one. The disadvantage is that patches are made for certain kernel versions and they may not be available for the latest ones. The patch is also not as well tested as the Linux kernel in general.

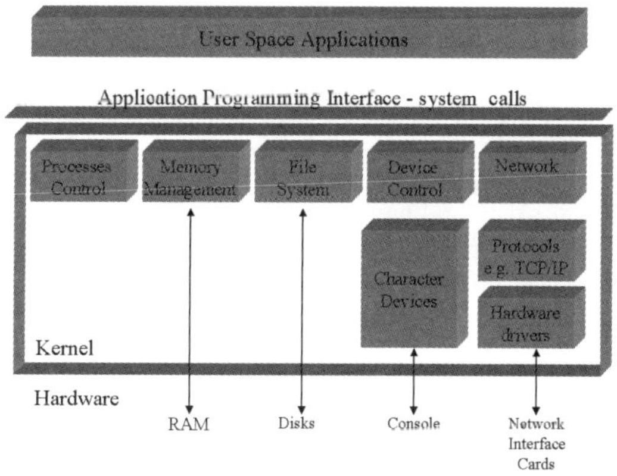

Real-time Linux approach

The second approach (RTLinux) overrides the Linux scheduler completely. Real-time tasks always run first and the kernel is scheduled to run only when there are no real-time tasks active. This makes it possible to run processes with very little overhead and with no interference from the kernel. Normal Linux applications can still be run on top of the kernel. Unfortunately, these real-time tasks cannot use normal Linux operating system services and must do everything by themselves. For example, normal device drivers will not work unmodified. It is also a little tricky to arrange communication between real-time and non-real-time tasks.

Security

Security is a crucial component in any computer or network system. It is well known that the system is as secure as its weakest link. Linux needs to be managed and configured appropriately, as with any operating system. However, there is one point that should be mentioned. One reason for Linux being more secure is that the source code is open and can be viewed by anyone competent enough in programming to verify that security is not compromised.

The other aspect to this is that the precompiled kernels need to come from a trusted party. Otherwise, they might contain parts that are not to the best interest of the system user. Also, one should be careful with modules that are loaded into the kernel, as their behaviour will have an effect to the primary functions of the system.

Networking

It is widely recognised that networking support in Linux is excellent. Networking support is handled in the kernel for efficienly and because it is not a function of a specific application. Packet reception and forwarding are asynchronous events, which is well illustrated with bursty Internet traffic.

There are two issues that need to be dealt with: packet and process handling. Packets need to be identified and delivered to correct interfaces be it a network interface or a user programming interface. Thus, the packet forwarding, routing and address resolution functions need to be implemented in the kernel. In some cases there are functions related to mobility and security that are implemented in the networking part of the kernel. These processes that interface with networking must be controlled accordingly: put to sleep when no packets are targeted for the process and awakend once there is a packet for the specific process.

Networking Functions

The versatility of Linux is best illustrated by naming the roles it can play in a network. It can carry out routing, bridging and firewalling functions as well as being a mail, Web, DNS and DHCP server.

Routing

The Linux kernel has built-in support for routing functions. A Linux device with proper network hardware interfaces can act as an IPv4 or IPv6 router for a fraction of the cost of a commercial router. A performance difference may be noticed if the criterion is based on delivered packets per second. However, it is more important to notice that the functionality exists and can serve a purpose for small and medium size corporations, for example. Recent kernels include special options for machines acting primarily as routers like IP multicast routing and policy routing.

Firewall

Different firewall toolkits exist for Linux. These firewall toolkits are very complete and when combined with other tools, allow blocking and redirection of all kinds of traffic and protocols. Different policies can be implemented via configuration files or GUI programs.

Development Cycle

Open source software is often claimed to be more reliable and faster to develop than proprietary software. This is because of the vast number of developers and ordinary users that participate in the process. Obviously, a large number of developers can produce new features faster than a small group, especially if the architecture supports this. As discussed earlier, all successful open source projects need such architectures. An interesting question is whether such development is also efficient, or is it only the huge amount of effort that speeds the development.

The argument for reliability is "Linus's Law" [7] as quoted by Eric S. Raymond: "Given enough eyeballs, all bugs are shallow." In other words, given a sufficient number of beta testers and those who correct the bugs, it is possible to essentially find all the defects. This means that the testing will be thorough if the feature or protocol is really required.

Most open source projects follow the model of stable and experimental releases. Stable releases are well tested and known to work, while experimental releases are just snapshots of what currently exists.Another distinction is between major and minor revisions that are reflected in version numbers. Major revisions are larger rewrites that can cause some incompatibilities to occur.

Licensing is one aspect needed to be considered once developing systems relying on Linux. Although Linux is free, it is copyrighted by Linus Torvalds. Its license is based on the Library GNU Public License (LGPL), which says that it is possible to create applications that run on top of Linux without revealing their source code. However, any modifications to the kernel are covered by the license and must be made public.

Kernel modules present an interesting case. Since they are dynamically loadable, they could be compared to normal application programs. However, they can potentially add new features to the kernel and for this reason they should be considered modifications and thus should be open source. In a statement, Linus Torvalds himself says that as long as kernel modules are used "as intended", they are covered by LGPL.

Conclusions

With all the features incorporated Linux has evolved into a feature rich Unix-like operating system with great importance in the Web domain to businesses today. It is provides an open development platform for reliable Internet services for both businesses and consumers.

References

[1] Tietoviikko, p.10, 20th September, 2001.

[2] Technical information, "Nokia Mediaterminal 511 S Technical information." http://www.nokia.com/multimedia/pdf/Mediaterminal.pdf . 2001.

[3] "The Tanenbaum-Torvalds Debate." In Chris DiBona, Sam Ockman, and Mark Stone: Open Sources. Voices from the Open Source Revolution. O'Reilly 1999. Pp. 221-251.

[4] Peter Wayner, "Free for All. How Linux and the Free Software Movement Undercut the High-Tech Titans." HarperBusiness 2000.

[5] "Out in the open," The Economist, April 12th, 2001.

[6] Linus Torvalds, "The Linux Edge." In In Chris DiBona, Sam Ockman, and Mark Stone: Open Sources. Voices from the Open Source Revolution. O'Reilly 1999. Pp. 101-111.

[7] Alessandro Rubini. "Linux Device Drivers," O'Reilly. p. 4. 1998.

[8] Ivan Bowman, Saheem Siddiqi, Meyer C. Tanyan, "Concrete architecture of Linux kernel," http://plg.uwaterloo.ca/~itbowman/CS746G/a2/, 1998.

[9] Eric S. Raymond, "The Cathedral and the Bazaar" at http://www.tuxedo.org/~esr/cathedral-bazaar/.

Java

This chapter presents the role of Java in the Mobile Internet. It gives an overview of the benefits and history of Java, and its role as an important software technology in the mobile Internet. Within this context, the chapter examines some key end-to-end issues.

A Short History

The development of Java started in project Green at Sun Microsystems in 1991. Sun first developed a programming technology called 'Oak' for interactive TV. The team, under the leadership of James Gosling, also developed a demonstrator including a real device. At that time, the contacted players in the interactive TV business was not interested. It is interesting to note, however, that Java-based technologies have now been standardized as a platform for the interactive digital TV services.

After realizing that interactive TV would not provide enough business for Java, the team had to try something else. As the enormous growth in the use of the Internet was emerging, the team decided to try integrating Java to the Internet. In 1994, they created a demo called 'WebRunner' based on a Mosaic browser and Java. The project was finally published in 1995 and the decision to integrate Java with the Netscape browser ensured that Java would become widely known in the computing industry.

In the early days of commercialization, Java was sold in two flavors: the Java Applet Environment for computers and browsers, and the Java OS for embedded devices. Java OS was a full operating system implemented in Java. The industry did not adopt JavaOS as most companies wished to keep their existing operating systems in place, and use Java for downloadable applications. This led to Java being repackaged in a new way, and Personal Java and Embedded Java were created.

At the same time, the Java platform for computers was expanded with new API libraries and new virtual machine technologies for speeding up the execution of Java bytecodes. However, only small changes have been made to the specification of the virtual machine and language.

The Benefits of Java™ Technology

Java is an application platform originally developed by Sun Microsystems. The popularity of Java has grown continuously since it was first introduced in late 1995. From the outset, Java was designed to be: "A simple, object-oriented, network-savvy, interpreted, robust, secure, architecture neutral, portable, high-performance, multithreaded and dynamic language."

The benefits of Java that have made it so popular are described below.

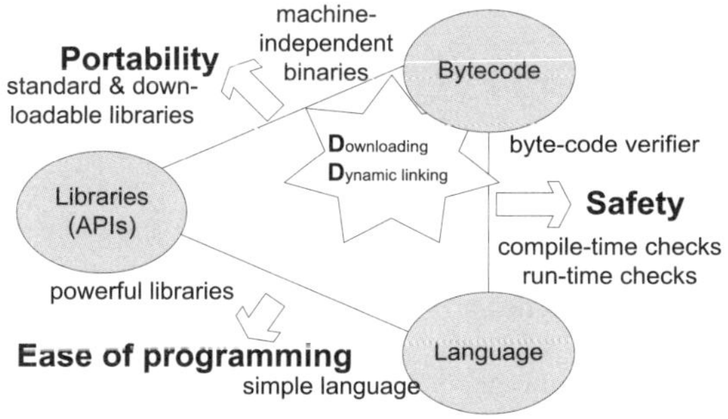

Technology cornerstones of Java.

Platform Independence

The most common justification for deploying Java is its platform independence and consequential application portability. The portability is based on standardized programming interfaces and the use of a device-independent format for representing executable programs. The programming interfaces are based on standardized libraries that are the same on different hardware platforms. Java applications are not distributed and installed in the form of machine code. Instead, Java programs are compiled into bytecode that is encapsulated in a standardized class file format. The bytecode is executed by an appropriate Java Virtual Machine (JVM). The virtual machine has been ported onto a wide variety of hardware and operating system platforms, enabling a high degree of portability for Java programs.

Safety

Compared to most other programming platforms, Java provides a certain degree of safety against programming mistakes and virus-like attacks.

Java helps programmers to avoid errors as it removes some error-prone aspects of C++, such as explicit pointers and multiple inheritance. It also uses strong typing and has several run-time checks such as array boundary checking.

At application execution, class files are run through a bytecode verifier to check that the bytecodes have not been modified to bypass the security mechanisms provided by the language.

Ease of Programming and Increased Productivity

Java is often compared to the C++ programming language and standard libraries, since Java can be used to replace C++ in many programming tasks. Although the syntax of Java is based on C++, there are several important differences.

A fundamental difference is that Java is a simpler language. C++ is a very complex language and very few programmers fully understand the use of all of its features. The simplicity of Java makes programming easier and programmers more productive. Also, compile and run-time checks reduce the time needed in the initial stages of program development for testing and debugging.

Java was designed to be a language that could be easily learned by the large existing pool of C and C++ programmers. The Java language and APIs were designed to be architecturally neutral, highly portable, network aware, robust and secure.

Dynamic Downloading and Late Binding

Java has been designed for dynamic class loading via a network. This allows it to be used for 'on-demand' applications. Although on-demand class loading is not always feasible in wireless networks, the dynamic class loading and late resolving of references still provides new types of dynamic behavior that have not been possible in embedded devices.

Java builds on existing infra-structure and market.

Java Core Platforms

The first published Java core platform was designed for desktop computers and to run together with a web browser. Since then, Java platforms have developed in two different directions: to fulfill the requirements of big desktop or enterprise computing systems, and to fit in consumer devices like set-top boxes and mobile phones.

As a consequence, the platform has been divided into three major categories. The current categorization of Java platforms is shown in a following figure.

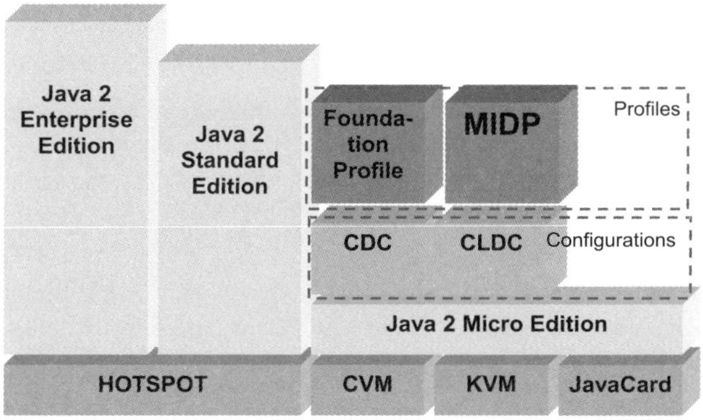

Java platforms.

The main categories are:

- **Java 2 Enterprise Edition (J2EE)** for Java-based enterprise and web servers,
- **Java 2 Standard Edition (J2SE)** for desktop and laptop computers, and
- **Java 2 Micro Edition (J2ME)** for resource-constrained consumer devices.

J2ME aims at a diverse set of devices and services and has been further divided into smaller entities called configurations and profiles. Configurations define the basic set of APIs and features of the virtual machine that are expected to be the same for devices with same amount of resources. There are currently two configurations:

- The Connected Device Configuration (CDC), which targets devices that Personal Java had earlier been targeted at. The underlying VM is assumed to have full VM support. In Sun's product portfolio the virtual machine supporting CDC is called CVM.
- The Collected Limited Device Configuration (CLDC) is for devices with very tight resource constraints. For the CLDC 1.0 VM, several features of the VM, such as support for floating point math, object serialization and reflection have been excluded. In Sun's product portfolio the virtual machine supporting CLDC is called KVM.

While configurations are targeted at devices with particular amounts of available resources, the domain specific sets of APIs are defined in profiles. Examples of profiles include:

- The Foundation Profile that is still a genetic profile and is usually assumed to be implemented by CDC devices.
- The Personal Profile that together with the Foundation Profile and CDC forms a successor of Personal Java.
- The Mobile Information Device Profile (MIDP) is an API set designed for devices like mobile phones and two-way pagers.

Building End-to-End Client/Server Applications Using Java

Java is currently used both on server (content) and terminal (consumption) sides. On the server side, Java is usually integrated with the business software and web servers. The client can be J2SE running on a desktop, J2ME running on a mobile terminal, or even a simple browser or applet.

Enterprise Java

Java plays an important role for enterprise software and applications. This is a very important part of providing content and services on the Mobile Internet. The Java 2 Enterprise Edition (J2EE) enables the building of highly robust, scalable and fault tolerant enterprise software. Among the underlying technologies used:

- **Servlets** enable building services that use HTTP/HTTPS communication. Servlets execute in a web server and respond to HTTP requests from web clients. The content of an HTTP request or response can be any form of textual or binary content required for the purposes of the desired service. Generally the content is some form of MIME media type such as text, HTML, XML, binary, image graphics, etc.
- **Enterprise Java Beans** (EJBs) provide a convenient mechanism for expressing business logic, typically related to transactions using databases.
- **Connectors** provide a standardized means for resource adapters to support legacy enterprise information systems, and integrate them into a J2EE based product.

In keeping with Java's ambition of "write once, run anywhere", there is a high degree of application portability across different J2EE containers. J2EE components can be easily deployed and upgraded on many J2EE servers and containers. This results in improved operational flexibility in managing the deployment and scalability of an application.

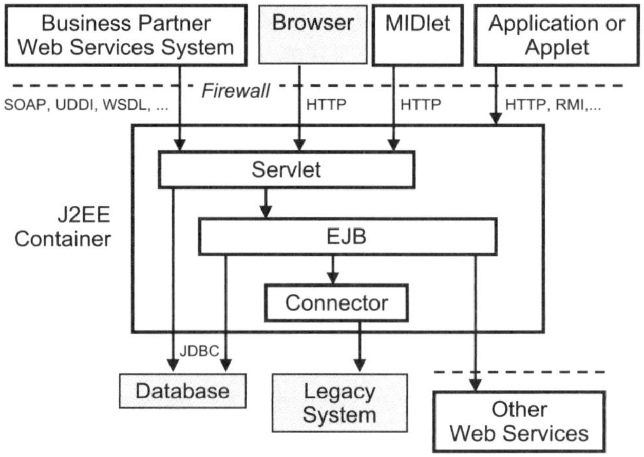

J2EE enables the building of services.

In addition, J2EE provides many other commonly needed application components and services, as well as standardizing their APIs [J2EE]. A few of these are briefly described here. The Java Message Service (JMS) supports reliable point-to-point messaging as well as a "publish / subscribe" messaging model. JavaIDL allows invocation of external CORBA objects. The Java Naming and Directory InterfaceTM (JNDI) is a standard API for naming and directory services. The Java API for XML Parsing (JAXP) provides support for the industry standard SAX and DOM APIs for parsing XML documents. The Java Authentication and Authorization Service (JAAS) enables services to authenticate and enforce access controls on users. The Java Database Connectivity (JDBCTM) defines a rich set of services for client access to databases. There is ongoing work in "Implementing Enterprise Web Services" standardization effort (JSR109), to define the role of J2EE in this context.

Standardization

Java standards are currently developed in the Java Community Process (JCP). The JCP is a process initiated, supported and guided by Sun. In this process, industry experts develop APIs in expert groups. Nokia has been an active member of these expert groups, creating Java Specifications, such as CLDC 1.0 and MIDP 1.0, for wireless mobile terminals and is currently a member of tens of expert groups in both J2ME - including next generation standards for wireless handsets, called CLDC NG, MIDP NG - and J2SE/EE space including specifications for operations support systems and intelligent networks.

Nokia is a member of the JCP Executive Committee for J2ME and is actively working on further development of the JCP process and participating decision making concerning Java Specifications.

MExE (Mobile Station Execution Environment) is a standardization effort originated by ETSI (European Telecommunications Standard Institute). Java is included in the classmarks of MExE:

- Classmark 2 is based on PersonalJava and the wireless profile of the JavaPhone API
- Classmark 3 is based on CLDC and MIDP

In addition to adapting existing Java technologies, MExE has defined security models for Java platforms and applications.

Current Status

Enterprise Java has a long history and is commonly used as the underlying technology in many mission-critical web servers, application servers and other enterprise servers. The scalability and reliability of such Java-based servers has been well proven by industry leading products such as Sun's iPlanet, IBM's WebSphere, BEA WebLogic, Oracle's and other J2EE-based servers.

Semantic Web

The World Wide Web has developed primarily as medium of content for human consumption. Auto-mating anything on the Web (information retrieval, synthesis of information, etc.) is difficult because human interpretation, in one form or another, is required to make Web content useful. Offering some relief, a new architecture for the Web is emerging, known as the "Semantic Web". In broad terms, it encompasses efforts to create mechanisms that augment content with formal semantics, thereby producing content suitable for automated systems [Berners-Lee et al 2001]. The Semantic Web will allow us to use more automated functions on the Web (such as reasoning, information and service discovery, and autonomous agents), easing the workload of humans.

The Semantic Web will also pave the way for true "device independence" and customization of information content for individual users. Information on the Web can now exist in a "raw form" and any context-dependent presentation can be rendered on demand (more generally, the Semantic Web represents a departure from the current "rendering-oriented" Web). It is important to note that the Semantic Web is not a separate Web but an extension of the current one, in which information, when given this well-defined meaning, better enables computers and people to work in tandem.

The World Wide Web essentially gives us an infrastructure for "pointing" (or, in other words, an infra-structure and mechanisms for linkages between documents). This infrastructure is quite general and consequently very powerful, but its problem is that the linking is indiscriminate: human interpretation (again) is required to understand what any particular link means. The Semantic Web aspires to change this by associating formal meaning with the linkages.

This brings us to the key aspect of the Semantic Web, something we will refer to as the "ontological approach": Instead of merely introducing new markup tags (via the use of XML, for example), the Semantic Web will focus on the meaning of the new markup, and on mechanisms that allow us to introduce, coordinate and share the formal semantics of our data, as well as reason (i.e., draw inferences) from the semantic data. In practice, computers will find the meaning of semantic data by following (semantic) hyperlinks to definitions of key terms and rules for reasoning about them logically. This aspect of the resulting infrastructure will enable the development of fully automated Web Services such as highly functional "intelligent" agents.

The Role of Ontologies

The ontological approach is predicated on the existence and use of ontologies: by this we mean a document or file that formally defines the relationships between terms for any particular domain of discourse. In this context (particularly since the Semantic Web community draws heavily from a branch of artificial intelligence known as knowledge representation) a widely cited definition of an ontology is Gruber's "a specification of a conceptualization" [Gruber 1993]; in that sense we depart from the abstract philosophical notion of ontology defined as "a branch of metaphysics concerned with the nature and relations of being" [Merriam-Webster 1998]. The most typical kind of ontology for the Web has a concept taxonomy and a set of inference rules.

Ontologies can enhance the functioning of the Web in many ways. They can be used in a simple fash-ion to improve the accuracy of Web searches by making the search program use precisely defined concepts (as opposed to ambiguous keywords possibly contained on a page). More advanced applications will use ontologies to reason from the relationships between concepts, and will be able to relate the information on a page to any associated knowledge and inference rules. This will enable conceptual reasoning across independently created web sites and their content.

People, as well as computational agents, typically have some notion or conceptualization of the meaning of terms. The specification a software program sometimes provides of its inputs and outputs could be used as a specification of the program itself. Similarly, ontologies can be used to provide a concrete specification of term names and meanings. Following this line of thought, though - where an ontology is a specification of the conceptualization of a term - there is much room for variation, and the spectrum of Web ontologies typically range from simple controlled vocabularies through informal concept hierarchies to something where arbitrarily complex logical relationships can be specified between defined concepts.

In practical terms, ontologies can be defined as finite yet extensible controlled vocabularies, they have unambiguous interpretation of classes and term relationships, and have strict hierarchical subclass relationships. In addition, ontologies typically allow property specification and value restriction on a per-class basis as well as the inclusion of individuals (class instances) in the ontology. Other features of ontologies, albeit not necessarily typical, are the specification of disjoint classes, inverse relationships, and part-whole relationships, as well as the specification of arbitrary logical relationships between terms [Lassila & McGuinness 2001].

Languages of the Semantic Web

In order to take the ontological approach, we need representation languages not only suited for expressing ontological information but also well matched with other Internet and World Wide Web standards. A layered approach is adopted:

For the purposes of serializing data and transporting it from one system to another, the Extensible Markup Language (XML) is used. XML is a very powerful formalism: it allows "customized" languages to be created for almost any purpose. In practical terms, XML lets everyone create their own markup tags that, for example, annotate Web pages or sections of text on a page. Programs could (and currently do) make use of these tags in sophisticated ways, but the programmer has to know the in-tended meaning attributed to the tag by the author. In short, XML allows users to add structure to their documents but says nothing about what these structures mean.

Resource Description Framework

Another language, called the Resource Description Framework (RDF) uses XML for serialization, but introduces a powerful abstract data model which allows systems to share the meaning of the aforementioned structures [Lassila 1998, Lassila & Swick 1999, Lassila 2000]. Originally, RDF is a standard for Web metadata developed by the World Wide Web Consortium (W3C) and some of its member organizations (including Nokia, IBM, OCLC, and others). Expanding the traditional notion of document metadata (such as library catalog information), RDF is suitable for describing any Web resource, and as such, provides interoperability between applications that exchange machine-understandable information on the Web.

Items that RDF expressions describe are called resources; broadly speaking, anything a Universal Resource Identifier or URI [Berners-Lee et al 1998] can name is a resource. Consequently, RDF can describe not just things on the Web (such as pages, parts of pages, or collections of pages) but also things not on the Web - as long as they can be named using some URI scheme (hypothetically we could describe any person with a "person" URI scheme, say, "person:US:123-45-6789"). The RDF description model uses object/attribute/value triples: we can view instances of the model as directed labeled graphs (resembling semantic networks), or we can take a more object-centric view and think of RDF as a frame-based representation system [Fikes & Kehler 1985, Lassila & McGuinness 2001]. In RDF, these triples are known as statements.

Descriptions in RDF can be limited to one resource (for example, a library catalog card that names a document's author and publisher), in which case the values of resource attributes (or properties, as they are called in RDF) are typically just strings. Descriptions can also span multiple resources: values of properties can be other resources - this way we can describe arbitrary relationships between multiple resources. URIs name properties, which are also resources, and can be described by asking, "What are a particular property's permitted values, which types of resources can it describe, and what is its relationship to other properties?"

Meaning in RDF is bound to specific terms and concepts that URIs define and name. Because URIs can be unique, two systems can define some concept (say, "person") and each use a different URI to name it to avoid clashes; however, two systems agreeing on a common concept will use the same URI and effectively share semantics.

The RDF model also defines some meta-level constructs (such as container types for describing collections of resources) and higher-order statements (statements about other statements). Higher-order statements are modeled in RDF itself, and allow the representation of modalities such as beliefs. Furthermore, an extensible, object-oriented type system known as RDF Schema is introduced as a layer on top of the basic RDF model [Brickley & Guha 2000]. Class definitions can be derived from multiple superclasses, and property definitions can specify domain and range constraints. We can think of RDF Schema as a set of ontological modeling primitives layered on top of RDF.

As already mentioned, RDF uses XML for the syntactic expression of model instances; this is a source of much controversy and confusion. RDF is essentially a data model and does not strive to replace XML. Instead, it builds a layer on top of it, making interoperable exchange of semantic information possible (for example, the object-oriented extensibility is intended to enable a partial understanding of data). The current version of RDF still lacks primitive data types (such as integer, float, etc.), so text strings are essentially the only literals available. Languages that build on top of RDF (such as the DARPA Agent Markup Language, described below) may expand on this, typically adopting the atomic data types from XML Schema [Biron & Malhotra 2001].

RDF's object-oriented extensibility lets developers take pieces of existing RDF schemata and extend them as needed. This is important, since it may lead to a type of "Darwinian evolution" of ontologies (and metadata) where strong solutions survive and evolve further (as opposed to the all-or-nothing situation that users of XML DTDs often face).

DARPA Agent Markup Language

The DARPA Agent Markup Language (DAML) program, launched in 2000, is a US Government spon-sored effort aimed at producing foundation technology for the Semantic Web. The program is funding critical research to develop languages, tools, and techniques that will allow the creation of more machine-understandable Web content. One of the program's results is DAML+OIL, a semantic markup language for constructing Web-based ontologies.

DAML+OIL is built on earlier RDF and RDF Schema, and extends them with richer modeling primitives (commonly found in frame-based representation languages). It also uses XML Schema to provide, for example, numeric data types. The language has clean and well defined semantics.

In comparison to RDF, DAML+OIL not only supports explicitly named classes, but also class expressions. These include classes whose instances are enumerated as well as boolean combinations of other class expressions. The notion of class equivalence is supported (i.e., it is possible to specify that two classes are equivalent and thus must contain the same instances). Furthermore, DAML+OIL also supports property restrictions; these allow properties to be tailored on a class-by-class basis (in RDF, characteristics of properties are global). DAML+OIL further clarifies the nature of RDF Schema domain and range constraints of properties (making them conjunctive, not disjunctive). It also supports notions of property equivalence and inverse relations.

Finally, DAML+OIL introduces a new container type which allows the definition of a collection in such a way that all its members are known (i.e., the collection cannot contain any fewer or more members than the ones explicitly specified).

The Semantic Web as a "Web of Services"

"Web Services" - remote functionality invoked over the Internet - are a step towards automating tasks on the Web. The Semantic Web, representing a more far-reaching form of Web automation, is closely related (and highly complementary) to Web Services.

Existing Infrastructure for Web Services

Some basic infrastructure for Web Services has recently emerged: Universal Description, Discovery and Integration (UDDI) is an industry initiative launched by Microsoft, IBM and Ariba to develop a standard online registry for publishing and dynamic discovery of Web Services offered by businesses. UDDI allows a business to locate potential business partners and form business relationships on the basis of the services they provide. The primary target is integration and some type of "semi-automation" of business transactions in B2B e-commerce. To achieve this, UDDI provides a registry for businesses and the services they offer, described using an XML schema defined by the UDDI specification. A Web Service provider registers its advertisements along with keywords for categorization; a Web Service user retrieves advertisements based on a keyword search. The UDDI search mechanism relies on predefined categorization through keywords and does not refer to any semantic content of the advertisements. The registry will function in a fashion similar to white pages or yellow pages, where businesses can be looked up by name or by a standard service taxonomy.

The Web Services Description Language (WSDL) specifies a protocol- and encoding-independent mechanism for Web Service providers to describe requests for offered services. WSDL is an XML grammar which describes network-reachable services and maps these to a messaging-capable collection of communication endpoints. WSDL separates the abstract definition of services and messages from their concrete binding to a network port and message format. This flexibility to redefine the deployed network bindings of a service (as more efficient message protocols and data formats emerge) makes the adoption of WSDL attractive. The current WSDL specification describes concrete bindings for SOAP, HTTP GET/POST, and MIME. Note that WSDL is a technology for only describing services, and does not address how a remote system would discover these services. WSDL can be used in UDDI as the language for describing interfaces to business services registered in a UDDI registry.

Completing the current Web Services architecture is the Simple Object Access Protocol (SOAP), a remote invocation protocol based on XML. It does not itself define a programming model or application semantics, yet it provides a packaging model for objects to exchange encoded data. The SOAP packaging model consists of three components: The envelope contains an expression of what is in the message, the target message handler, and whether the message can be considered optional. The encoding rules describe both simple and compound data types. Finally, the body describes the RPC-related data: the URI of the target object, method name, method signature, and method parameters.

Web Services Using DAML

On the Semantic Web, to make use of a Web Service, a software agent needs a computer-interpretable description of both the service, and the means by which it is accessed. An important goal within the DAML program is to establish a framework within which these descriptions are created and shared. Web sites will then be able to employ a set of basic classes and properties for declaring and describing services.

Part of the ongoing work of the DAML program is DAML-S [Ankolenkar et al 2001, Burstein et al 2001], a Web Service ontology expressed in DAML+OIL. It gives Web Service providers a core set of markup language constructs for describing the properties and capabilities of their Web Services in an unambiguous, computer-interpretable form. DAML-S markup of Web Services will facilitate the automation of the following Web Service tasks:

1. Automatic discovery involves the automatic location of Web Services that provide a particular service and that adhere to requested constraints.
2. Automatic invocation involves the execution of a discovered Web Service by a computer program or agent. Execution of a Web Service can be thought of as a collection of function calls. DAML-S markup of Web Services provides a declarative, computer-interpretable interface for executing these function calls. A software agent should be able to interpret the markup to understand what input is necessary to the service call, what information will be returned, and how to execute the service automatically.

3. Automatic composition and interoperation involves the automatic construction of a plan to use a number of Web Services to perform some task, given a high-level description of an ob-jective. With DAML-S markup of Web Services, the information necessary to select and com-pose services will be encoded at the service Web sites. Software can be written to manipulate these representations, together with a specification of the objectives of the task, to achieve the task automatically.
4. Automatic execution monitoring: Individual services and, especially, compositions of services, will often require some time to execute completely. Users (human or artificial) may want to know during this period what the status of their request is. Their plans may also have changed requiring alterations in the actions the Web Service provider takes.

In comparison to service profiles - the DAML-S characterization of services - the other industry standards are limited, first and foremost, in that they cannot express logical statements (e.g. pre- and postconditions, or rules to describe dependencies between profile elements). Input and output types are supported to varying extents. From the DAML-S standpoint, services can be simple (or primitive); they invoke only a single Web-accessible computer program, sensor, or device that does not rely upon another Web Service; and there is no ongoing interaction between the user and the service, beyond a simple response. Alternately, services can be complex, composed of multiple primitive services, often requiring an interaction or dialogue between the user and the services, so that the user can make choices and provide information conditionally. DAML-S is meant to support both categories of services, but complex services have provided the primary motivation for the features of the language.

DAML-S provides an upper ontology for services, including the properties normally associated with all kinds of services. The upper ontology does not address what the particular subclasses of the base service class should be, or even the conceptual basis for structuring this taxonomy. This structuring may take place according to functional and domain differences, or market needs. A service profile provides a high-level description of a service and its provider, and is used as a request, or as an advertisement, within a service discovery process.

Service profiles consist of three types of information: a human readable description of the service, a specification of the functionalities that are provided by the service (represented as a transformation from the inputs required by the service to the outputs produced), as well as a host of functional attributes which provide additional information and requirements about the service that assist when reasoning about several services with similar capabilities (these include guarantees of response time or accuracy, as well as the cost of the service).

Furthermore, a more detailed perspective on services is to view them as processes. The DAML-S representation of processes draws upon established work in a variety of fields, such as automated planning and workflow automation, and will support the representational needs of a very broad array of services on the Web.

Ultimate Form of Interoperability

Given the automation aspect, the Semantic Web enables a radical departure from the current browsing paradigm of the Web. Indeed, we can foresee applications and usage situations where browsing is not a viable alternative. On the other hand, one might argue that browsing still takes place, but by automated agents and not by human users. If we can represent information about Web-based services in a formal manner, the discovery of additional functionality needed by agents becomes easier. Generally, the process of matching one agent's description of "missing" functionality with descriptions of services offered by other agents we call "service discovery", and resembles the act of advertising and querying. It is becoming an increasingly important aspect of distributed information systems.

In addition to the Web Service standards described earlier, a number of mechanisms for low-level service discovery are emerging - examples include Sun's Jini and Microsoft's Universal Plug and Play [e.g., Richard 2000] - but these mechanisms attack the problem at a syntactic level, and - just as their higher-level relatives - rely heavily on the standardization of a predetermined set of functionality descriptions.

Standardization, unfortunately, can only take us half way towards the goal of automated agents on the Web, as our ability to anticipate all possible future needs is limited. The Semantic Web offers a possibility to elevate the mechanisms of service discovery to a semantic level. Here a more sophisticated description of functionality is possible, and the shared understanding between the consumer and the producer can be reached via the exchange of ontologies which provide the necessary vocabulary for discussion.

More importantly, it will also be possible to reach a "partial match" between a request and an advertisement [e.g., Sycara et al 1999], yet be able to take advantage of the discovered service. An agent could enhance its functionality in the following manner:

1. Exchange ontologies to establish partial understanding,
2. Find a partial match of needs with services offered, and
3. Compose exact required functionality from services discovered (using, for example, automated planning or configuration techniques).

This approach takes information system interoperability well beyond what mere standardization or simple interface sharing enables, since it is based on "deeper" descriptions of service functionality and can be performed ad hoc and on demand.

Eventually the Semantic Web will "break out" of the virtual realm and extend into the physical world. As we have observed, URIs can name anything, including physical entities, which means we can use the RDF and DAML+OIL languages to describe people, their personal preferences, and various devices such as cell phones and home appliances. Such devices can advertise their functionality - what they can do and how they are controlled - much like software agents. Being more flexible than low-level schemes such as Universal Plug and Play, the semantic approach opens up many exciting possibilities: So-called home automation today requires careful configuration for appliances to work together, but semantic descriptions of device capabilities and functionality will allow us to achieve such automation with minimal human involvement.

Concrete steps have already been taken here, with work on developing a standard for describing functional capabilities of devices (such as screen sizes) and user preferences. Built on RDF, this standard, called Composite Capability/Preference Profile (CC/PP), will initially be used to allow cell phones and other nonstandard Web clients to describe their characteristics so that Web content can be tailored for them on the fly [Klyne et al 2001]. With the full versatility of languages for handling ontologies and logic, devices could eventually automatically seek out and employ services and other devices for added information or functionality. Given all this, we can now imagine a Web-enabled microwave oven consulting the popcorn manufacturer's Web site for optimal popping parameters.

Conclusions

The power of the Semantic Web will be realized when we are automatically able to collect Web content from diverse sources, process the information and exchange the results with other automated systems. The effectiveness of these systems (e.g., software agents) will dramatically increase as more machine-readable Web content and automated services (including other agents) become available. The Semantic Web promotes this synergy: even agents that were not expressly designed to work together can transfer data between each of them when these data come with semantics. This serendipity represents the ultimate form of system interoperability, and is a departure from a strict standardization approach towards mechanisms that are more flexible and enable operation in unanticipated situations.

References

[Ankolenkar et al 2001] Anupriya Ankolenkar, Mark Burstein, Jerry R. Hobbs, Ora Lassila, David Martin, Sheila A. McIllraith, Srini Narayanan, Massimo Paolucci, Terry Payne, Tran Cao Son, Katia Sycara, and Honglei Zeng: "DAML-S: A Semantic Markup Language for Web Services", Proceedings of the First Semantic Web Working Symposium (SWWS'01), Stanford University, July 2001

[Berners-Lee et al 1998] Tim Berners-Lee, Roy Fielding, and Larry Masinter: "Uniform Resource Identifiers (URI): Generic Syntax", Internet Draft Standard RFC 2396, August 1998

[Berners-Lee et al 2001] Tim Berners-Lee, James Hendler, and Ora Lassila: "Semantic Web", Scientific American 284(5):34-43 (May 2001)

[Biron & Malhotra 2001] Paul V. Biron & Ashok Malhotra: "XML Schema Part 2: Datatypes", W3C Recommendation 02 May 2001, World Wide Web Consortium, Cambridge (MA); http://www.w3.org/TR/xmlschema-2/

[Brickley & Guha 2000] Dan Brickley & R.V.Guha, "Resource Description Framework (RDF) Schema Specification 1.0", W3C Candidate Recommendation 27 March 2000, World Wide Web Consortium, Cambridge (MA); http://www.w3.org/TR/rdf-schema/

[Burstein et al 2001] Mark Burstein, Jerry Hobbs, Ora Lassila, David Martin, Sheila McIlraith, Srini Narayanan, Massimo Paolucci, Terry Payne, Katia Sycara, and Honglei Zeng: "DAML-S 0.5 Draft Release", DARPA Agent Markup Language Program draft document, May 2001; http://www.daml.org/services/daml-s/2001/05/

[Dublin Core 1999] "Dublin Core Metadata Element Set, Version 1.1: Reference Description", Dublin Core Metadata Initiative, 1999; http://purl.org/dc/documents/rec-dces-19990702.htm

[Fikes & Kehler 1985] Richard Fikes & Tom Kehler, "The Role of Frame-Based Representation in Reasoning", CACM 28(9): 904-920 (1985)

[Gruber 1993] Tom R. Gruber: "A translation approach to portable ontologies", Knowledge Acquisition 5(2):199-220, 1993

[Hendler & McGuinness 2000] James Hendler & Deborah McGuinness: "The DARPA Agent Markup Language", IEEE Intelligent Systems Trends and Controversies, November/December 2000

[Klyne et al 2001] Graham Klyne, Franklin Reynolds, Chris Woodrow & Hidetaka Ohto: "Composite Capability/Preference Profiles (CC/PP): Structure and Vocabularies", W3C Working Draft 15 March 2001, World Wide Web Consortium, Cambridge (MA); http://www.w3.org/TR/CCPP-struct-vocab/

[Lassila 1998] Ora Lassila: "Web Metadata: A Matter of Semantics", IEEE Internet Computing 2(4): 30-37 (1998)

[Lassila 2000] Ora Lassila: "The Resource Description Framework", IEEE Intelligent Systems 15(6): 67-69 (November/December 2000)

[Lassila & McGuinness 2001] Ora Lassila & Deborah L. McGuinness: "The Role of Frame-Based Representation on the Semantic Web", Knowledge Systems Laboratory Report KSL-01-02, Stanford University, 2001; http://www.ksl.stanford.edu/KSL_Abstracts/KSL-01-02.html

[Lassila & Swick 1999] Ora Lassila & Ralph Swick: "Resource Description Framework (RDF) Model and Syntax Specification", W3C Recommendation 22 February 1999, World Wide Web Consortium, Cambridge (MA); http://www.w3.org/TR/REC-rdf-syntax/

[Merriam-Webster 1998] "Merriam-Webster's Collegiate Dictionary", Merriam-Webster, 1998

[Richard 2000] Golden G. Richard III, "Service Advertisement and Discovery: Enabling Universal Device Cooperation", IEEE Internet Computing 4(5): (2000).

[Sycara et al 1999] Katia Sycara, Matthias Klusch, Seth Widoff & Jianguo Lu: "Dynamic Service Matchmaking Among Agents in Open Information Environments", in: A. Ouksel, A. Sheth (eds.): ACM SIGMOD Record (Special Issue on Semantic Interoperability in Global Information Systems) 28(1): 47-53 (March 1999)

[van Harmelen et al 2001] Frank van Harmelen, Peter F. Patel-Schneider, and Ian Horrocks (eds.): "Reference description of the DAML+OIL (March 2001) ontology markup language", DARPA Agent Markup Language draft document, March 2001; http://www.daml.org/2001/03/reference.html

MOBILE DOMAIN

WCDMA Technology

This chapter is intended to briefly cover the main technical aspects of UMTS Terrestrial Radio Access Network (UTRAN), specifically from the WCDMA technology point of view. As the UMTS radio interface, WCDMA technology is going to be the most widely used technology for IMT-2000. The WCDMA radio technology within the Universal Terrestrial Radio Access Network is standardised by the 3rd Generation Partnership Project (3GPP). The 3GPP organisation and working methods are introduced in Chapter 6 of this book.

Air Interfaces for IMT-2000

Within ITU (International Telecommunications Union), several different air interfaces have been defined for third generation systems, based on either CDMA or TDMA technology. WCDMA (Wideband Code Division Multiple Access) is the main third generation air interface in the world and will be deployed in Europe and Asia, including Japan and Korea, in the same frequency band of around 2 GHz. The wide deployment of WCDMA will create large markets for manufacturers and for the providers of content and applications. Note that in second generation systems different air interfaces are used in Europe, in Japan and in Korea: GSM in Europe, PDC in Japan and IS-95 in Korea.

In North America the frequency spectrum around 2 GHz has already been auctioned to operators using second generation systems, and no new spectrum is available for the new systems. Thus, third generation services must be implemented within the existing bands by replacing part of the spectrum with third generation systems. This approach is referred to as re-farming.

In addition to WCDMA, the other air interfaces that can be used to provide third generation services are EDGE and multi-carrier CDMA (cdma2000). These solutions are targeted mainly at providing third generation services within existing frequency bands, for example, in North America. EDGE (Enhanced Data Rates for GSM Evolution) can provide third generation services with bit rates up to 400 kbps with a GSM carrier spacing of 200 kHz. EDGE can be used flexibly together with GSM in all existing GSM bands, such as GSM900 and GSM1800. EDGE is expected to be used by most of the GSM operators to improve spectral efficiency of the existing GSM frequencies.

Multi-carrier CDMA is designed to be deployed together with IS-95 in the same frequency band. The name for the multi-carrier CDMA comes from the downlink transmission direction, where instead of a single wideband carrier, multiple (up to 12) parallel narrow-band CDMA carriers are transmitted from each base station. The bandwidth of each narrow-band carrier is equal to the bandwidth of IS-95. Similar to the GSM evolution to GPRS/EDGE there is evolution in the IS-95 side for the higher data rates or new performance improvement methods.

The expected frequency bands and geographical areas where these different air interfaces are likely to be applied are shown in the next figure.

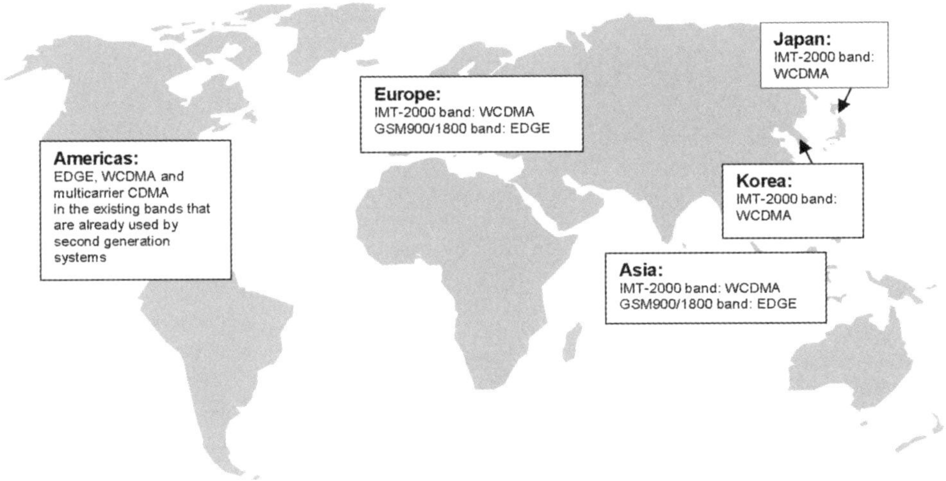

Expected air interfaces and spectrums for providing third generation services.

The Spectrum for UMTS

Work to develop third generation mobile systems began when the World Administrative Radio Conference (WARC) at its 1992 meeting, identified the frequencies around 2 GHz that were available for use by future third generation mobile systems. The spectrum allocation in Europe, Japan, Korea and the USA is shown in the next figure. In Europe and in most of Asia the IMT-2000 bands of 2 × 60 MHz (1920-1980 MHz plus 2110-2170 MHz) will be available for WCDMA FDD. The availability of the TDD spectrum varies: in Europe it is expected that 25 MHz will be available for licensed TDD use in the 1900-1920 MHz and 2020-2025 MHz bands. The rest of the unpaired spectrum is expected to be used for unlicensed TDD applications (SPA: Self Provided Applications) in the 2010-2020 MHz band. FDD systems use different frequency bands for uplink and for downlink, separated by the duplex distance, while TDD systems utilise the same frequency for both uplink and downlink.

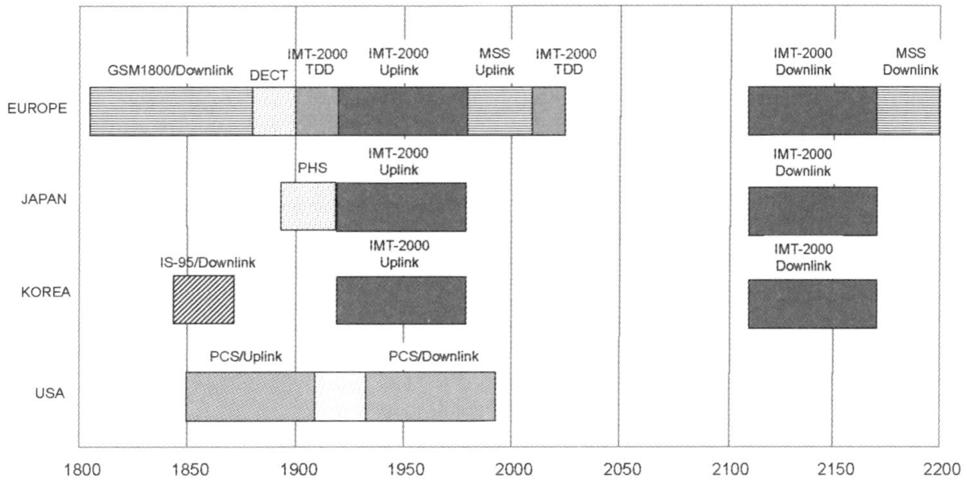

Spectrum allocation in Europe, Japan, Korea and USA.

A few example UMTS licenses are shown in the next table in Japan and in Europe. The number of UMTS operators per country is between 4 and 6.

Examples of the first phase UMTS licenses allocated

Country	Number of operators	Number of FDD carriers (2 x 5 MHz) per operator	Number of TDD carriers (1 x 5 MHz) per operator
Finland	4	3	1
Japan	3	4	0
Spain	4	3	1
UK	5	2-3	0-1
Germany	6	2	0-1

UMTS licenses are given either based on the operator's capabilities for providing services (beauty contest) or based on auction. Finland, Japan and Spain used beauty contest, while UK and Germany applied the auction approach.

More frequencies have been identified for IMT-2000. At the WARC-2000 meeting of the ITU in May 2000 the following frequency bands were identified for IMT-2000 use:

- 1710-1885 MHz
- 2500-2690 MHz
- 806-960 MHz

It is worth noting that some of the bands listed, especially below 2 GHz are partly used by other systems like GSM. What the exact duplexing arrangements will be is under discussion at the moment. The feasibility studies for the different duplex arrangements with the new bands have been initiated as part of the 3rd Generation Partnership Project (3GPP) standardisation activities in connection with WCDMA evolution work.

WCDMA Radio Access Network Architecture

WCDMA Technology has shaped WCDMA radio access network architecture due to the requirements of the basic features of CDMA, such as soft handover. In the following figure the basic UTRAN architecture is shown indicating the connections to the core network for both circuits switched traffic (Iu-CS) and packet switched traffic (Iu-PS). The existence of the open, standardised Iur interface is essential for proper network operation including soft handover support in multi-vendor environment. The same architectural principles as with WCDMA are also considered in GSM/GERAN standardisation, paving the way for multi-radio networks. Multi-radio aspects, such as radio resource management between UTRAN and GERAN, are also being worked on as part of 3GPP standardisation for high performance multi-radio networks.

Also as an important trend on the architecture side, IP based technology making its way to the radio access side in addition to the core network. An example of this is IP based transport, which is currently in the final stages of 3GPP standardisation. After this UTRAN internal interfaces (Iub/Iur/Iu-CS) will provide support for IP in addition to ATM technology in Rel'99/Rel'4. IP already exists on Iu-PS.

WCDMA Basics

The WCDMA air interface is based on CDMA technology. All the users share the same carrier with the shared resource being the power on the carrier. The characteristic feature is the wide 5 MHz carrier bandwidth over which the signal for each user is spread, as illustrated in the next figure. The transmission bandwidth is the same for all data rates with the processing gain being larger for the smaller data rates than for higher rates. The achieved processing gain provides protection against the interference from other users active on the same carrier. In the receiver despreading separates the transmitted and spread signal for data detection.

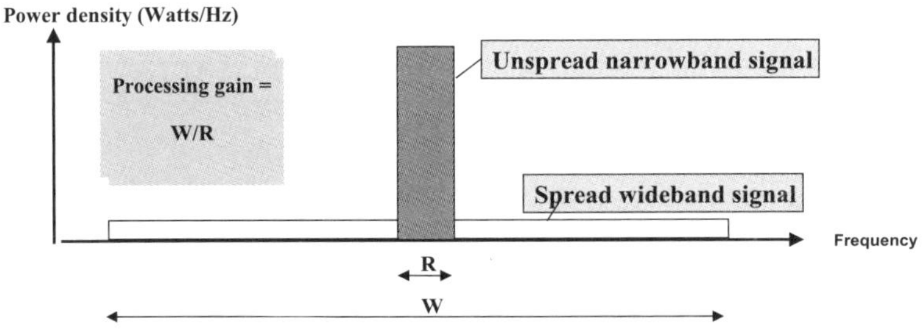

The CDMA principle used in WCDMA.

There are two key features resulting from the use of CDMA technology, namely, fast power control and soft handover. They both contribute to WCDMA system capacity but are also requirements for proper system operation.

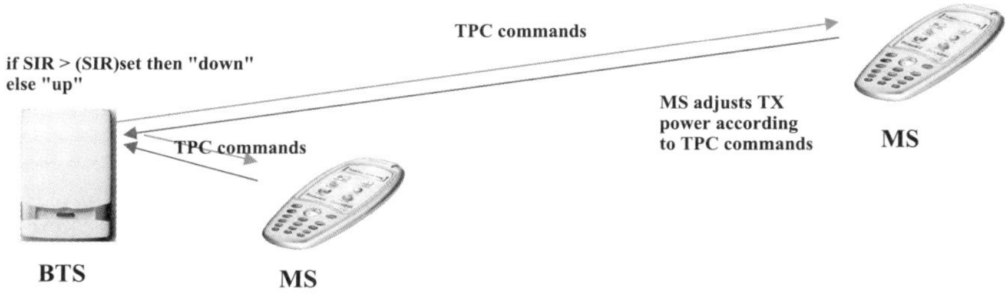

The Fast power control principle.

Fast power control, especially in the uplink, is required so that users do not generate extra interference and do not block the reception of signals for other users. Without power control a terminal transmitting near the base station would block the reception of other users further away when exceeding the achieved processing gain. This is also known as the near-far problem in CDMA. The power control command rate in WCDMA is set to 1500 Hz with a typical 1 dB step either up or down.

Soft handover is required for similar reasons. In soft handover, a terminal is connected simultaneously to two or more cells on the same frequency. Especially in the uplink this is vital as otherwise a terminal between two cells could cause problems for the cell to which it is not connected. In soft handover, all cells provide power control information to the terminal and the near-far problem is avoided.

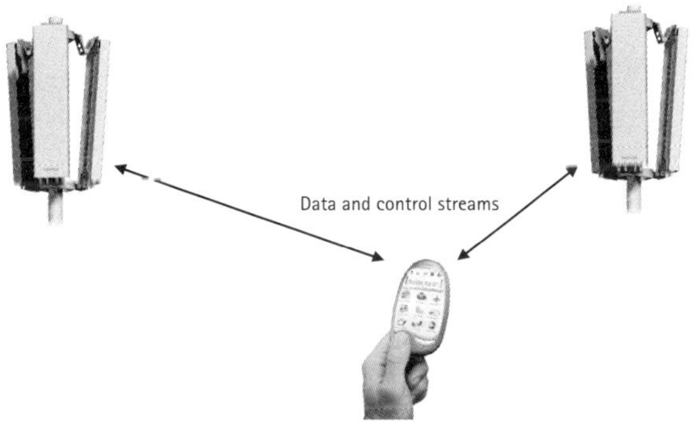

Soft handover.

WCDMA Physical Layer and Terminal Capabilities

In this section the main differences between the third and second generation air interfaces are described and the key physical layer parameters of WCDMA are explained. Also, the terminal capabilities from the air interface point of view are addressed on how they impact the services each type of terminal can support.

GSM and IS-95 (the standard for cdmaOne systems) are the second generation air interfaces considered here based on TDMA and CDMA respectively. Other second generation air interfaces are PDC in Japan and

US-TDMA mainly in the Americas; these are based on TDMA (time division multiple access) and have more similarities with GSM than with IS-95. Second generation systems were built mainly to provide speech services in macro cells.

To understand the background of the differences between second and third generations systems, we need to look at the new requirements of third generation systems which are listed below:

- Bit rates up to 2 Mbps
- Variable bit rate to offer bandwidth on demand
- Multiplexing of services with different quality requirements on a single connection, e.g. speech, video and packet data
- Varying delay requirements
- Quality requirements from a 10% frame error rate to a 10-6 bit error rate
- Coexistence of second and third generation systems and inter-system handovers
- Support for asymmetric uplink and downlink traffic
- High spectrum efficiency
- Coexistence of FDD and TDD modes

The WCDMA air interface has been defined to provide first phase data rates up to 2 Mbps in Release - 99 and Release 4. For Release 5, higher data rates up to 10 Mbps and even beyond are under discussion in 3GPP standardisation.

The WCDMA way of sharing resources with the main resource - the power shared amongst users - being the key when dealing with variable bit rate services. With no fixed allocation of the physical resources that would be used by a terminal even when there is no data to transmit, the capacity sharing on the carrier can be done very fast, even on a 10 ms basis. Power based resource sharing is fully valid in the uplink but for the downlink, more attention needs to be paid to code resource usage when the peak and average data rates differ. Methods have been defined for efficient use of downlink resources, for example, with a shared channel concept for downlink packet data. The fundamental principle of WCDMA resource sharing is illustrated in the next figure.

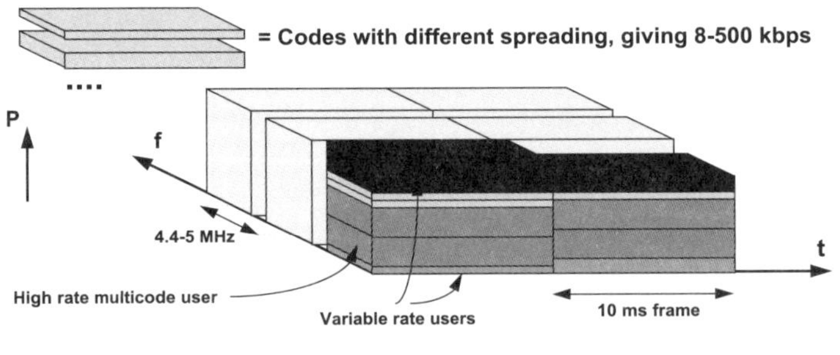

Dynamic resource sharing with WCDMA.

For other service requirements, such as different delay requirements, WCDMA provides different Transmission Time Intervals, with values being configurable to 10, 20, 40 or 80 ms. Multi-service capability has already been provided for fairly simple terminals by means of multiplexing different services with different QoS requirements on the same connection. The simplified WCDMA service multiplexing principle is illustrated in the next figure. In the actual multiplexing chain there are additional steps such as rate matching or interleaving functions that are necessary for proper system operation.

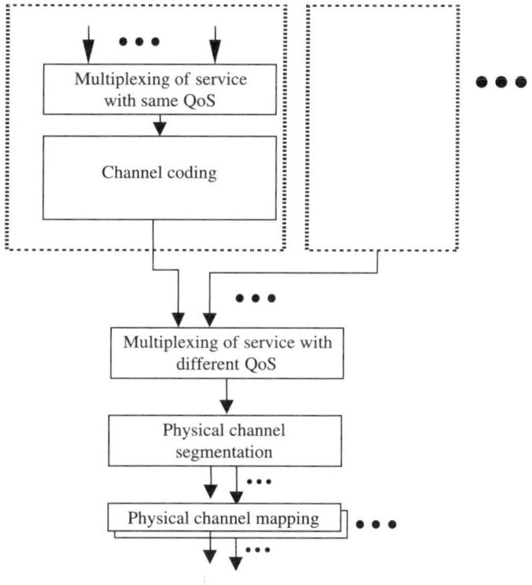

The WCDMA principle of providing different QoS on a single physical channel.

For co-existence with GSM, the handover and necessary measurements have been specified to allow the continued use of GSM in WCDMA or vice-versa for the service set specified on the GSM side.

WCDMA can provide uplink and downlink data rates independently of each other, thus facilitating uplink and downlink asymmetry per a connection basis. With variable duplex spacing it is also possible to configure less spectrum for uplink if this is considered necessary especially with some of the extension bands thought to be available in the future. As current IMT-2000 frequency allocation also provides an unpaired spectrum, a TDD mode of operation has been specified, where users share the same carrier in both uplink and downlink transmission directions. Here, the TDMA principle has been combined with CDMA with users sharing a time slot by spreading codes as illustrated in the next figure. Each time slot can be allocated either for uplink or downlink direction with some limitations. Also, the (wideband) TDD mode of operation uses a 5 MHz bandwidth and the same solutions as with WCDMA in FDD mode have been applied where applicable for the TDD mode as well for maximal commonality. As part of Release 4 there is also a 1.28 Mcps TDD option.

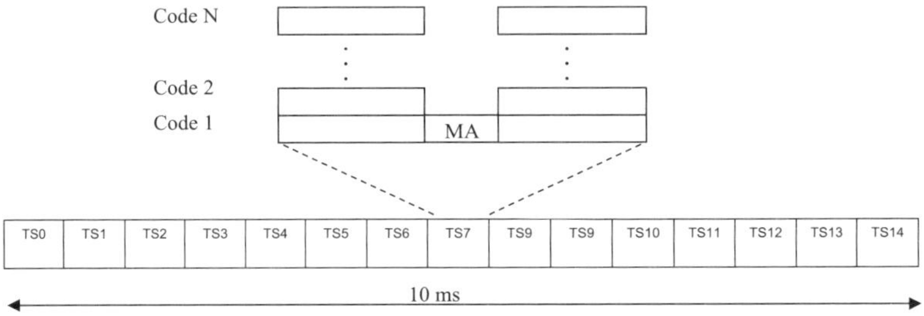

The WCDMA TDD mode of operation with the TDMA/CDMA principle applied.

To achieve high spectrum efficiency, besides using diversity methods due to the bandwidth, WCDMA also incorporates methods not used in 2G standards, such as transmit diversity in the downlink direction. Also, the advanced channel coding methods provide coding gain to achieve low SIR requirements for the desired quality level.

The GSM Terminal class mark is not used in WCDMA. WCDMA terminals inform the network, upon connection, of parameters, which indicate the capabilities of the particular terminal. It is worth noting that terminal capabilities can be given independently for the uplink and downlink direction. To provide guidance for which capabilities should be applied together, example capability classes have been specified in WCDMA standardisation in 3GPP. The following capability classes have been defined:

- 32 kbps class: This is intended to provide basic speech service, including AMR speech as well as some limited data capabilities but not together with speech services.
- 64 kbps class: This is intended to provide a speech and a data service, with simultaneous data and speech capability e.g. AMR speech and 32 kbps data simultaneously.
- 128 kbps class: This class has the air interface capability to provide video telephony and various other data services
- 384 kbps class is being further enhanced from 144 kbps and has multicode capability, which points to support for advanced packet data methods in WCDMA. This class can provide e.g. AMR speech simultaneously with packet data up to 384 kbps
- 768 kbps class has been defined as an intermediate step between the 384 kbps and 2 Mbps class.
- 2 Mbps class: This is the state of the art class and the 2 Mbps capability has been defined for the downlink direction only.

The intention is also that each example class can provide the services that are provided by the classes with lower data rates. A terminal can provide the capability with FDD or with both FDD and TDD. In WCDMA there are several performance improvement features that are mandatory on the terminal side. This allows to take these features into account in initial network capacity/coverage planning as they are supported in all terminals from the beginning . An example of such a feature is downlink transmit diversity.

As part of WCDMA evolution, terminal capabilities are going to develop together with new features. As part of Release 5 work, peak data rates up to 10 Mbps are being discussed.

WCDMA Air Interface Performance

From the end user perspective the quality of a cellular service may be limited by a weak signal, i.e. by the coverage, or by too many users trying to access the service at the same time. In other words, by the capacity of the system. The coverage and capacity of WCDMA will be covered in this section.

Coverage is important when the network is not limited by capacity, such as at the time of initial network deployment, and typically in rural areas. Macro cell coverage is determined by the uplink range from the mobile station to the base station, because the transmission power of the mobile terminal is much lower than that of the base station. The output power of a mobile terminal is typically 21 dBm (125 mW) and that of the macro cell base station 40-46 dBm (10-40 W) per sector.

To understand the typical coverage areas of WCDMA cells, we can compare the WCDMA cell area to existing GSM cells. This comparison is important as GSM operators will utilise their existing cell sites for their UMTS networks. With existing GSM1800 sites WCDMA offers bit rates of 64-144 kbps in uplink traffic with the same coverage probability as for GSM1800 speech. In downlink traffic, the offered bit rates are higher, typically up to 384 kbps. With GSM900 sites WCDMA offers bit rates of about 32 kbps in the uplink direction with the same coverage probability as for GSM900 speech assuming that simple smart antenna solutions are used in WCDMA. The cell size in GSM900 can be larger than in GSM1800 because the signal attenuation is lower in the lower frequency band. Therefore, the coverage of WCDMA on the 2 GHz band will not be as good as for GSM900 speech on the 900 MHz band without smart antenna solutions. On the other hand, the GSM site density is higher in most urban areas to provide more capacity, and therefore, reusing existing GSM sites will provide a high quality WCDMA network.

Typical site areas for GSM1800 and WCDMA will be 2 km^2 in urban areas, 5-10 km^2 and 20 km^2 in rural areas. A 3-sector site is assumed in here. The exact maximum site area depends on the environment and on the required coverage probability.

When traffic increases in the network, a higher base station density may be needed. The typical capacity of a WCDMA air interface is 1 Mbps per carrier per sector per 5 MHz carrier in 3-sector macro cells and 1.5 Mbps micro cell per 5 MHz carrier. This value represents a maximum total throughput per sector which can be shared between the users who request service at the same time. The radio resource management algorithms in the radio network controller allocate the capacity between users. The capacity allocation can be based on the requested services and their priorities.

Providing high capacity will be challenging in urban areas where the amount of traffic can be very high.

Typical capacities of a WCDMA air interface

	Macro cell layer	Micro cell layer
Capacity per site per carrier	3 Mbps with 3 sectors	1.5 Mbps
Capacity per site per operator with 3 UMTS FDD carriers	9 Mbps	4.5 Mbps
Maximum site density in an urban area	5 sites / km^2	30 sites / km^2
Maximum capacity per operator	45 Mbps / km^2	135 Mbps / km^2

The figures do not show the potential of the WCDMA enhancements currently being finalised in 3GPP, such as High Speed Downlink Packet Access (HSDPA). HSDPA is a result of the 3GPP feasibility study, which has shown up to 100% performance improvements for downlink packet data operation. Also, when compariing with other radio access technologies the support of e.g. transmit diversity in all WCDMA terminals gives WCDMA a competitive advantage on the market place.

WCDMA is designed to be deployed together with a GSM network, and handovers are supported from WCDMA to GSM and from GSM to WCDMA. In the initial deployment phase the WCDMA network coverage will not be as large as the GSM coverage and handovers from WCDMA to GSM can be used to provide continuous service even if the user leaves the WCDMA coverage area in connected mode. Also idle mode parameters are supported for inter-system cell reselection.

The inter-system handovers are triggered by the source radio access network, and the handover triggers are vendor specific. Inter-system handover can also be triggered due to high load to balance loading between WCDMA and GSM to fully utilise GSM and WCDMA networks together.

Summary

This chapter briefly covered air interfaces, the spectrum, WCDMA basics, physical layers and terminal capabilities, as well as WCDMA performance. WCDMA technology offers a wide range of possibilities for service introduction, including means for providing a different quality of service as well as means for providing multiple services simultaneously. The radio access network architecture has been shaped by WCDMA requirements and standardisation in 3GPP has produced the specifications to enable proper system operation in a multi-vendor environment with open internal interfaces in the access network. WCDMA networks can use advanced performance enhancement schemes from the beginning and the evolution of WCDMA technology during 2001/2002 is going to further boost the packet data capacity, making WCDMA especially suited to carry IP based traffic. The WCDMA radio access technology of UMTS is covered in more depth in [1] and the references therein. The latest WCDMA radio specifications developed by the 3GPP TSG RAN can be found in [2].

REFERENCES

[1] H.Holma & A.Toskala, "WCDMA for UMTS", Revised Edition, John Wiley & Sons, 2001
[2] www.3gpp.org

GSM/EDGE Radio Access Network (GERAN)

The GSM/EDGE standard is taking steps to define a fully capable 3G mobile radio access network. 3GPP Release 5 will specify a GSM/EDGE radio access network (GERAN) which can connect to a 3G mobile core network through the Iu interface and is capable of offering the same set of services - conversational, streaming, interactive and background - as the UMTS terrestrial radio access network (UTRAN). With Iu-ps interface, GERAN can connect to the IMS (Internet Multimedia Services) domain of the 3G core network for the efficient support for IP based services.

The GSM standard has evolved from a basic voice service to a wide variety of speech and data services over the cource of a decade. The latest completed evolution step, Release 1999, introduced 8-PSK modulation providing a threefold increase in the peak data rates. The addition of 8-PSK modulation was made without modifications to the established GSM/GPRS architecture. Two new data features, Enhanced GPRS (EGPRS) [1][2][3][4] and Enhanced Circuit Switched Data (ECSD) [5] were introduced as was the Compact mode for the narrowband deployment (1 MHz bandwidth at the minimum) [1][2]. EGPRS is an enhancement to GPRS and uses the Gb interface in communication with the packet switched core network. ECSD is an enhancement to HSCSD (High Speed Circuit Switched Data) and uses the A interface to the circuit switched core network.

The next major step in the GSM/EDGE evolution path is Release 5 from 3GPP (to be completed in March 2002). In this step, the functional split between the GSM/EDGE Radio Access Network (GERAN) and the core network (CN) will be harmonized with UTRAN, enabling GERAN to connect to the same 3G core network and provide the same set of services as UTRAN. The new architecture implies significant modifications to the GERAN radio protocols. Also the efficient support of IP based services requires new functionalities in the radio protocols.

In Release 1999, only the data services were enhanced with 8-PSK modulation. In the GERAN Release 5, the voice capacity will be increased as well by defining 8-PSK modulated speech channels. The narrowband AMR codec from GSM Release 1998 is assumed to be the speech codec, and two new channel types - 8-PSK half rate channel and 8-PSK quarter rate channel - are defined.

The section is organized as follows: The 3G traffic classes are described in part two (3G Mobile Services) and the GERAN architecture in part three (GERAN Ar-chitecture). Part four describes the GERAN radio protocols. The GERAN physical layer is described in the fifth part (Physical Layer). Part six presents a few key GERAN performance aspects and the Chapter VII the conclusions.

3G Mobile Services

As a part of the radio access network harmonization, services provided by GERAN are aligned with the UMTS traffic classes. There are four different classes of traffic:

- Conversational,
- Streaming,
- Interactive, and
- Background.

In the conversational class, speech or video is carried between end-users. The quality of service (QoS) requirements are mainly set by human perception. Failure to meet a requirement causes degradation of the user-perceived quality of service, often to a level that is un-acceptable.

Streaming traffic is one-way user data transfer (control data is also transferred in the other direction, typically uplink) having no absolute delay requirement. However, there is a strict requirement to preserve time relation between the information entities. The stream can be time aligned in the receiving end by buffering the data. During the transmission phase relatively large delay variations can be accepted.

Interactive traffic consists of sessions where a person or machine is requesting data from a remote computer. In this traffic type, it is characteristic that a request-response pattern exist in data transfer. Another characteristic feature is that it is often important to guarantee the correct delivery of the requested data. As an example: web browsing, data base enquiry and the polling of measurement data.

Background traffic has the most relaxed delay requirements. It is assumed that the end user does not expect the data for some time. Therefore, data transfer can be carried out in the background when there is no delay critical data to be transferred. Typical traffic belonging to this category is SMS (short message service), e-mail and file transfer.

IP-based multimedia is a new service category in GERAN standard. This means both conversational speech and video. In the circuit switched domain, these services already exist in Release 1999. However, GERAN Releases 5 and 6 make it possible to provide all services (both real-time and non-real-time) in the packet domain, which is attractive to operators, as only one (packet) core network needs to be maintained [6]. Consequently, GERAN will offer the user the freedom to have multiple traffic flows in parallel, e.g. a speech call, an email transfer and web browsing at the same time.

GERAN Architecture

The basic principles that are used for the design of GERAN architecture for the next standard release (Release 4) are:

- Separation of the radio related and not radio related functionalities between the core network and the radio access network. This would enable development of the platform that would allow for provisioning of services independent of the access type.
- Support of services for pre-Release 5 terminals must be ensured.
- Maximize commonalties between GERAN and UTRAN but maintain the backward compatibility for GERAN. The key alignment point between GERAN and UTRAN is a common interface to the 3G core network (Iu interface). Consequently, GERAN should provide the same set of services as UTRAN, e.g. support the same radio access bearers and QoS mechanisms as in UTRAN.
- Standardization of the GERAN Release 5 should support a true multivendor environment.

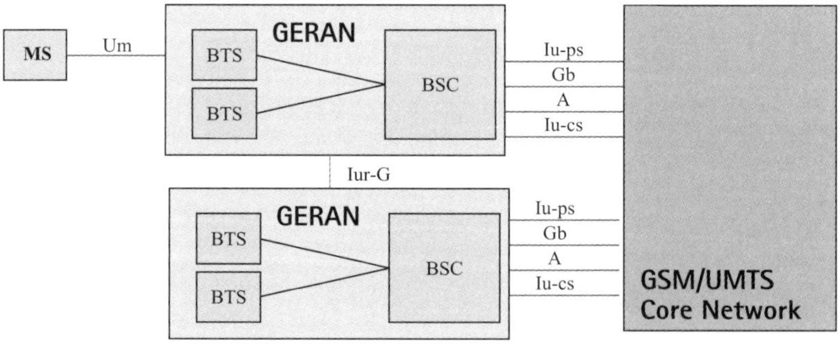

The proposed new GSM/EDGE radio access network architecture.

Legacy Interfaces

The protocol model for legacy interfaces is based on Release 4 for GSM/EDGE and the functionality split between the radio access and the core network remains unchanged.

The Gb interface is used in GSM/GPRS to connect a Serving GPRS Support Node (SGSN) and Base Station Subsystem (BSS). This interface is needed in GERAN for supporting pre-Release 5 terminals.

The A interface which is a traditional interface in GSM that is used to connect BSS and MSC (mobile switching center) will be supported by GERAN. This interface is needed to support pre-Release 5 terminals, and provide new enhanced services like e.g. wideband telephony (with WideBand Adaptive Multi Rate - WB-AMR - speech codec) and support for 8-PSK modulated speech channels.

New Interfaces

The general protocol model for new interfaces is based on the protocol model used in UMTS and is shown in the next figure. The layers and planes are logically independent of each other, which offers additional flexibility. For example, if needed, transport network protocol layers may be changed in the future with virtually no impact on the radio network layer.

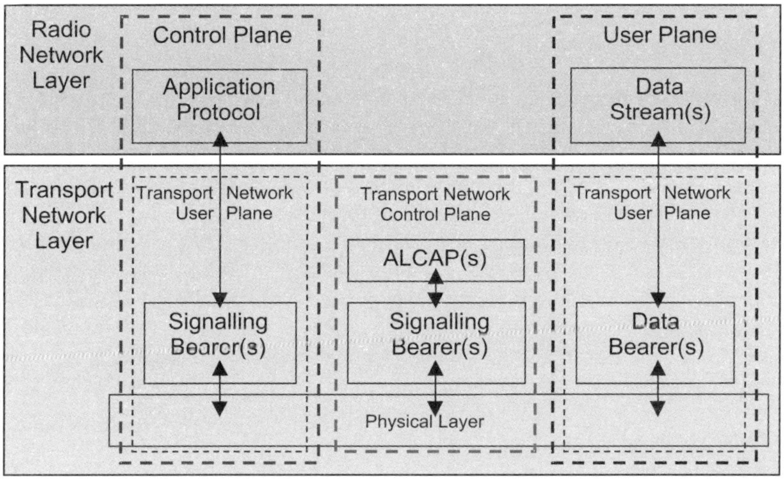

General UMTS protocol model.

The Iu interface is common for UTRAN and GERAN. It is specified at the boundary between the core network (CN) and the radio access network (RAN). The Iu interface towards the PS domain of the core network is called Iu-ps, and the Iu interface towards the CS domain is called Iu-cs. Both interfaces are foreseen to be supported by GERAN specifications. The protocol model for Iu-ps interface is shown in the next figure.

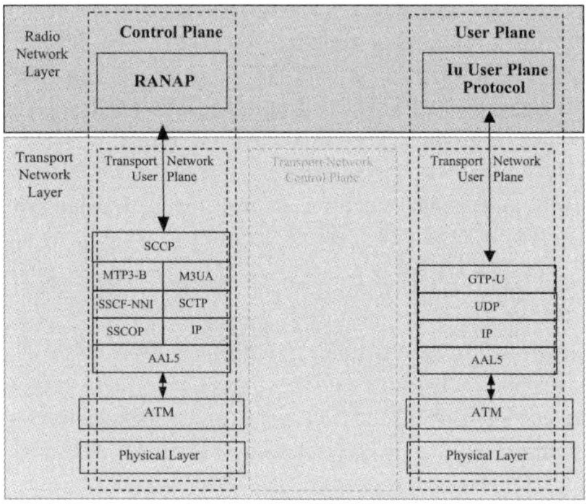

Protocol model for Iu-ps interface.

The Iur-g interface provides a logical connection between any two GERANs and supports the exchange of signaling information between two GERANs. The Iur-g interface is based on the UTRAN Iur interface [7]. However, only a subset of the functionalities is needed for the GERAN of Release 5, mainly related to cell-level mobility management. Therefore, it is proposed that only Radio Network Subsystem Application Part (RNSAP) Basic Mobility Procedures and RNSAP Global Procedures be adopted from UTRAN and modified as necessary. These procedures do not require the user plane, as illustrated in the protocol model in figure below.

The functionality split across these interfaces is aligned with the UTRAN-CN split. For GERAN, this means that functionalities like ciphering and header adaptation are moved from CN and are part of the radio access network. In addition, the level mobility management tasks are solely a part of the radio access network.

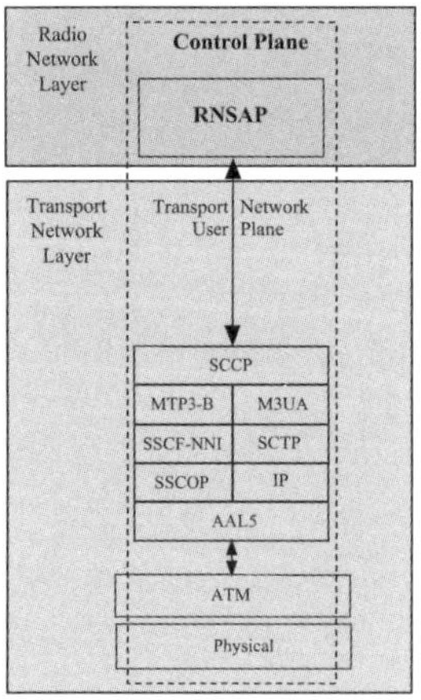

Protocol model for Iur-g interface (subset of UTRAN Iur).

The main benefit of the harmonization is that an operator deploying both radio access technologies (GERAN and UTRAN) may operate one single core network, and both access technologies can provide the same set of services. This includes provision for all the UMTS traffic classes through GERAN. For the network and terminal manufacturers, the harmonization gives benefit in the form of synergies in protocol and interface implementation.

The harmonization of the GERAN and UTRAN architectures simplifies the future development of these radio access technologies. The IP transport is expected to give benefit for the operators by bringing down the transmission costs.

GERAN Radio Protocols: PDCP, RRC, RLC AND MAC

Following figures show the user plane and the control plane protocol stacks in a packet-switched domain (up-to layer 2 only), when connected through Iu-ps to the 3G CN. The Protocol stacks when connecting to the circuit switched core network (CN) domain with Iu-cs, are not shown. Neither are the protocol stacks, when connecting to the CN through legacy interfaces, A and Gb.

Packet Data Convergence Protocol - PDCP

The PDCP allows for the transfer of user data using services provided by the RLC, and header adaptation of the redundant network PDU control information in order to make the transport over the radio interface spectrum efficient. Two different header adaptation mechanisms may be used by the PDCP, for GERAN, header compression and header removal.

Header Compression consists of compressing transport and network level headers so that the decompressed headers are semantically identical to the original uncompressed headers. Header compression is suited for generic Internet applications, especially multimedia applications.

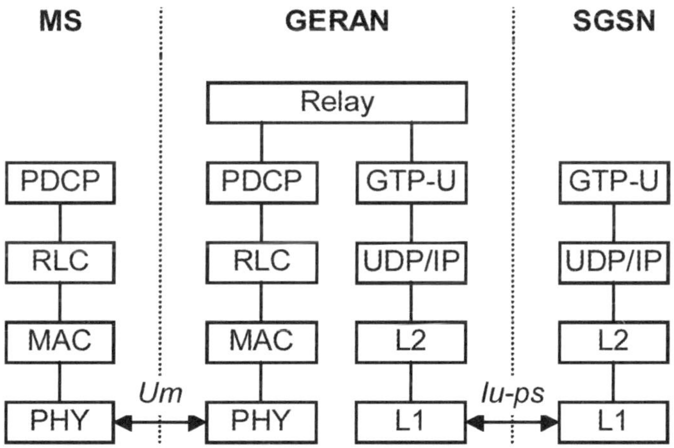

User plane protocol stack towards PS CN domain through Iu-ps

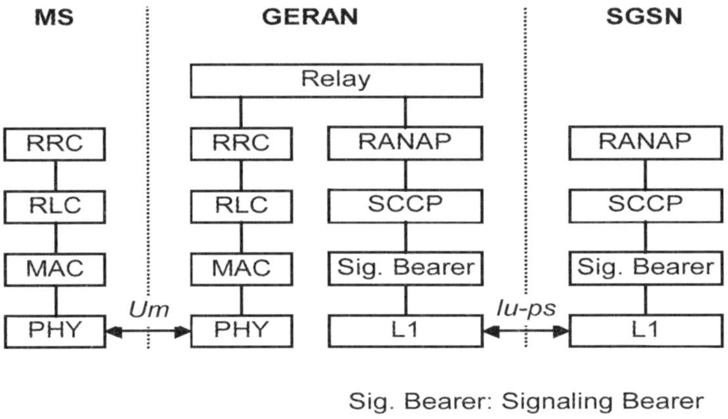

Control plane protocol stack towards PS CN domain through Iu-ps.

Header Removal consists of completely removing transport and network level headers (e.g. RTP/UDP/IP). The receiving entity can, however, regenerate the headers based on information submitted at the call set-up and from the link and physical layers. Semantically, the original and regenerated headers may be different. Header removal targets an optimized speech bearer for which the transmission over the radio interface is circuit-switched like.

Radio Resource Control - RRC

The RRC protocol is based on both GSM RR [8] and UTRAN RRC [9] specifications. Many new RR concepts introduced to UTRAN (compared to GSM Release 4), will be adopted from UTRAN RRC to GERAN RRC, e.g. the Radio Bearer, the RRC connection and the RRC connection mobility management concepts. Other RRC functions are include the broadcast of information related to the non-access stratum (core network) and the access stratum, paging/notification, routing of higher layer PDUs, control of requested QoS, initial cell selection and re-selection in idle mode, support for location services, control of security functions and MS measurement reporting and control of the reporting.

RRC realizes the QoS requirements of the RAB by establishing a Radio Bearer (RB) between MS and GERAN. For each RB RRC allocates either a dedicated or a shared channel. In the case of dedicated channels, RRC also takes care of radio resources allocations. In the case of shared channels, MAC takes care of radio resource allocations.

RRC connection mobility means that in RRC connected mode, the GERAN tracks the location of MS without CN involvement. When there is no RRC connection, i.e. in idle mode, MS is identified by non-access stratum identities (e.g. IMSI, TMSI and P-TMSI) and the GERAN has no own information about the individual MS. RRC connected mode has two levels of accuracy for RRC connection mobility: Cell and GERAN Registration Area (GRA). On the cell level, MS makes cell updates after each GERAN cell reselection and the network tracks the MSs location on the cell level. For inactive users, MS may fall back to the GRA level of accuracy. In this state the MS location is known only with GRA accuracy.

Radio Link Control - RLC

The RLC protocol allows for data transfer in transparent, acknowledged or unacknowledged modes, and notifies unrecoverable errors to the upper layer.

When in transparent mode, RLC has no functionality and does not alter the data units of the upper layer.

In non-transparent mode, the RLC is responsible for ciphering RLC PDUs in order to prevent any unauthorized acquisition of data. The RLC non-transparent mode is built based on (E)GPRS if the Radio Access Bearer (RAB) is mapped onto a Packet Data Traffic Channel (PDTCH). Alternatively, the RAB may be mapped onto an (E-)TCH (Enhanced Traffic CHannel). Signaling is performed on a Packet Associated Control Channel (PACCH) when PDTCH is used, and on a Fast Associated Control CHannel (FACCH) and Slow Associated Control Channel (SACCH) when (E-)TCH is used. The mapping of a RAB onto the proper traffic channel is dependent on the MAC mode and the targeted traffic class, as described in the next part. In acknowledged mode, Backward Error Correction (BEC) procedures are provided that allow error-free transmission of RLC PDUs through two possible selective ARQ mechanisms: Type I ARQ and Hybrid Type II ARQ (Incremental Redundancy). In unacknowledged mode, no BEC procedure is available.

Medium Access Control - MAC

The MAC sublayer allows the transmission over the physical layer of upper layer PDUs in dedicated or shared mode. A MAC mode is associated with a physical subchannel for use by one or more mobile stations (dedicated or shared mode respectively). The MAC layer handles the access to and multiplexing onto the physical subchannels of mobile stations and traffic flows.

GERAN Physical Layer

The physical layer of GERAN interfaces the MAC and the RRC protocols. It offers logical channels and associated transmission services. Logical channels are divided in two categories:

- Traffic channels, and
- Control channels.

Traffic channels are intended to carry either encoded speech or user data, while control channels carry signalling or synchronization data. Logical channels are multiplexed either in a fixed predefined manner, or dynamically by the MAC on physical subchannels. Physical SubCHannels (PSCH) are the units of the radio medium. Some are reserved by the network for common use, others are assigned to dedicated connections with MSs (Dedicated Physical SubCHannel - DPSCH), or are assigned to a shared usage between several MSs (Shared Physical SubCHannel - SPSCH). The types of logical channels that are allowed on a PSCH depend on the mode of the PSCH. In the following, emphasis is given to traffic channels.

Traffic channels of type TCH are intended to carry either encoded speech or user data on a physical subchannel in dedicated MAC mode only, i.e. for one MS only. The TCH use acircuit-like connection over the radio interface. The TCH can be either GMSK or 8-PSK modulated. One main characteristic of the channel coding for the TCH is the use of diagonal interleaving. The two categories of TCH are Speech TCH and Data TCH.

The channel coding of Speech TCH is optimized to maximize the perceived speech quality. Output bits from the speech codec are divided into different classes based on their subjective importance and the bits of the different classes are protected unequally in channel coding. Earlier releases of the GSM standard already introduced a Full Rate (FR) and Half Rate (HR) speech on GMSK channels, an Enhanced Full Rate (EFR) speech on GMSK channels, and finally the AMR codec on both full rate and half rate GMSK channels.

The next GERAN releases will introduce the wideband AMR codec for high quality speech services, and will introduce narrowband AMR on half rate 8-PSK channels to increase the speech capacity [10]. The definition of the QR channels in GERAN standard is for further study.

Table below lists the average code rates for the different AMR codec modes and codec rates. Diagonal interleaving over eight bursts is used on the full rate channel, over four bursts on the half rate and over two bursts on the quarter rate channel.

Unlike Speech TCH, the channel coding of Data TCH includes equal error protection for the payload bits. Furthermore, due to the more relaxed delay requirements, the channel coding is enhanced with longer interleaving.

Codec Mode	*Bits per Speech Frame*	*Average Code Rate*			
		GMSK		*8PSK*	
		FR	*HR*	*HR*	*QR*
12.2	244	0.54	x	0.36	*0.72*
10.2	204	0.45	x	0.30	*0.60*
7.95	159	0.35	0.70	0.24	*0.47*
7.4	148	0.33	0.65	0.22	*0.44*
6.7	134	0.30	0.59	0.20	*0.40*
5.9	118	0.26	0.52	0.18	*0.35*
5.15	103	0.23	0.46	0.15	*0.31*
4.75	95	0.21	0.42	0.14	*0.28*

Speech frame sizes and code rates for different AMR codec rates (both in GMSK and 8-PSK channels).

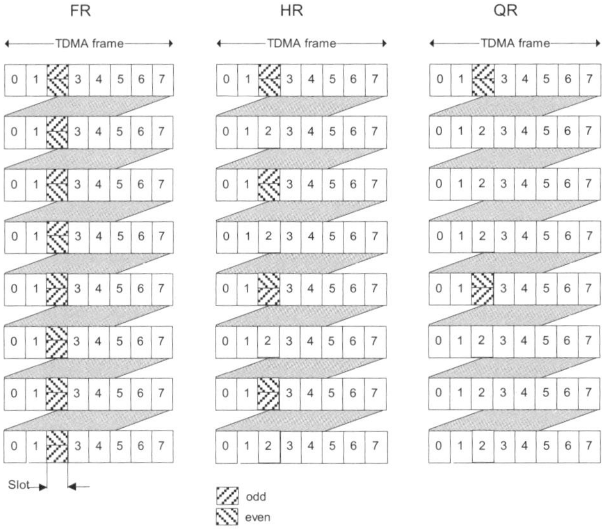

Channel mapping and interleaving for full rate, half rate and quarter rate channel modes.

A wide range of data TCHs already exist in the GSM/EDGE standard, using both GMSK and 8-PSK modulation, and offering data rates from 9.6 kbps to 43.2 kbps in a single slot configuration. In E-TCH/F32.0 multislot configuration, e.g. the 64 kbps data rate can be provided with two time slots. The GERAN physical layer will reuse these Data TCHs. Typically GERAN will use Data TCHs for services with relaxed delay requirements (e.g. streaming services) in dedicated MAC mode.

TCH	*Modulation*	*Code rate*	*Data rate (kbps)*
TCH/F9.6	GMSK	0.53	9.6
TCH/F14.4		0.64	14.4
E-TCH/F28.8	8PSK	0.42	28.8
E-TCH/F32.0		0.47	32.0
E-TCH/F43.2		0.64	43.2

Data TCH modulation and coding schemes.

MCS	*Modulation*	*Code rate*	*Data rate (kbps)*
MCS9	8PSK	1.0	59.2
MCS8		0.92	54.4
MCS7		0.76	44.8
MCS6		0.49	29.6
MCS5		0.37	22.4
MCS4	GMSK	1.0	17.6
MCS3		0.80	14.8
MCS2		0.66	11.2
MCS1		0.53	8.8

EGPRS modulation and coding schemes

Packet Data Traffic CHannels (PDTCH) are intended to carry user data on physical subchannels in either dedicated or shared MAC mode. PDTCH is temporarily dedicated to one MS but one MS may use multiple PDTCHs in parallel for individual packet transfer (multislot configuration). PDTCH allows several MSs to be multiplexed on the same SPSCH. PDTCH also allows several traffic classes from the same MS to be multiplexed on the same PSCH (shared or dedicated).

The channel coding of PDTCH uses short rectangular interleaving in order to ensure efficient multiplexing. In GPRS, four different coding schemes based on GMSK modulation were standardized. In EGPRS an additional set of 9 coding schemes, using both GMSK and 8-PSK modulation were introduced. The available data rates run from 8.8 kbps to 59.2 kbps in a single slot configuration. In multislot configuration the maximum data rate can be brought up to 473.6 kbps. PDTCH can be used by GERAN for any traffic class.

The fast block level power control technique was introduced in Release 1999 for the ECSD (Enhanced Circuit Switched Data) to increase the spectral efficiency, especially for the transparent circuit switched service. In the Release 5, the fast power control technique is defined for the speech.

GERAN PERFORMANCE

EGPRS Link Adaptation

The MCS scheme can be dynamically selected to maximise the throughput depending on the radio link conditions. Link Adaptation (LA) algorithms can be based on the various available link quality indicators (e.g. BLER, C/I, BEP and size of IR memory). One of these indicators, the Bit Error Probability, including both mean and deviation, is a new MS channel quality measurement that has been introduced in EDGE for more accurate DL quality reporting. The user data rate for a single TS with ideal LA can be seen in the next figure (TU3 and ideal FH).

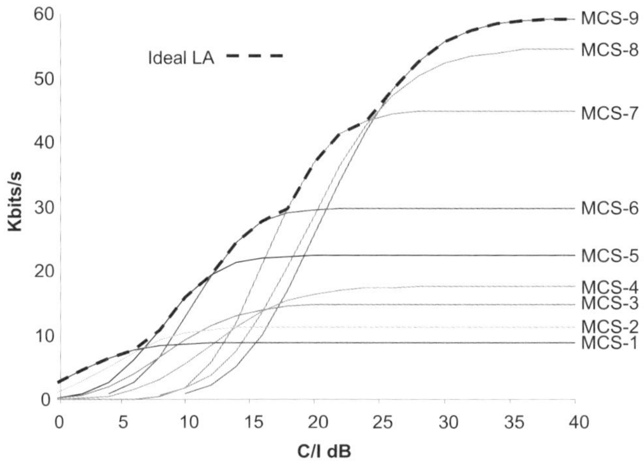

Throughput experienced for the different MCS (TU3 channel and ideal FH conditions).

The use of LA increases the overall network throughput (better spectral efficiency) compared with the case of a fixed coding scheme. LA performance benefits can be envisaged from the GPRS studies carried out (only four coding schemes considered). The benefits of link adaptation have been evaluated in different network scenarios (different traffic load and reuses), as depicted in Figure 10. Throughput is increased by more than 170% in particular scenarios. LA network performance is very sensitive to initial coding scheme selection due to the speed of the LA algorithm.

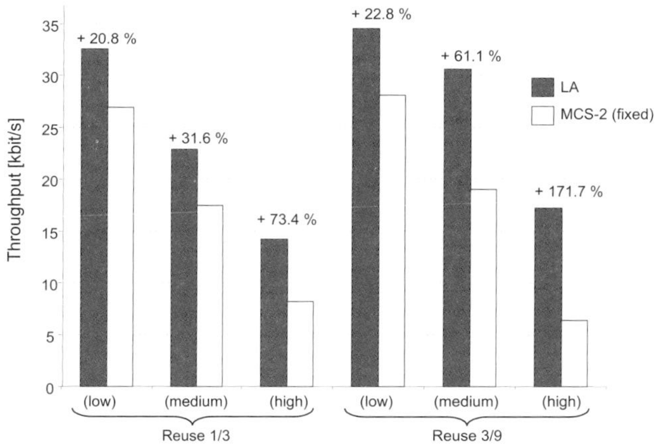

Performance of LA with MCS1-4
LA based on C/I, initial coding scheme is MCS-2.

Incremental Redundancy

Incremental Redundancy (IR) automatically adjusts the code rate to actual channel conditions by incrementally transmitting redundancy information until decoding is successful. IR does not require any channel quality measurements and can operate based solely on acknowledgement messages sent by the receiver. Typically the same MCS is used for the initial transmission and also for all the retransmissions, but with different puncturing schemes. Combining a retransmitted block with the previous transmissions of the same block significantly reduces the block error probability of the retransmitted block. This combining reduces the effective coding rate and therefore high MCSs can be used even in relatively poor radio conditions.

Link level simulations for MCS-8, MCS-6 and MCS-2 with and without IR combining can be seen in figure below. Appreciable performance increases are envisaged from the link level results. LA and IR are similar in the sense that they adapt the coding rate to the radio environment. In order to maximise efficiency both techniques can be efficiently combined.

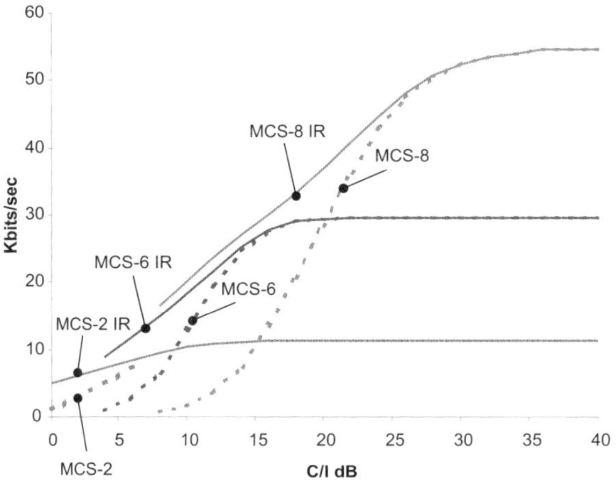

Incremental Redundancy (IR) performance in EGPRS. TU3 channel, ideal FH.

The IR gain versus C/I is illustrated in the next figure. It shows that the highest gain is progressively obtained for lower C/I.

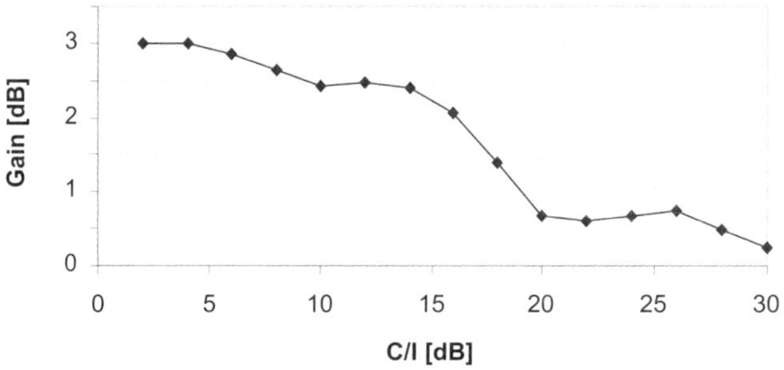

IR Gain (ideal LA) versus C/I

The network level impact of IR is available from EDGE Compact studies. The results presented in the next figure. Up to 50% capacity gain has been obtained compared to fixed MCS-6. However, higher delays are typically experienced with IR.

For boosting the EGPRS performance further, network vendor specific enhancements like the use of smart antennas in the base station - including the downlink delay diversity - can be used.

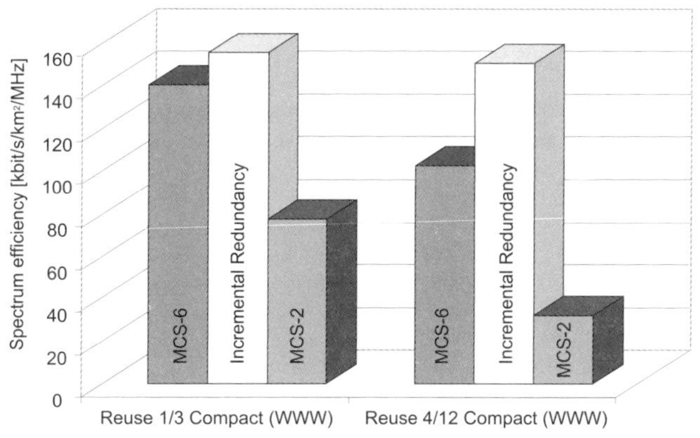

Spectrum efficiency for fixed MCS-6, MCS-2 and IR in EDGE Compact. The initial coding scheme for IR is MCS-8.

ECSD Performance

Link level results and network level performance simulation results for ECSD are presented in the following sections.

ECSD Link Level Performance

The link level results in terms of throughput versus C/I conditions are depicted in the figure below [11].

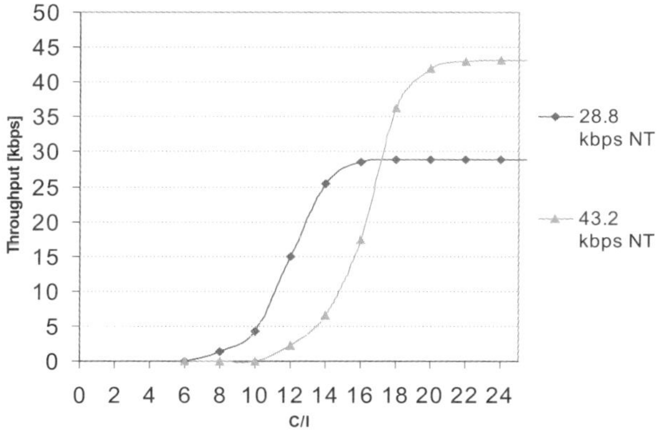

Performance of ECSD/NT at TU3/iFH

ECSD Network Level Performance (T and NT Services)

For the T service spectral efficiency analysis, three different connection types are compared: 2-slot GSM 14.4 kbps, 1-slot 28.8 kbps ECSD/T, and 2-slot 32 kbps ECSD/bit. Such types could be used as an example for video calls. For E-TCH, fast power control is used, while the 14.4 kbps GSM channel uses the existing power control scheme. The next figure shows the results, spectrum efficiency for ECSD/T is doubled when compared to the one achieved by GSM [12].

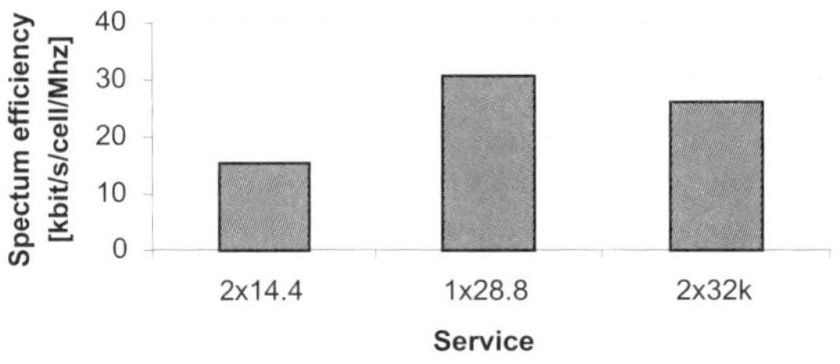

Spectral efficiency for T service in microcell environment. Frequency reuse 3

For the NT service, two different cases have been studied [12]: in Case 1 (ECSD NT) all users used 1-slot 43.2 kbps ECSD/NT and in Case 2 (HSCSD NT) the users requested a maximum of three 14.4 kbps GSM channels in order to get the same user data throughput.

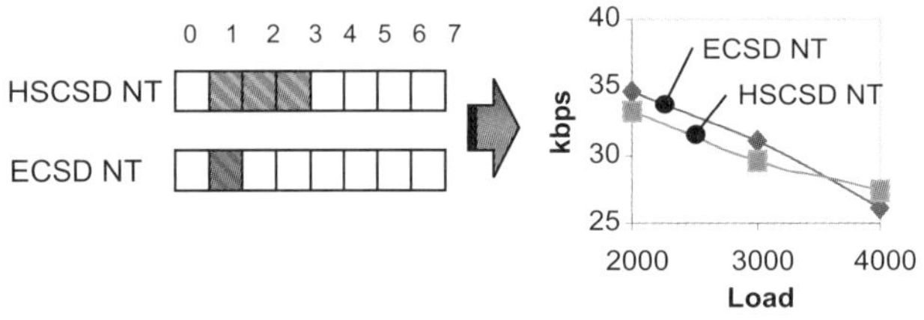

Performance for NT service in macrocell environment

The results show that 2-3 times less radio resources are needed with ECSD in order to satisfy the same required throughput. ECSD link adaptation functionality will ensure that the coding scheme is dynamically selected in order to ensure maximum performance.

GSM/EDGE Speech Evolution and Performance

The speech quality of a connection is defined by the speech codec in use and the radio network performance. Since the development of the first GSM FR speech codec the enhancement of these aspects has been under continuous development. In addition to plain voice quality aspects, the drivers behind the evolution have also been coverage and capacity related. The latest AMR narrowband codec introduced an important improvement as the speech quality, coverage and capacity can be dynamically maximised according to the radio channel conditions.

Currently the standardisation forums are working on the next generation AMR speech codec, which can provide significantly higher speech quality through the introduction of a wideband AMR codec. Next figurepresents the evolution path of the current and foreseen speech codecs.

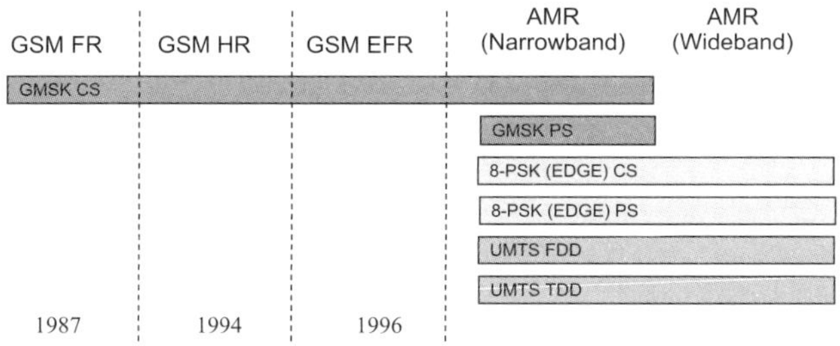

Evolution path of GSM speech codecs

The figure below presents some performance results for full rate (FR) 8-PSK speech channel, half rate (HR) 8-PSK speech channel, quarter rate (QR) 8-PSK channel and octal (1/8) rate (OR) 8-PSK speech channel for TU3 channel with iFH. The table below shows the channel coding and interleaving parameters, that were used in the simulations. Note that quarter rate and octal rate channels are not part of GERAN standard.

	CC rate	K	Interleaving Depth
14.4 TCH/FS	1/5	7	8F
15.7 TCH/FS	1/5	7	8F
7.4 TCH/HS	1/4	7	4F
12.2 TCH/HS	1/3	7	4F
14.4 TCH/HS	1/3	7	4F
5.9 TCH/QS	1/3	7	2F
7.4 TCH/QS	1/3	7	2F
5.9 TCH/OS	1/3	7	1F
7.4 TCH/OS	1/3	7	1F

Channel coding and interleaving parameters of example simulations.

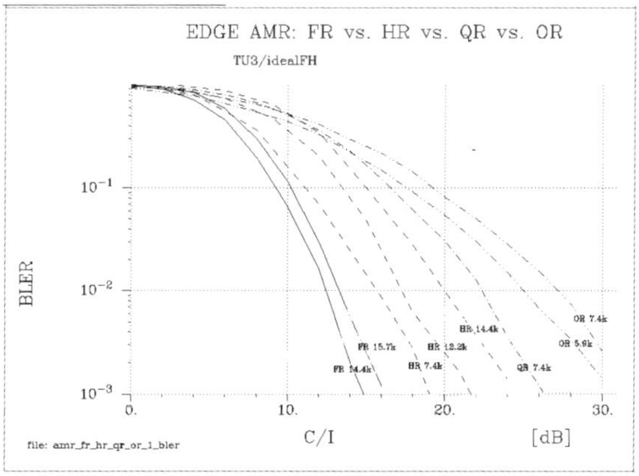

Performance of 8-PSK FR, HR, QR and OR speech channels.

The link level simulation results presented above give a rough estimate of the required C/I values in order to provide speech services with acceptable QoS. In high interference conditions, AMR link adaptation can be used to switch to the more robust codec or channel mode.

Conclusion

The alignment of the GSM/EDGE radio access network (GERAN) architecture with the UTRAN architecture allows GERAN to

- connect to the 3G core network through the Iu interface, and
- to provide the same set of 3G services as UTRAN.

Further, the alignment makes the future development of the both radio access technologies (GSM/EDGE and WCDMA) towards an IP based multiradio networks smoother.

The other significant developments in the GERAN Release 5 include the support for IP based services and spectral efficiency and quality enhancements for voice services - in terms of half rate 8-PSK speech channels, fast power control and Wideband AMR speech codec.

References

[1] R. Pirhonen, T. Rautava, J. Penttinen, TDMA Convergence for Packet Data Services, IEEE Personal Communications, 1999.

[2] A. Furuskär et al, EDGE: Enhanced Data Rates for GSM and TDMA/136 Evolution, IEEE Pers. Commun., Vol. 6, No. 3, June 1999, pp. 56-66.

[3] Y. Zhao, G. Sébire, E. Nikula, Advanced Measurements for EGPRS Link Quality Control, to appear in the proceedings of APCC 2000, November 2000.

[4] K. Balachandran et al, A Proposal for EGPRS Link Control Using Link Adaptation and Incremental Redundancy, Bell Labs Technical Journal, July-September 1999.

[5] S. Hamiti, M. Hakaste, M. Moisio, N. Nefedov, E. Nikula, H. Vilpponen, Enhanced Circuit Switched Data for Real Time Services over GSM, VTC'99 fall, Sep. 1999, pp. 578-582.

[6] X. Qiu, K. Chawla, L.F. Chang, J. Chuang, N. Sollenberger, J Whitehead, RLC/MAC Design Alternatives for Supporting Integrated Services over EGPRS, IEEE Pers. Commun., Vol. 7, No. 2, April 2000, pp. 20-33.

[7] 3GPP TS 25.420, UTRAN Iur Interface General Aspects and Principles, Release 1999.

[8] ETSI GSM 04.18, Radio Resource Control Protocol, Release 1999.

[9] 3GPP TS 25.331, RRC Protocol Specification, Release 1999.

[10] T. Bellier, M. Moisio, E. Nikula, R. Pirhonen, H. Rantala, Quarter Rate AMR Channels - A Speech Capacity Booster in GSM/EDGE, to appear in the proceedings of WPMC'00, November 2000.

[11] Shkumbin Hamiti et al. "Enhanced Circuit Switched Data for Real Time services over GSM". Nokia Research Centre.

[12] ETSI SMG2 "ECSD-concept evaluation", Tdoc SMG2, 670/99, 31 May-4 June, 1999, Tucson.

Bluetooth

The Framework

In early 1998 five companies, Ericsson, IBM, Intel, Nokia and Toshiba, funded an industry consortium to develop a new wireless technology to connect different portable devices. The newly created 'Special Interest Group' (SIG) specified a system for short-range radio communication suitable for small handheld devices with low current consumption. The system operates in the almost globally available, unlicensed 2.4 GHz ISM band. A fundamental objective of the five promoter companies was to produce an open specification, publicly available and royalty free. This was achieved by a cross-license agreement of all intellectual properties within the specification beyond all members of the SIG. In order to participate in this agreement, companies have to sign the SIG agreement. Today, the SIG comprises more than 2800 companies. In 1999, the promoter group was extended by 3COM, Lucent, Microsoft, and Motorola.

Based on early studies by Ericsson and Nokia, and based on many contributions from other members of the 'Bluetooth Special Interest Group', the 1.0 specification of the wireless technology was released on the 26th of July 1999. The specification is being improved with successive new releases, including improved and new functionality. The specifications consist of two documents, the core specification and the profile specification. The core specifies components such as the radio, baseband, link manager, service discover protocol, transport layer, and interoperability with different communication protocols, while the profile specifies the protocols and procedures required for different types of Bluetooth applications.

Interoperability is seen as one of the most critical success factors for the technology. To foster interoperability, the SIG has established a qualification program to test Bluetooth products for their compliance with the specification. The actual IP license is tied to the result of the qualification tests.

Technical Overview

Bluetooth Lower Layers

Bluetooth wireless technology defines a system specification for short-range wireless voice and data communications. The system operates in the 2.4 GHz Industrial, Scientific, Medical (ISM) band. It uses a fast frequency hopping scheme with 79 frequency channels of each 1 MHz bandwidth. The following table summarizes the main parameters of the Bluetooth radio and baseband specification:

Frequency	2401.5 – 2480.5 MHz
Modulation	GFSK, BT = 0.5, Index: 0.28 – 0.35
Bit Rate	1 Mbps
Hopping Rate	1600 hops/s (typical) 3200 hops/s (inquiry, paging)
Output power	Class 3: 0 dBm (typical) Class 2: 4 dBm (optional) Class 1: 20dBm (optional)
Receiver sensitivity (@ 0.1% BER)	-70 dBm
Topology	Star (Master-Slave)
Duplexing	Time Division Duplexing (TDD)
Connection Types	Synchronous, Asynchronous
Encryption	1 bit stream cipher with key up to 128bit

The star topology of the system requires one master device. This device can handle active connections with 7 slave devices simultaneously. However, it is possible to temporarily suspend an active communication with a slave device (park mode). Parked devices are also part of the piconet, thus the number of devices in a piconet can exceed seven.

The Radio is specified to be simple, resulting in a low cost implementation, as well as having relaxed requirements for sensitivity and implementation accuracies, which allows low current consumptions in implementations. A typical current consumption for a state-of-the-art implementation is in the area of 15-30 mA for a 64 kbps connection. Furthermore sophistiacted low-power modes allow to reduce the current consumption during idle times.

The Baseband defines different packet types for both synchronous and asynchronous transmissions, as well as packet types supporting different error handling techniques (error correction or detection, encryption, etc.). Also, data whitening is used to reduce any DC offsets in the receiver due to special payload characteristics.

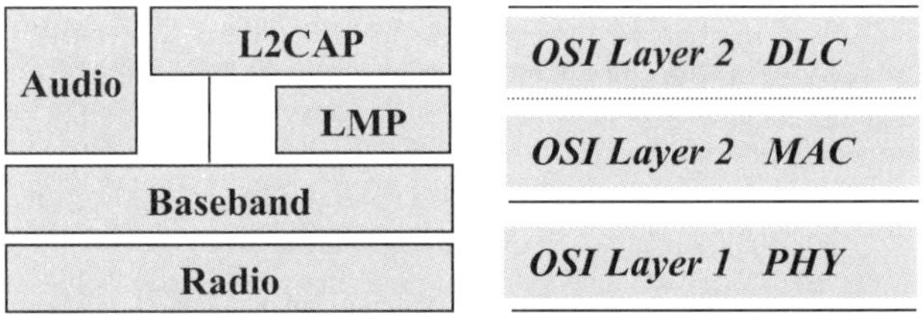

The figure shows the lower layers of the Bluetooth protocol stack in comparison with the ISO/OSI reference model. The Bluetooth specifications do not strictly follow the ISO/OSI model, but the Bluetooth functions can be related to the functions in that model as shown in the figure.

The Logical Link Control and Adaptation Protocol (L2CAP) layer shields the middleware and applications protocols from the complex lower layer Bluetooth functions. It offers transparent transmission of higher layer data over the Bluetooth air interface. L2CAP supports functions like protocol multiplexing, assembly/reassembly of large data packets received from higher layers and the control of quality of service.

The Link Manager Protocol (LMP) has the responsibility to control the connections of a device. This includes functions like connection establishment and link detachment, security management (authentication, encryption) and power management of different low power modes. There might be multiple instances of Link Managers if a device maintains multiple connections.

Bluetooth Middleware Protocols

The middleware protocols comprise references to exisiting middleware protocols, modified exising protocols as well as protocols explictly developed for Bluetooth wireless technology. One of the important protocols is the Service Discovery Protocol (SDP), as Bluetooth focuses on ad-hoc networks of different devices and different capabilities. This dynamic environment is managed by the SDP, which defines a standard way for devices to describe supported services offered to other devices.

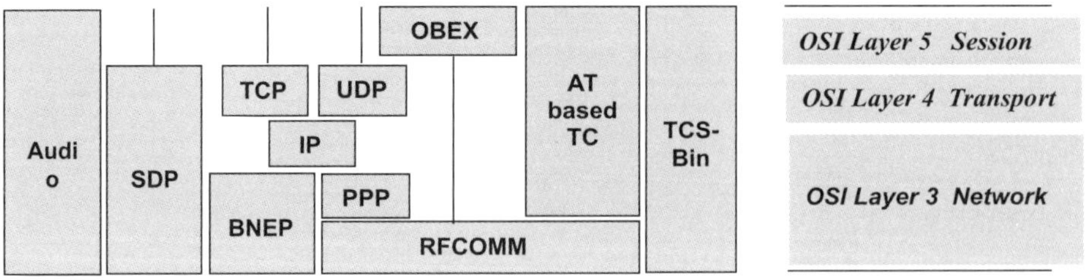

The RFCOMM protocol defines a virtual serial port for applications that makes the reuse of existing applications relatively simple. Especially for the usage model of cable replacement, the serial port is an important functionality of the Bluetooth specifications. The definition of RFCOMM incorporates major parts of the ETSI TS 07.10 standard, which defines multiplexed serial communications over a single serial link. As many of the other transport protocols in Bluetooth, RFCOMM uses an L2CAP connection to establish a link between two devices.

Another key protocol in the specifications is the OBject EXchange (OBEX) protocol, which provides an easy and spontaneous mechanism to transfer objects over the channel provided by RFCOMM. This session protocol is based on a specification defined in the Infrared Data Organization (IrDA), called IrOBEX used for object exchanges over infrared channels. In principle, the OBEX protocol can also run over TCP/UDP, but this is not cleary specified in the Bluetooth specifications. Additionally, the OBEX specifications also define four object formats to be transmitted over Bluetooth. These are:

- **vCard** format to describe electronic business cards, managed by the Internet Mail Consortium
- **vCalendar** format to define electronic calendar and schedule entries
- **vNote** short electronic note format defined by the IrDA organisation
- **vMassage** electronic message and electronic mail format by IrDA.

The OBEX protocol and OBEX applications are supported by the Service Discover Protocol, which offers information about the supported OBEX services in the SDP's service records.

Although RFCOMM and OBEX based applications are important for ad-hoc networking and non-reprogrammable small devices, the industry trend is moving towards IP transport based applications. While in the 1.1 core specifications the Bluetooth IP specification was based solely on IP on top of PPP, during summer 2001, a new profile for Bluetooth was released, which allows IP packets to be encapsulated into a protocol running directly over L2CAP. This new protocol, Bluetooth Network Encapsulation Protocol, BNEP, allows to exploit the full benefits of IP as the restrictions of an underlying Point-to-Point Protocol (PPP) no longer exist.

Bluetooth wireless technology also supports audio and telephony applications. The Telephony Control System (TCS) specifications define how audio is transferred as well as how telephony functions are controlled over a Bluetooth link, i.e. how call control functions like dialling or call acceptance can be controlled by a remote device. While the control functions are defined in the TCS specifications, the audio is routed directly to the baseband layer, without being routed through L2CAP or other layers. In addition to TCS, the specifications also define AT commands between two devices for AT based telephony control.

Bluetooth Profiles

Although the Middleware and Lower Layer functionality are defined in the core specifications of Bluetooth, profiles are defined in a separate profile specifications document. Profiles determine the mandatory and optional parts of the specification for the use cases. The definiton of profiles fosters interoperability between devices from different manufacturers.

Bluetooth profiles can be classified into 4 categories:

- Generic Bluetooth profiles
- Serial Port Profiles
- Telephony Profiles
- Networking Profiles

Two generic profiles exist, which are generic as they cannot be directly related to a special use case, therefore they are mandatory profiles for all Bluetooth devices. The Generic Access Profile (GAP) is the root of all other profiles and defines the functions needed for basic Bluetooth connectivity. The other generic profile, Service Discovery Application Profile (SDAP) maps to the SDP defined in the core specifications and is mandatory in all Bluetooth devices.

The serial port profiles are used for applications, which can be described as cable replacement usage scenarios, therefore mapping directly to the RFCOMM protocol. The profiles are Generic Object Exchange (GOEP), Synchronization (SP), Object Push (OPP), File Transfer (FP) and Serial Port (SPP).

Telephony profiles include profiles which define typical telephone use cases, like cordless telephony or the use of a headset. The profiles are Cordless Telephony (CTP), Headset (HSP) and Intercom (IntP).

For data applications the networking profiles are of interest. They include profiles for Dial-Up Networking (DUNP), Fax (FaxP) and LAN Access (LAP). Just recently a new profile was approved, which defines the how IP is used directly on top of L2CAP. This Personal Areas Networking profile is going to be a new, interesting alternative for Mobile Internet Applications.

Future Developments

Bluetooth wireless technology has become a mature technology, both from a standardization point of view and for the developed technology and products. The Bluetooth SIG has an outrageous number of industry members and many companies are working on products and technical solutions. The key for Bluetooth to become a real volume business are products that are easy to use, without any interoperablity problems between devices of different manufacturers and for a low price. The material cost for Bluetooth functionality is expected to go down dramatically, when the volumes of products increases to tens of millions. From current cost levels of 10-20 USD it is expected that in 2004 Bluetooth functionality will add less than 5 USD to the material cost of a device, if not even less.

While the majority of current Bluetooth products support functionality, which concerns cable replacement solutions, we will see a transition into the area of Personal Area Networking applications. Low current consumption and low cost allows integration to many personal devices, enabling the creation of your own short range network. The Bluetooth SIG is focusing in currently ongoing standardization work, heavily on the definition of personal area networking profiles.

Summary

Bluetooth wireless technology is an integral part of the Mobile Internet. Emerging from a widespread use in personal ad-hoc networking and cable replacement scenarios, the technology will migrate into usage models of Internet access. Especially local services will be provided by the use of Bluetooth wireless technology.

The Symbian Platform

In the past, most mobile phone manufacturers implemented their own, proprietary operating systems for their devices. The next generation of mobile terminals is supposed to provide a wealth of features and capabilities, and it will become increasingly difficult to extend proprietary operating systems to support these new features. Because of this development, an increasing effort has been made by different mobile terminal manufactures to develop a standard operating system that would be suitable for a wide variety of mobile devices.

Motivation

In recent years, the integration of computing and communication devices has increased enormously. Phones and computers have become not only wireless, but are also more integrated into one device. Such devices can be divided into two categories:

- communication oriented, so-called "feature phones" or "Smart phones", with limited means to input, process, store or transfer data (mobile phones with add-on applications and PC connectivity) or
- computing oriented devices, so-called "communicators" with enhanced data input capabilities, like a keyboard, more processing power and memory (handheld computers with connectivity to or built-in mobile phones).

However, these two categories are constantly merging, resulting in one wireless device with capabilities for communication as well as computation.

For such devices, Symbian has developed the Symbian platform.

The Symbian consortium was founded in June 1998 and consists of the leading mobile phone manufacturers Nokia, Motorola, Ericsson and Matsushita and the PDA manufacturer Psion. Furthermore, the Symbian platform is supported and licensed by an increasing number of mobile terminal manufacturers.

Derived from Psion's PDA operating System EPOC, the Symbian OS is one of the longest coordinated efforts to develop an operating system for mobile devices. The current version of the operating system is release 6.0, which is used for example in the new 9210 Communicator from Nokia [FoNo01].

Overview of the Symbian platform

The Symbian platform consists of

- the operating system
- a set of applications
- support for different means of communications (voice, messaging and browsing)
- support for a PC connection (file browsing and conversion as well as synchronization and installation of additional software)
- a framework for graphical user interfaces for different device families
- an application architecture framework and application development support

To cope with the many different form factors of mobile devices, Symbian has defined different device families with different input / output capabilities (screens, pointing devices, data input devices). These so called "Device Family Reference Designs" (DFRD) enable the reuse of components within the same product family, the sharing of system resources and binary compatibility between different devices that belong to the same device family [EPOC5], [Task00].

In addition to the operating system, the Symbian platform provides an application suite for the most common PDA applications: Personal Information Management, messaging, browsing, office applications and more.

The user interface framework eases the development and porting of applications and enables an application to have the same look and feel as native applications [Symb01]. At the same time the application architecture supports the development of document-centric and/or task-oriented applications. To support the development of further applications, the Symbian platform also contains Software Development Kits for applications written in C++ or Java as well as tools for the development of applications that allow the connection to a PC.

The Symbian OS

In beginning of November 2000 Symbian launched their sixth version of the operating system, which is optimised for pen and keyboard based Communicators supporting color displays, Bluetooth, WAP, HTML, SMS, Java and email [Symb01].

The Symbian OS is designed for devices with limited battery capacity, memory and processing power, especially Smartphones and Communicators.

Current hardware requirements are a 32 bit CPU, like the StrongARM processor, and a processor frequency or 36 MHz or more [EPOC5]. A typical device is the 9210 Communicator from Nokia.

The main features of the Symbian OS are

- a component-based operating system that was designed from the "ground up" to be fully object-oriented
- a robust, modern and well-architectured 32-bit microkernel with pre-emptive multi-tasking,
- which allows its processes to provide their own memory protection
- support for the client-server architecture, which ensures the stability of the system even in the case of a problem on the client side of an application

Due to these features, as well as a flexible object oriented application architecture, the Symbian OS can satisfy the need for an operating system that does not lose user data and provides a consistent and simple User Interface.

Generic Technologies

Symbian's Generic Technology (GT) is the common core of APIs and technology, which is shared between all Symbian reference designs. The use of the same Generic Technology for different reference designs enables the simultaneous development of the designs for different wireless devices. Furthermore it eases the porting of applications between reference designs, since those parts of an application, which use Generic Technology APIs, do not need to be changed [SyCr01], [SyQu01].

The major components of the Symbian's Generic Technology are

- Base: The fundamental runtime system, the tools required to build it, and lowest-level security.
- Application framework: Middleware APIs for data management, text, clipboard, graphics, internationalization, and core GUI components.

- Communication: Wide-area communications stacks including TCP/IP, GSM, GPRS and WAP. Personal-area communications stacks including infrared, Bluetooth and serial.
- Messaging: Internet email, SMS text messages and fax.
- Browsing: WML and HTML browsing engines.
- Java runtime: Personal Java 3.0 specification JVM-based Java runtime system, with JavaPhone 1.0 APIs.
- Connectivity: Converters and viewers for foreign data formats and communications framework for connecting with a PC running Symbian Connect.
- Media server: Audio recording and playback, and image related functionality.
- Application engines & services: Engines for contact management, schedule and to-do list management, and other applications

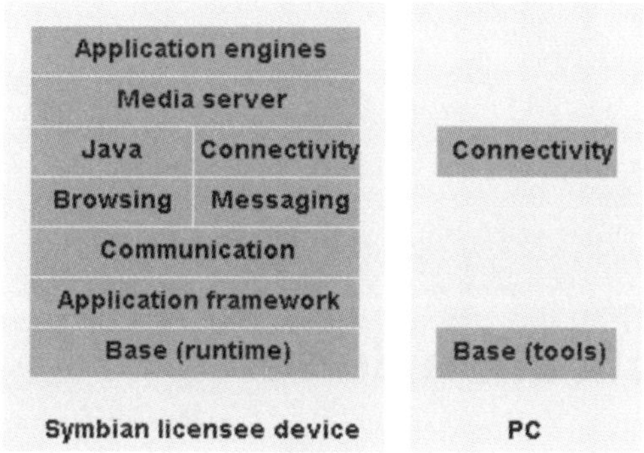

Symbian Generic Technologies

Device Family Reference Designs

In order to deal with the wide variety of mobile devices, Symbian developed the concept of Device Family Reference Designs (DFRDs). The idea is the availability of similar resources within the devices that are running the same DFRD, thus allowing the binary compatibility and therefore the transfer of application between devices of the same family as well as the use of shared components within one device [SyCr01], [SyQu01].

Version 6 of the Symbian OS includes three different DFRDs (see previous figure):

- Crystal is Symbian's reference design for large-screen keyboard-oriented communicators. The recommended screen size is about half VGA (640x200 pixels) with options available for colour or grey scale, depending on manufacturer. Crystal supports a full QWERTY keyboard and options are built-in to have either a pen interface, which is similar to traditional Psion devices, or a button-oriented interface
- the Quartz DFRD is designed for devices with a quarter-VGA portrait colour touchscreen, 240 pixels wide and 320 pixels high, that are pen operated and have no keyboard but hardware keys for cursor functionality. Furthermore it supports devices with a built-in phone, with speaker, microphone, and perhaps additional keys

- the Pearl DFRD (Smartphone UI), which is intended for Smartphones and other devices with a limited user interface, like a keypad and only a small screen, but that still have reasonable processing power. For such devices Symbian does not develop the user interface, which is done instead by the licensee

Symbian OS History and DFRDs

The Symbian Application Suite

The Symbian platform is supplied with an application suite that is optimized for the needs of wireless information devices [EPOC5]. The applications can be categorized into

- Communications applications: email, fax, WAP and WWW and SMS
- Personal Information Management applications: contacts, calendar, world map and alarm
- Office applications: word, spreadsheet, presentation viewer, database, spell checker and jotter
- Utilities: calculator, imaging, voice notes, video player and games
- System: control panel, file manager, print and installer
- Connectivity: PC connect, server synchronization, file conversion and application installation

The Symbian OS is an open operating system, and there are many more applications from the vast developer community available on the Internet.

Application Development

Together with the Symbian Application Framework, Symbian also offers Software Development Kits (SDKs) for C++ and Java. For previous versions of the Symbian OS there is also a SDK for developing applications in OPL. Furthermore, there is a separate development kit for the development of connectivity applications that allow data transfer between a mobile device and a PC [SDN01].

The C++ and Java SDKs contain debug and release versions of emulators for one DFRD and development, debugging and deployment tools as well as tools for building installation images and localizing applications [SyCr01], [SyQu01]. All that is necessary to start developing applications is an Integrated Development Environment, and in the case of Java development, a Java compiler (version 1.1).

The Java Runtime Environment for the Symbian OS is a port of the Java Runtime Environment from SUN, version 1.1.6, which passed the SUN certification suite. It offers a full enterprise and Javaphone API set and full support for Java Native interfaces to offer C++ APIs to Java applications.

To add additional devices and use their functionality in an application, it is necessary to write a device driver for that device. In the Symbian OS, it is necessary to obtain a device Driver Development KIT from Symbian [SDN01] for this task.

As the Symbian OS uses so called UIDs, unique 32-bit numbers, to identify file types and the structure of a file (an executable, dynamically-loadable library or a type of storage or some other file format), it is necessary during the build process to provide such a UID for an application. For experimental purposes it is possible to use UIDs in the range from 0x01000000 to 0x0FFFFFFF. However, to release an application it is necessary to allocate a UID through the Symbian Developer Network [SDN01], [Uid01]. The release of applications with experimental UIDs is not recommended, since it might cause complications on devices where the application is to be installed.

Conclusions

The Symbian platform represents an operating system and an application suite that is extremely well suited for mobile devices with limited computational capabilities and battery power. Supported by the major players in the mobile market, it has a good potential to become the most important operating system for wireless terminals.

The next version of the Symbian platform, version 6.1 will extend the current version with additional technologies such as WAP 1.2.1, GPRS-based packet data and additional Bluetooth functionality according to the Bluetooth 1.0b standard [Symb01].

References

[Task00] Martin Tasker & al.: Professional Symbian Programming. Wrox Press, 2000
[Symb01] http://www.symbian.com
[SDN01] http://www.symbiandevnet.com
[EPOC5] EPOC R5 SDK documentation and sample applications
[SyQu01] Symbian Platform version 6 Quartz DFRD SDK
[SyCr01] Symbian Platform version 6 SDK for Nokia 9210 (based on the Crystal DFRD)
[Uid01] uid_ew@symbian.com
[FoNo01] http://www.forum.nokia

SyncML - Universal Data Synchronization

An important factor in the increasing popularity of mobile computing is the possibility to use several copies of the same information. Personal information (Calendar, Contacts, Tasks) is a prime example of this kind of important shared data. Several users access the same information, make updates, and act based on the information. The majority of modifications to data is being made on personal copies of information, to be updated to the general copy available to everybody. The process of making the databases identical is called data synchronization.

Unfortunately, the very same factor has created a major challenge. The root of the challenge lies in data synchronization protocols, definitions of the workflow for communication during the synchronization. There have been a myriad of proprietary synchronization protocols, each of which have had only a limited number of supported applications, operating systems and transport protocols. As a result, users have not been able to share their information with others in a practical and efficient way.

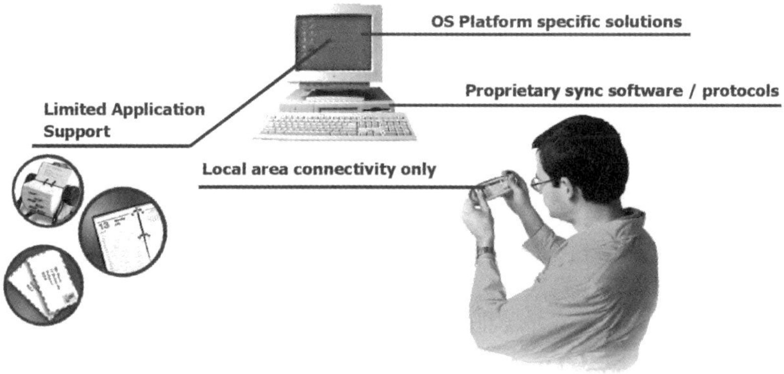

The synchronization challenge

SyncML is the next generation in synchronization protocols. It is an open standard for universal data synchronization that supports a variety of transport protocols and arbitrary networked data. It is easily expandable and based on XML, which is widely considered a future-proof choice. SyncML effectively addresses the resource limitations of mobile devices and has been optimized to work with a variety of applications and devices. SyncML is the only available synchronization protocol that can deliver this kind of universal connectivity.

SyncML technology has firm market acceptance, and is becoming the protocol of choice for universal data synchronization. At the moment, there are over 650 companies supporting SyncML technology worldwide. Ericsson, IBM, Lotus, Matsushita, Motorola, Nokia, Openwave, Starfish Software, and Symbian are actively contributing in the development of the technology. In addition, major communications industry bodies - 3GPP and WapForum have adopted SyncML as the preferred remote synchronization protocol.

SyncML Technology

SyncML is based on the server-client architecture, where a SyncML server handles the synchronization session between tOwo databases. As illustrated in a following figure, a user connects with a SyncML enabled handset to a SyncML synchronization server. The SyncML server is connected to the remote database, which may also be updated via a WWW-calender by other users. The SyncML server handles the resolution of conflicting entries and updates the information both on the handset, and the remote database. The information in the handset and the remote database are synchronized and, thus, identical.

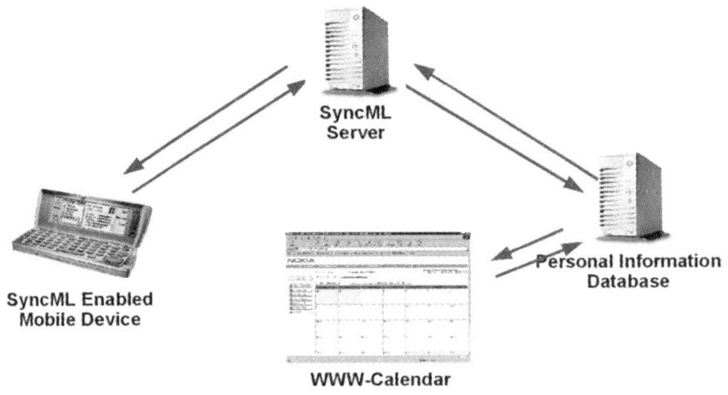

A SyncML synchronization session in practice

The SyncML is transport independent and, thus, capable of synchronizing data over many different data protocols used by wireless and networked applications. It works smoothly and efficiently over the Internet (HTTP), the Wireless Session Protocol (WSP), OBEX (Bluetooth, infrared etc.), e-mail standards (SMTP, POP3 and IMAP), TCP/IP networks and other wireless and wireline communication protocols. The most commonly used protocols will be HTTP, WSP and OBEX.

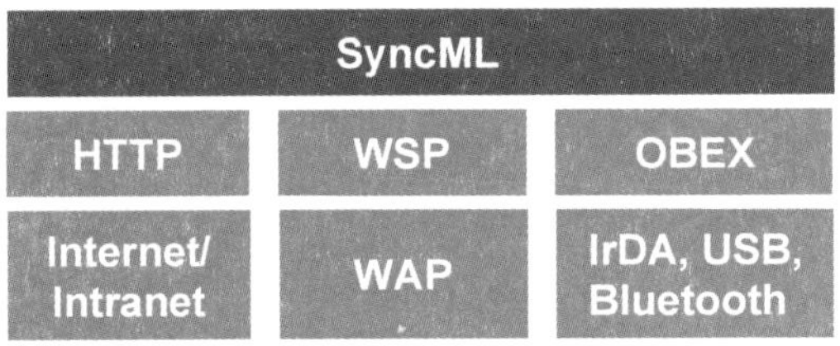

Current SyncML transport bindings

In short, SyncML offers a versatile tool for developing data synchronization capabilities into any service concept. The protocol does not mandate how data is represented or structured on the device or within the networked data repository after synchronization is complete. As SyncML enables mobile devices to communicate with a vast range of data, the protocol also supports simultaneous synchronization with multiple network data stores.

In the first phase, SyncML devices synchronize PIM-data (contacts/phone numbers, calendars, to-do lists). This provides operators and service providers with an opportunity to offer new sticky services, where the personal information of customers is stored and synchronized with their system. By combining SyncML and time-critical content, calendaring can be turned into a valuable service where interesting events and time-critical information are fed into mobile phones. The users will receive the information they need and want at any time and any place.

Security

The SyncML specification uses the security mechanisms of underlying transports. Thus, no restricting security schemes have been defined. SyncML authentication is based on basic or digest access authentication defined by IETF: Base64 coding and MD5 are used, respectively. Authentication can be performed on several levels: between synchronization server and a client, on the database level and on operation levels. Other authentication schemes can be specified by prior agreement between the originator and the recipient.

Interoperability

To ensure interoperability, the protocol describes how common data formats are represented over the network. To ensure extensibility, the protocol permits the definition of new data formats as needs arise. In addition, the protocol allows implementors to define experimental, non-standard synchronization protocol primitives. The most commonly used data formats that the protocol supports from the outset include vCard for contact information and vCalendar for calendar, to-do list and journal information.

Interoperability is ensured by a specific interoperability testing procedure determined by the SyncML Initiative, a non-profit organization developing SyncML technology. Currently the SyncML Initiative has made available a self-test for testing specification conformance and held a Syncfest event for interoperability testing with other vendor's and manufacturer's products. The First SyncFest was held at the end of February 2001 in Europe, and it has been followed by tri-monthly events in different geographical locations.

SyncML Evolution

SyncML is rapidly evolving as the de-facto standard in universal data synchronization. The SyncML Initiative, an open industry organization for SyncML technology has firm industry backing. Ericsson, IBM, Lotus, Matsushita, Motorola, Nokia, Openwave, Starfish Software, and Symbian are actively contributing to the activities of the Initiative. In addition to these companies, there are over 650 companies supporting the technology worldwide. Major industry bodies, 3GPP and Wap Forum have also adopted SyncML as the preferred remote synchronization protocol.

The market is about to experience explosive growth within the next 6 months. In June 2001, Nokia introduced the first commercially available SyncML enabled mobile device, the Nokia 9210 Communicator. Since then, other SyncML enabled handsets and servers have also appeared on the market. Several mobile operators, service operators and corporates are ramping up production on their first SyncML enabled services.

The first services will concentrate on offering personal information synchronization, and combining the information with interesting content. In the future, SyncML is going to be expanded to support new data types (images, files, database objects). New expansions are used to build and offer vertical applications for corporate users (company files, sales data, SCM, fleet management, etc.), and consumers (information services, interactive games, etc.).

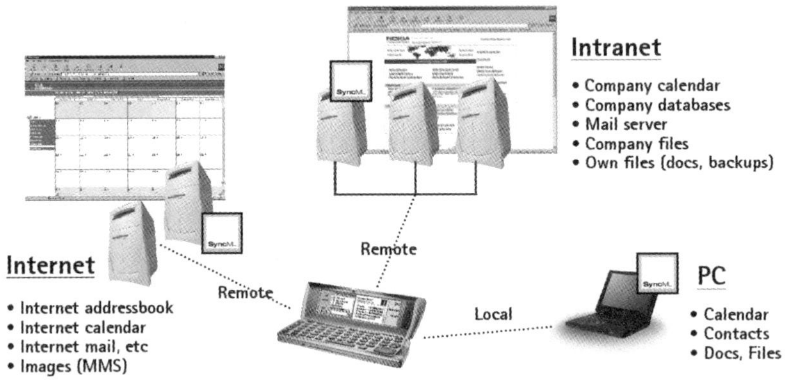

SyncML enabled services

Currently the SyncML initiative is also looking into opportunities to leverage the forthcoming SyncML terminal base in remote device management, which would, for example, allow wireless operators to configure mobile devices over the air.

Device Management

The nature of mobile terminals is dramatically changing as their complexity increases with the addition of new and more powerful and richer applications. At the same time, these increasingly complex terminals remain in mass-market environments to be used by end users with minimal technical interest or skill. This is similar to the history of personal computers, as current computer users are mostly persons without a technical background. This has lead to a situation, where external parties locally or remotely maintain or configure the computers for end users.

To hide the complexity of mobile terminals from the end users and to conveniently provide the seamless functionality of all mobile applications, Device Management (DM) has been created. Device Management is the term used for a technology that allows third parties to carry out the difficult procedures for configuring mobile terminals on the behalf of an end user. Third parties would typically be wireless operators, service providers or corporate information management departments.

DM, as a mechanism, can be used to manage very different data objects. Some of the data objects are simple numeric or textual parameters, while others are binary. Numeric objects may include connectivity parameters such as access point addresses and proxy configurations. Binary objects may be security keys or even displayable content such as animation, screen savers and caller identification animations. In the future, the DM system could even be used to distribute and install executable software.

Technology

The Device Management technology can be divided into two separate technical areas, bootstrap technology and continuous DM. Bootstrap technology is only used to set the initial settings for continuous DM. After the bootstrap phase, only continuous DM is needed between a mobile terminal and a DM server. In general, the bootstrap is done over unsolicited message functionality such as WAP Push over SMS.

Continuous DM is called SyncML DM, as it uses the SyncML technology as a basis and is being standardized by the SyncML initiative. The SyncML DM protocol closely resembles One-way Synchronization defined by the SyncML initiative for mobile data synchronization. Instead of databases being synchronized, management objects are synchronized or transferred. A management object might reflect a set of configuration parameters for a device. Actions that can be performed on this object might include reading and setting parameter keys and values. Another management object might be the run-time environment for software applications on a device.

A Device Management environment is depicted in the following figure. The SyncML DM and bootstrap components are illustrated.

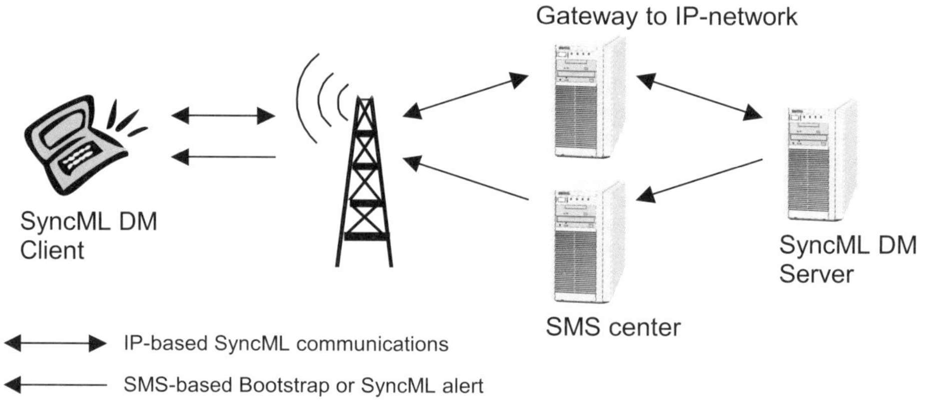

Regular Device Management (DM) Environment.

The standard-based approach, such as the SyncML DM technology, ensures compatibility between a large number of devices and management servers. The use of SyncML DM has substantial advantages for relatively small and cost effective devices. It is wireless-friendly and transport independent. It also provides an extensible, mature and flexible security model and the independency of a run-time environment.

An important characteristic related to the nature of DM, is the differentiation of the paying party, depending on service. Some management operations are initiated and desired by the service provider, and they should not typically cause costs to the end user, while the end user directly triggers other operations and should thus be charged for them.

After all, Device Management technology offers services to other applications, e.g. provides them with configuration data. It cannot be classified as a standalone technology or application but a very important and vital enabler for mobile terminals with a large set of mobile applications.

Middleware

The current trend in developing forthcoming telecommunication networks is to utilize Internet protocols. An immediate implication is that IP is the layer 3 protocol and that the addresses are IP addresses. However, this is not sufficient. Other solutions—both above and below the IP protocol—are also needed to meet the requirements for the next generations of telecommunication networks. Issues under study in the Internet community and in various standardization bodies, forums and consortia of telecommunications include mobility, Quality-of-Service, security, management of networks and services, discovery, ad-hoc networking, dynamic configuration, and geospatial location.

Another significant trend is the requirement of ever-faster service development and deployment. The immediate implication has been the introduction of various service/application frameworks/platforms. Middleware is a widely used term to denote a set of generic services above the operating system. Although the term is popular, there is no consensus for a definition [1].

Typical middleware services include directory, trading and brokerage services for discovery, transactions, persistent repositories, different transparencies such as location transparency and failure transparency. Examples of middleware include Common Object Request Broker Architecture (CORBA) [2], Java 2™ Enterprise and Micro Editions (J2EE and J2ME) [3, 4], Distributed Common Object Model (DCOM) [5], and Wireless Application Environment (WAE) [6]. Characteristically, the competing middleware specifications provide many similar but slightly different services. In order to overcome the problems due to different specifications, the Parlay Group [7] has specified a set of UML models and corresponding APIs that can be implemented in CORBA, Java and DCOM environments.

This article examines ways to incorporate existing middleware solutions and Internet protocols in the Mobile Internet.

Application Requirements

The next figure depicts a highly abstracted vision of how a service application is distributed among various application servers, network elements and terminal or end-user systems. It should be noted that, for simplicity, the figure only shows a single terminal device although multi-party applications will be much more important and challenging than one-party applications. In addition, we must also be ready to cope with end-user systems based on body area networks and home communication systems.

The execution environments or the service stratum consist of middleware, operating systems and protocol stacks that should support fast service development and deployment.

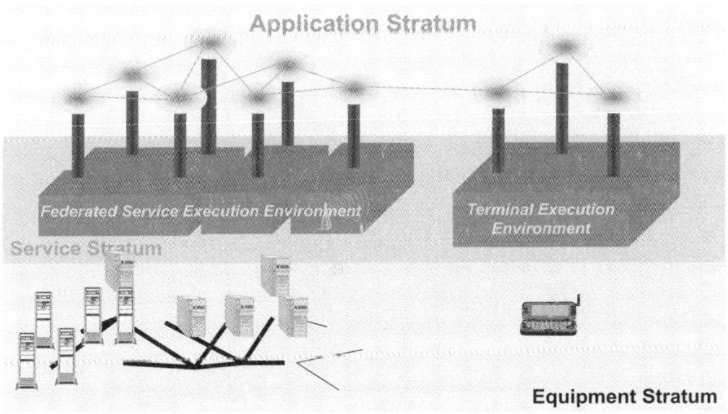

Mobile Distributed Application

The service stratum should make it easy to divide the application logic into co-operating parts or components, to distribute and configure these components as well as to redistribute and reconfigure them. Additional requirements for future mobile applications include adaptability to changes in the execution and communication capabilities, efficient use of available communication resources, dynamic configuration of end-user systems as well as ultimate robustness, high availability and stringent fault-tolerance.

The requirements for data accessed by these applications are quite similar. The execution environment should provide a consistent, efficiently accessable, reliable and highly available information base. This implies a distributed and replicated world-wide "file system" that also supports intelligent synchronization of data after disconnections.

Internet Protocols and Middleware

The next figure decomposes the abstract notion of the execution environment in the previous figure into a layered architecture. It should be noted that the operating system is not shown in the next figure because it would introduce another dimension. In most operating systems the layers below the middleware are encapsulated into program libraries that are available through system calls, file descriptors, etc.
The main stream seems to be the "IP-TCP-HTTP-XML" protocol stack, in spite of the facts that

1. TCP behaves poorly on transmission paths containing both wireless and wired links [8, 9, 10],
2. HTTP is one of the best examples of using TCP capabilities in an extremely inefficient manner, and
3. XML is verbose and expensive to process.

In TCP transport over a wireless path, there has been significant enhancements due to the proposals from the IETF working groups on transport area (tsvwg [11]), on performance implications of link characteristics (pilc [12]), and on endpoint congestion management (ecm [13]). The HTTP transport over a wireless link has not yet been addressed in IETF but practical proposals are available [14, 15]. For XML the situation is even worse. The WAP forum has proposed a binary XML format [16] but that only addresses the efficient presentation of tags. The real problem is the efficient presentation of the tag values on the wire. The XML Protocol Activity in W3C [17] has laid the basis of a solution but not yet provided a one to this problem.

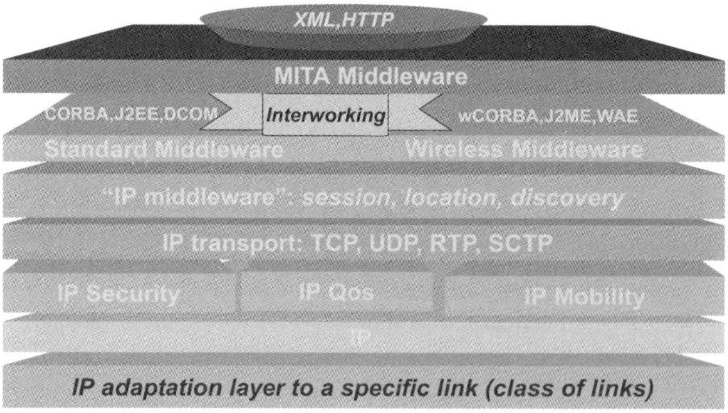

Layered Internet Protocol Architecture for Middleware

Programming Models

The programming model of the Internet is based on sockets, protocol specific APIs and state machines of application protocols. The state machines need to be separately implemented for each protocol. Although the Internet approach is efficient for a single protocol, it is very cumbersome and expensive for application development that involves several operating systems and protocols. Moreover, many Internet protocols do not explicitly define the presentation format on the wire, which has introduced serious interoperability problems between implementations. Another concern is that the Internet approach is prone to feature interactions since many protocols are developed independently from each other.

In middleware solutions, particularly in CORBA, a considerable amount of effort has been put on interoperability in heterogeneous operating environments. This has resulted in a uniform baseline protocol that also specifies the presentation format. In the case of CORBA, the unifying protocol is the Generic Inter-ORB Protocol (GIOP) and the presentation format is the Common Data Representation (CDR) that specifies how different data types and complex data structures are put on the wire. Therefore, many details that need to be handled by applications in the Internet solution are handled by the platform in the middleware solution.

When the CORBA approach—as well as the Java and DCOM approach—is compared to the Internet approach, the only possible conclusion is that the middleware solutions provide a much more convenient programming model than the Internet solution does. The middleware solutions are usually based on object-oriented programming and method invocations. The invocations are based on strongly typed interfaces that provide both compile and run time error checkings. They also hide many implementation details. Therefore, middleware-based application development is much faster than the Internet based one.

The fundamental problem of the current middleware specifications is that they only take advantage of a narrow subset of useful Internet protocols. The current middleware specifications were born in a time when the Internet protocols were a synonym of the TCP/IP transport. Later they have developed solutions of their own for Quality-of-Service, directory, discovery, and so on, independently from each other and from the IETF specifications.

Next Steps

From the software architecture point of view, the fundamental challenge is the question of how the developments in Internet protocols and in different middleware solutions could be harmonized. This requires both an overall vision on a broad frontier of standardization activities and a deep insight into various specifications. In particular, we do not want to discard existing middleware solutions since they are well-established and provide obvious benefits for application development.

By harmonization we mean two things. Firstly, we need to solve the problem of incorporating evolving Internet solutions of QoS, mobility, discovery, and security into the existing middleware specifications without breaking those specifications. Secondly, we need to discover how different middleware solutions can become interoperable, in the sense that the components of an application can be executed on different middleware platforms.

Instead of separate, stand-alone protocols, we are advocating a base protocol approach. A good candidate as the base protocol would be the GIOP specified in the OMG. The primary advantage of GIOP over the COPS (Common Open Policy Service) Protocol and the Simple Network Management Protocol (SNMP), which are specified in the IETF, is its usability in the user plane. In principle, the COPS - perhaps also the SNMP - could be used in the user plane but there is no tradition of such use in the community of developers of applications for end-users. HTTP together with XML - through the W3C SOAP (Simple Object Access Protocol) initiative for example - might be an alternative to the GIOP but the problems of HTTP / XML over wireless links are much more serious than those of GIOP. The same is also true for Java RMI (Remote Method Invocation) over wireless [18].

In middleware standardization we have focused on interoperability and interworking, particularly on aspects of mobile or nomadic users and wireless links. By active involvement we try to reduce the footprint of run-time software as well as the memory requirements and computational complexity of system software and application components in the end-user system.

The increasing diversity of devices—terminals, network elements, and application servers—imply that different middleware solutions will also be used in the future. This heterogeneity requires interoperability on two levels: between middleware platforms and between parts of an application running on different middleware platforms.

The interoperability between different platforms is quite mature. In particular, the OMG has been the leading forum in specifying interoperability bridges between CORBA and other middleware platforms. In contrast, the interoperability between parts of an application running on different middleware platforms is still immature. There are practically no tools available to support this kind of interoperability. The burden of interoperability is totally left to application developers.

The OMG, however, has started a comprehensive development of a new architecture denoted as Model Driven Architecture (MDA) [19]. The objective is to interrelate IDL specifications, UML modeling [20], Meta-Object Facility [21], and XML Metadata Interchange (XMI) [22]. The forthcoming MDA might provide a useful starting point for tools supporting interoperability between parts of an application running on different middleware platforms.

Conclusion

The second (or third) generation of middleware products begins to be mature enough for deployment in telecommunication networks. However, opportunities for improvements still exists and both ORB and Java communities are working towards developing specifications that are missing.

The main implication of Internet protocols on the middleware is the increasing richness of available functionality and features. This implies that the middleware standardization bodies need to take another look at their solutions for Quality of Service, mobility, discovery, and security. They should also consider alternative Internet transport mechanisms to TCP/IP, which is a good protocol for bulk data transfer but not for messaging. When the middleware specifications are based on the IP solutions for QoS, mobility, discovery, and security, interoperability and interworking will be much more easy to arrange.

References

[1] Network Policy and Services: A Report of a Workshop on Middleware. RFC 2768, February 2000.

[2] OMG CORBA v2.4.2 Specification. OMG document formal/2001-02-01, February 2001.

[3] Sun Microsystems. Java™ 2 Platform, Enterprise Edition Specification, v1.3, February 2001.

[4] Sun Microsystems. Java™ 2 Platform, Micro Edition Specification: http://java.sun.com/j2me/

[5] Eddon, G. and Eddon, H. Inside Distributed Com (Microsoft Press, 1998).

[6] WAP Forum. Wireless Application Environment Overview. Document id 195-WAEOverview-20000329-a. March 29, 2000.

[7] Parlay Group. Parlay 2.1 Specification. 2000-2001: http://www.parlay.org/

[8] Alanko, T.; Kojo, M.; Laamanen, H.; Liljeberg, M.; Moilanen, M.; and Raatikainen, K.: "Measured Performance of Data Transmission over Cellular Telephone Networks," Computer Communications Review 24(5): 24–44, October 1994.

[9] Cáceres, R. and Iftode, L.: "Improving the Performance of Reliable Transport Protocols in Mobile Computing Environments," IEEE Journal on Selected Areas in Communications 13(5):850–857, June 1995.

[10] Kojo, M.; Raatikainen, K.; Liljeberg, M.; Kiiskinen, J.; and Alanko, T.: "An Efficient Transport Service for Slow Wireless Telephone Links", IEEE Journal on Selected Areas in Communications 15, 7 (Sep. 1997), pp. 1337-1348.

[11] IETF TSVWG home page. http://www.ietf.org/html.charters/tsvwg-charter.html

[12] IETF PILC WG home page. http://pilc.grc.nasa.gov/pilc/.

[13] IETF ECM WG home page. http://www.ietf.org/html.charters/ecm-charter.html.

[14] Liljeberg, M; Helin, H; Kojo, M and Raatikainen, K: Mowgli WWW software: Improved usability of WWW in mobile WAN environments. In Proc. IEEE Global Internet 1996 Conference, 1996, 33-37.

[15] Hausel, B. and Lindquist, D. B.: "WebExpress: A System for Optimizing Web Browsing in a Wireless Environment", in Proceedings of MobiCom'96, ACM SigMobile, 1996, 108-116.

[16] WAP Forum. WAP Binary XML Content Format. Document id WAP-192-WBXML-20000515, Version 1.3, May 15, 2000

[17] W3C. XML Protocol Activity: http://www.w3.org/

[18] Campadello, S., Helin, H., Koskimies, O., and Raatikainen, K. Wireless Java RMI._http://www.cs.Helsinki.FI/research/monads/papers/edoc2000/rmi.html

[19] OMG. Model Driven Architecture, A Technical Perspective. OMG Document Number ab/2001-02-01, February 14, 2001.

[20] OMG. UML Specification. OMG document formal/2000-03-01, March 2000.

[21] OMG. MOF Specification. OMG document formal/2000-04-03, April 2000.

[22] OMG. XMI Specification. OMG document formal/2000-11-02, November 2000.

Java

As a programming platform, Java brings the benefits explained above to terminal programming. More importantly, it enables application development for a large new market of mobile phone owners.

From a user's point of view, Java brings new types of services with rich graphics and interaction, for example games. Java will also increase the personalization of the phone.

Java enables personalization, entertainment and new applications for the Mobile World.

Technically speaking there are two Java platforms for mobile terminals. PersonalJavaTM together with JavaPhone™ APIs provide a rich programming environment and allows software compatibility with J2SE. The Nokia 9210 Communicator is shipped with the PersonalJavaTM platform and JavaPhone™ APIs.

Later, Nokia will also ship phones with MIDP support. MIDP has been optimized for small devices like mobile phones, and the special characteristics of small devices were considered to be more important than compatibility with J2SE. The core components of MIDP include:

- Core java.lang and java.util libraries defined by CLDC,
- Timer APIs for delayed and periodic execution of events,
- A Networking API based on the Generic Connection Framework of CLDC. The idea of the Generic Connection Framework is to provide an abstraction layer over several communication mechanisms, such as TCP/IP sockets. An application using the Generic Connection Framework needs only provide the destination URL and the system returns a connection object for the application. For most communication, the application does not need to know the detailed connection class and thus the application can encapsulate the connection-dependent issues in the URL string.
- MIDP 1.0 requires that the HttpConnection class must be implemented. HttpConnection is defined in such a way that it can be implemented either by using a WAP protocol stack or by using TCP/IP.
- The user interface library (lcdui) has been designed for mobile devices. The API design is based on two layers. The high-level API is designed for business applications whose client parts run on wireless terminals. For these applications, portability across devices is important. To achieve this portability, the high-level API employs a high level of abstraction and provides very little control over look and feel. Also, by using the high-level API, the application can utilize the native look and feel of the system and the user is immediately familiar with the look and feel.
- The low-level API, on the other hand, provides very little abstraction. This API is designed for applications that need precise placement and control of graphic elements, as well as access to low-level input events. Some applications also need to access special, device-specific features. A typical example of such an application would be a game.

- The Record Management Storage API (RMS) was defined for storing small data items in persistent storage.

In order to maintain the integrity of the system, all MIDP applications are run under the control of the Application Management System (AMS). The AMS is responsible for the installation of applications, starting and stopping of applications - according to user requests - and ensuring that the basic functionality of a phone is not compromised by the Java applications.

End-to-End Java Applications

Client-server communication in MIDP is enabled by an API that allows HTTP-type of communication. The API has been specified in such a way that both HTTP and WTP can be used as the underlying transport protocols.

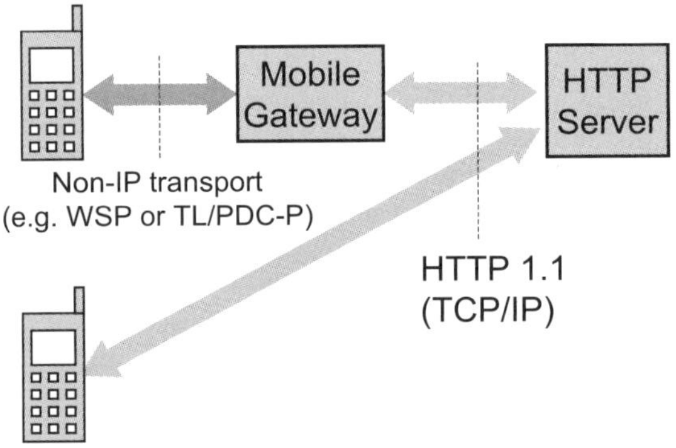

MIDlet networking possibilities.

The MIDP API specification does not specify the means for the discovery and installation of MIDP applications into a phone. There are several possible ways to do this, applications can be pre-installed at the factory, downloaded over a serial cable from a PC, or downloaded over the air (OTA) by an application such as a browser. In the future, over the air downloading is expected to be the most important means for downloading new MIDlet applications to phones. A recommended practices document which describes over the air provisioning is available.

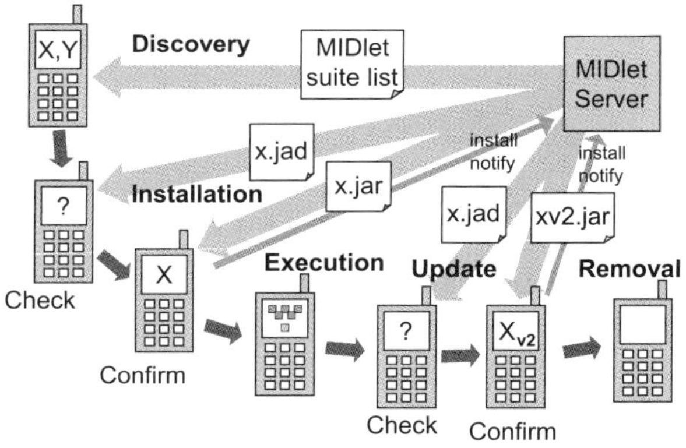

MIDlet deployment and lifecycle.

For OTA downloads, the deployment and lifecycle of the MIDlet is described roughly as follows (see the previous figure):

1. Discovery

   A person uses a browser on their mobile device to browse web pages on a remote server. The user finds a page containing a link to an interesting MIDP application. Actually the link points at a Java Application Descriptor (application_name.jad) which contains the core attributes of the application. The MIME type of a JAD descriptor is: text/vnd.sun.j2me.app-descriptor.

2. Installation

   When the user clicks on the link, the browser downloads the descriptor and provides it to the Application Management Software (AMS).

   The AMS uses the attributes in the descriptor to check if the terminal is able to run the application, and also checks if there are other reasons that would prevent downloading. If application downloading can be done, the AMS uses the URL given in the descriptor to locate and download the Java Archive (JAR) file from a MIDlet Server.

   Next, the AMS installs the MIDlet on the terminal, perhaps after asking for permission from the user. At this point, the AMS may also send a confirmation notification to the server. The server uses this confirmation for billing and customer care purposes.

3. Execution

   The application is then executed and used as a separate user action. During its normal course of execution, a networked MIDlet may contact a HTTP Server. The terminal establishes a 'client role' HTTP or WTP connection for such cases. This depends on the type of network and terminal used. While a connection is open, the client application may converse with the server by means of performing simple HTTP GET and PUT operations.

4. Update

   An application may be updated in a similar fashion, with a newer version of the MIDlet.

5. Removal

   Eventually, the user may wish to delete the application. The AMS will provide services to allow the user to do this.

Current Status

Deployments of Java in mass market terminals has recently begun. Since its introduction in June 2000, MIDP has gained wide interest. Some initial products implementing Java/MIDP are already available on the market. Also, several operators are starting or piloting Java-based services. In the JavaOne Conference 2001 Nokia promised to ship 50 million phones supporting Java by the end of 2002. The mass market of such terminals, coupled with the enormous interest by developers and operators is expected to yield many new areas for application and service growth.

Java's Role in the Mobile Domain

Although powerful, browsing does not fulfill all the needs of client applications. Downloadable Java MIDlets (or Personal Java applications in high end terminals), allow the opening of the terminal to third party software developers, operators and others. This is expected to create a number of new business opportunities and application types. In particular, many types of games and dynamic graphical content (toys and entertainment programs) will developed. Task-oriented client applications for existing enterprise services can also be expected.

Multimedia Messaging – The evolution of SMS to MMS

SMS has been a great success story. The service provides a considerable source of revenues for telecom operators and service providers. However the content that can be transmitted over SMS is rather limited: text, ringtones and small pictures. With recent advances in technology, creating rich multimedia content has become easier and more affordable. For instance, digital cameras are offering new means of instant photography; point, shoot and the picture is ready to be transmitted. However, there is a lack of an application allowing users to instantly share those images. Today, they would typically have to transfer their images to a PC and send them to a friend through e-mail, hoping that he will read the message soon. Unfortunately, the magic of the moment often passes.

It would be much better if the image could be sent directly and displayed instantly on the colour screen of the friend's mobile terminal. It would allow the moments of life to be shared when they occur; when the magic is still there.

The Multimedia Messaging Service (MMS) is the next evolutionary step from the Short Message Service (SMS). For a user, it works similarly to SMS. However, MMS provides the opportunity to utilise a wide variety of richer content types than SMS. These include pictures, audio, music and video clips. MMS is expected to become as much a mass-market service as SMS is today.

The following presents the various aspects of the new Multimedia Messaging Service from the technology and challenges to the applications MMS brings.

The Short Message Service

Originally launched in 1992, SMS has become the most successful wireless data service. SMS has mainly been used for terminal-to-terminal messaging. In SMS, a user typically composes a text message on his mobile terminal and sends it to another person using the recipients phone number as the address. The message is received by an SMS Center (SMSC), which acts as a store-and-forward unit. It is responsible for delivering the message to the destination phone. It may have to wait until the recipient terminal becomes reachable (powered on or within GSM range) for that to be possible. The message is delivered without any intervention from the recipient; it appears naturally on his terminal screen and a sound alerts him of the arrival of the new message.

Another important aspect of SMS, from a usability point of view, is that once the destination terminal has received the message, a delivery confirmation is conveyed to the sender. Since most people carry their phone with them, it provides confidence that the message reached or will soon reach the person.

On the other hand SMS does have several limitations. Each SMS message is limited to quite a short length of 160 characters. Methods exist to concatenate multiple SMS messages to create longer messages. There are also ways to convey other types of (binary) content such as pictures and ringtones using SMS. However, SMS is transmitted over signalling channels of a wireless network. This makes it impractical to use SMS to convey multimedia messages, which are several kilobytes in size.

The Multimedia Messaging Service

The Multimedia Messaging Service concept is built on the familiar and successful SMS service. From the user's perspective, the logic and behaviour of the service is similar to SMS.

Like SMS, MMS is based on a store and forward service model. Receiving multimedia messages in a mobile terminal works similarly to SMS: messages are delivered directly to the mobile device with no need for user interaction. Furthermore, the sender receives delivery confirmation.

However, there are important differences between SMS and MMS. A major one is that MMS provides the opportunity to utilise richer content types than SMS. The content will include pictures (in formats such as JPEG and GIF) as well as audio, music and video clips. Because MMS uses data traffic channels and not signalling, it is well suited to deliver much larger multimedia content. The following figure shows how a multimedia message composed of text and an image might look like.

An example of a MMS message composed of text and an image.

SMS was mainly used for messaging between mobile terminals. In addition to this, MMS will be used for messaging from Web applications to mobile devices and between the Internet and mobile terminals as shown in the figure below. These exchanges will lead to the development of numerous media-rich applications: picture messages, electronic postcards, instant pictures and video clips, audio messages, etc.

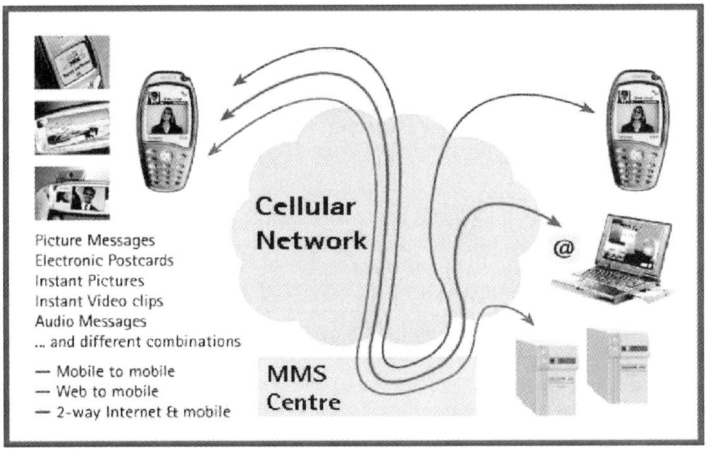

Exchange of MMS messages between mobile devices, the Internet (e-mail, instant messaging), and Web applications.

In sending multimedia messages, the address used can be either the recipients phone number (MSISDN, familiar from SMS), or email address (RFC 822 [12]). Messages addressed to MSISDN are routed to the recipient terminal whereas email-addressed messages are routed directly to the SMTP network that handles the delivery.

MMS messages are routed via both the senders' and the recipients' MMSC. The routing mechanism enables fast introduction of services, and, more importantly, messaging between networks of different technologies, such as GSM, 3G, TDMA and CDMA. This interoperability makes MMS a truly global service.

MMS and Standards

The Multimedia Message Service architecture and its overall concepts have been standardised in 3GPP. Based on the work and requirements of 3GPP, the WAP-based implementation specifications are the responsibility of the WAP Forum.

WAP 2.0, released in July 2001, contains the specifications for MMS. They are composed of three documents:

- WAP-205-MMSArchOverview: WAP MMS Architecture Overview [1].
- WAP-206-MMSCTR: WAP MMS Client Transactions [2]. Defines the exchange of messages (protocol) between MMSC and client to send/receive messages.
- WAP-209-MMSEncapsulation: WAP MMS Encapsulation Protocol [3]. Defines the composition of messages based on MIME. More details in the section on Messaging Encapsulation.

Those specifications address the MMS concepts and requirements presented in 3GPP Release 99 [5][6].

Work is under way in WAP and 3GPP to address future MMS functionality.

MMS Network Architecture

The overall architecture of the MMS implementation in cellular networks is shown in the following figure. The role of the MMSC (Multimedia Messaging Service Center) is analogous to the role of SMSC: it provides access to the cellular network with WAP, store-and- forward service logic and an application interface towards applications that provide value-added services on top of MMS.

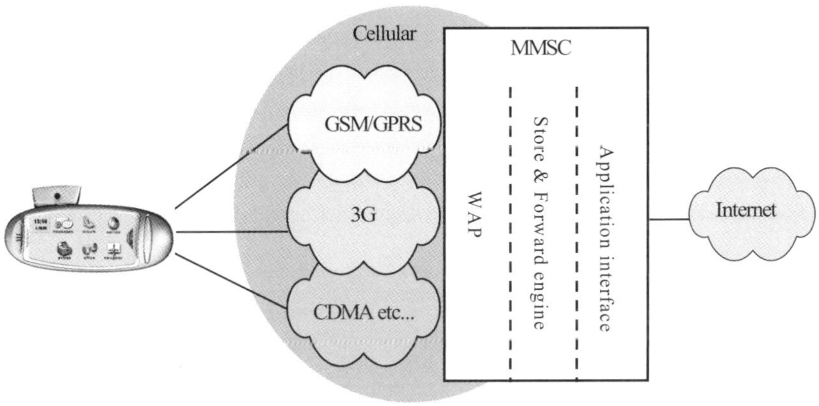

The MMS Network Architecture

The mechanisms employed to deliver messages in the cellular network are different from the ones used in SMS. In MMS the message sizes in the first phase are estimated to be around 30 kilobytes, depending on the content types. This is considerably more than the allowed size of an SMS message, so a mechanism different from SMS was required in MMS. The WAP protocol suite was selected as the mechanism because it provides data transport services that are optimised for cellular networks. WAP also provides uniform transport services regardless of the underlying network (GSM, GPRS, TDMA, 3G).

The WAP protocol WSP layer [4] is used as the basis of the transport mechanism. WSP can be viewed as a binary version of the HTTP protocol [11]. The following figure shows the message flow for multimedia messages delivery over a GPRS network. Message notification at the recipient terminal uses the WAP push mechanism. A WAP push Service Indication (SI) is delivered to the recipient terminal on top of the SMS bearer. This notifies the recipient's terminal that a new message has arrived. The service indication triggers a WAP WSP get operation that fetches the message to the terminal. After the message has been fetched, a delivery notification will be issued to the sender.

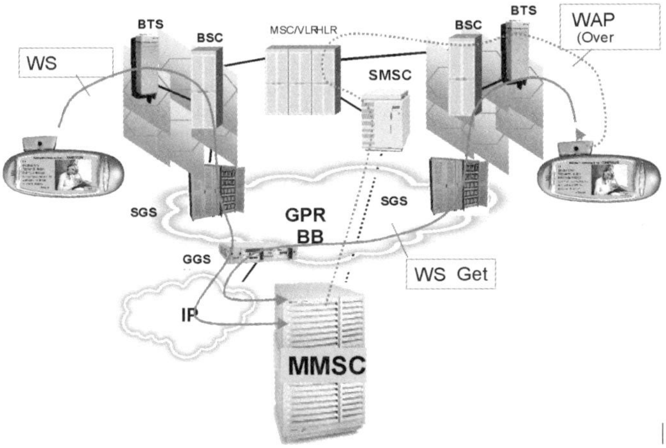

Multimedia message delivery in a GPRS network.

Inter-Network Routing

Delivery between networks in MMS is based on an architecture different from SMS. Short Messages are delivered from the sender's SMSC directly to the recipient whereas in MMS the messages are routed via both the senders' and recipients' MMSC.

Reasons behind the selected architecture and routing mechanism are:

- Introducing recipient services is easy and fast – no need for agreement between operators regarding the services.
- Messaging between different network technologies (GSM, 3G, TDMA and CDMA) is made feasible. This interoperability makes MMS a truly global service.

The MMS network architecture in a two-network case is illustrated in the following figure.

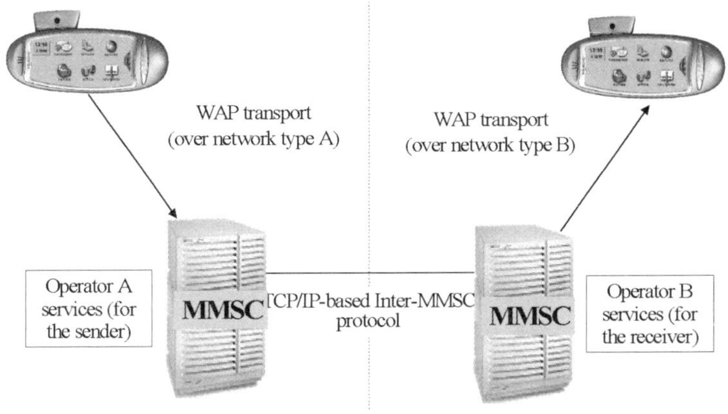

Inter-MMSC Architecture

Being able to route messages between networks of different technology, such as between GPRS and CDMA networks, is a step forward from the SMS world where messages are transmitted only on similar networks. This means that the messaging volume potential of MMS exceeds that of SMS. The benefits, however, do not materialise free of charge: MSISDN–based (or other) routing requires routing tables to be set up between MMSCs.

An added complexity is handling Mobile Number Portability, which requires routing decisions to be based on individual MSISDNs instead of prefixes. However, the benefits brought by the adopted network architecture outweigh the disadvantages.

Inter-MMSC protocol standardisation work is progressing in 3GPP.

Message Encapsulation

A multimedia message is encapsulated in a WAP WSP protocol data unit (PDU) consisting of MMS headers and a message body. These two entities form a PDU of type application/vnd.wap.mms-message. This is illustrated on the next page.

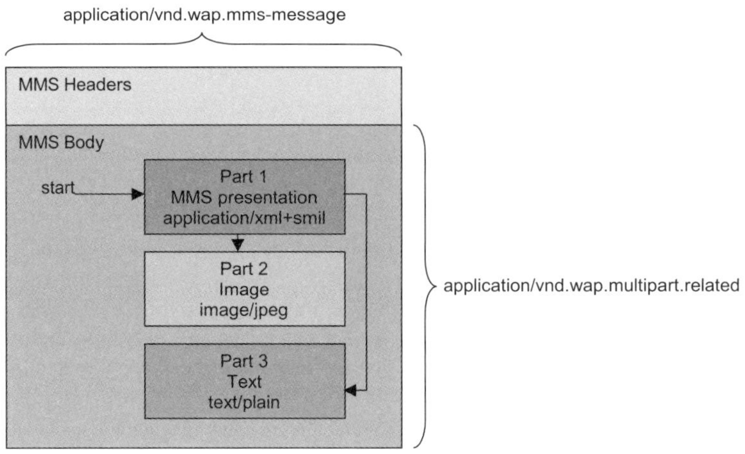

Message encapsulation

The MMS headers contain information related to the transfer of the message from the sender to the recipient. Some headers originate from RFC 822 (from, to, date, subject, etc.) while others are MMS-specific (such as X-Mms-Transaction-ID and X-Mms-Expiry). In MMS the message body typically has the MIME multipart/related structure of RFC 2387 [10]. It is composed of distinct multimedia objects and an optional presentation part. The formats of choice for the presentation part are SMIL [13] and WML [14].

The body can also follow the MIME multipart standard of RFC 2045-7 [7-9]; providing compatibility with e-mail systems. In any case, conversion from/to multipart mixed/related and multipart can be easily performed.

The message body is present only for posting and retrieving messages. For other transactions, such as message notification and delivery reporting, only headers are sent.

MMS Applications

MMS introduces a generic mechanism to encapsulate and transport multimedia content without restricting the formats used. Because of this, MMS can be the foundation of numerous and very diverse applications:

- Mobile to mobile: sending/receiving photographs, voice mail, business cards, etc.
- Web applications to mobile terminal: electronic postcards, greeting cards, advertisements, news of the day (video/audio clips), screen savers, animations, maps, photo albums.
- Internet to/from mobile device: receive selected e-mails, send e-mails, etc.

For instance, although postcards are very popular, it is troublesome to search for a post office, or a mailbox, find correct stamps, etc. It would be more convenient to take a picture from a camera attached to, or integrated in the phone and send it to a friend through MMS. The message reaches the destination almost instantly and can be personalised (e.g. a picture of yourself in Paris).

Message Adaptation

Multimedia Message Adaptation (MMA) will be a key technology in ensuring the success of multimedia messaging. Its role is to seamlessly link (from the user's perspective) different services to MMS and ensure interoperability between MMS phones having diverse capabilities. MMA is achieved by modifying, in the network, the multimedia message to make it suitable for the destination device and application. This section presents the motivation behind MMA.

Adaptation for Interoperability with Internet Applications

Interoperability with Internet applications is important for the success of multimedia messaging; especially in the early stages of introduction of the service. Initially, the MMS users will represent a relatively low number and will start using the service only if it allows them to reach many users. The way to access those users is to link MMS with existing technologies such as e-mail, SMS, and Web.

To support these scenarios, messages will need to be adapted within the delivery system. Such adaptation includes media format conversion and media modification (e.g. image scaling) to match mobile terminal characteristics. Also, adaptation and routing should be hidden from the end-users. When sending a message, the user should not have to consider the destination application, e.g. MMS, Web, or e-mail. He should only need to provide the recipient's (destination) address and the message delivery system should handle the rest automatically and perform the required adaptations.

Adaptation for Interoperability Between Mobile Terminals

Most GSM phones support the short text message service. SMS is bundled with most phones regardless of their product category. Similarly, we expect that phones of several different product categories will include MMS in the future. However, MMS only refers to the message exchange mechanisms over which a wide variety of content can be transmitted without specifying or restricting it. This is an important aspect since it allows the evolution of the service and its support by a wide variety of phones. We foresee a very pervasive environment under which MMS will operate and evolve.

Both the composition and presentation of multimedia messages will involve phones with very different capabilities: display size, colour or grey scale display, processing power, support for content formats, available memory to store messages, regional differences, etc. The technological advancement in phone technology further adds to the diversity of phones that will soon be involved in MMS.

We also foresee that MMS content will evolve from text and images, to audio and motion picture clips. Therefore, support of previously deployed phones will always remain an issue as the MMS content and services evolve. Since MMS content needs to remain flexible and still ensure interoperability, message adaptation will play a key role in the service.

From these arguments it becomes clear that MMS profiles or interoperability guidelines on how to use MMS are very useful as long as they only define minimum requirements that can be exceeded. Otherwise product and service differentiation and evolution may be severely compromised.

For instance, some high-end mobile terminals may support VGA resolution images while lower-end devices may support images of to 80x60 pixels. It would be a shame to limit phones to sending images with a resolution corresponding to the lowest common denominator. Content adaptation in the network would enable the exchange of VGA resolution images between high-end phones while reducing the resolution when the destination is a low-end phone. Network-side content adaptation encourages phones to send messages with maximum content while ensuring interoperability.

We believe it is not an acceptable option to have multimedia messaging only work satisfactorily between similarly capable phones. Nor it is an option that the sender of a message needs to tailor the message so that it fits the receiver's phone capabilities. In general, he will not know how to do this and will probably not be able to predict which phone the recipient is using to receive the message. When the service works seamlessly for him, using MMS will provide a much better user experience because he can concentrate on his communication needs. By contrast, requiring the user to consider technical constraints will lower the quality of his experience and using MMS will not be perceived as "fun".

Automatic network-side message adaptation will improve the end-user's experience of delivered content. Without adaptation, delivering the multimedia message may even fail completely in some cases, e.g. if the size of the message is too large for the receiving phone.

Mobile Terminal Capability Negotiation

In order to properly adapt messages for the destination terminal, its capabilities and characteristics must be reported to the server. There are two ways to achieve this. First, the terminal, upon requesting the download of the message, may indicate in the User-Agent header field of WSP the type of terminal it is. It may also use the User Agent Profile (UAProf) specification [15]. Servers could also maintain databases containing the terminal characteristics corresponding to each mobile number (a user profile). These solutions will help the server perform proper adaptation.

Conclusions

The new Multimedia Messaging Service has been presented, and major differences between SMS and the Multimedia Messaging Service described:

- Use of data traffic channels for increased message size.
- Generic encapsulation mechanism allowing usage of rich multimedia content.
- Inter-network traffic architecture providing a global messaging service between different networks such as 3G, TDMA, CDMA and GSM.
- Need for content adaptation to ensure interoperability between applications and mobile devices of different capabilities, and to enhance the user experience.

With all these features, MMS is expected to be a successful mass-market service.

References

[1] "Wireless Application Protocol, WAP MMS Architecture Overview Specification", WAP-205-MMSArchOverview. WAP Forum. URL: http://www.wapforum.org.

[2] "Wireless Application Protocol, WAP MMS Client Transactions Specification", WAP-206-MMSCTR. WAP Forum. URL: http://www.wapforum.org.

[3] "Wireless Application Protocol, WAP MMS Encapsulation Protocol Specification", WAP-209-MMSEncapsulation. WAP Forum. URL: http://www.wapforum.org.

[4] "Wireless Application Protocol, Wireless Session Protocol Specification", WAP-203-WSP. WAP Forum. URL: http://www.wapforum.org.

[5] "Multimedia Messaging Service: Service aspects; Stage 1", 3rd Generation Partnership Project TS 22.140 Release 1999. URL: http://www.3gpp.org/ftp/Specs.

[6] "Multimedia Messaging Service: Functional description; Stage 2", 3rd Generation Partnership Project TS 23.140 Release 1999. URL: http://www.3gpp.org/ftp/Specs.

[7] "Multipurpose Internet Mail Extensions (MIME) Part One: Format of Internet Message Bodies", Freed N., November 1996. URL: ftp://ftp.isi.edu/in-notes/rfc2045.txt.

[8] "Multipurpose Internet Mail Extensions (MIME) Part Two: Media Types", Freed N., November 1996. URL: ftp://ftp.isi.edu/in-notes/rfc2046.txt.

[9] "MIME (Multipurpose Internet Mail Extensions) Part Three: Message Header Extensions for Non-ASCII Text", Moore K., November 1996. URL: ftp://ftp.isi.edu/in-notes/rfc2047.txt.

[10] "The MIME Multipart/related content type", Levinson E., August 1998. URL: ftp://ftp.isi.edu/in-notes/rfc2387.txt.

[11] "Hypertext Transfer Protocol HTTP/1.1", Fielding R., Gettys J., Mogul J., Frystyk H., Masinter L., Leach P., Berners-Lee T., June 1999. URL: ftp://ftp.isi.edu/in-notes/rfc2616.txt.

[12] "Standard for the Format of ARPA Internet Text Messages", Crocker D., August 1982. URL: ftp://ftp.isi.edu/in-notes/rfc822.txt.

[13] "Synchronized Multimedia Integration Language (SMIL) 2.0 Specification", W3C Recommendation, August 2001. URI: http://www.w3.org/TR/smil20/.

[14] "Wireless Application Protocol, Wireless Markup Language Specification, Version 1.3", WAP-191-WML, WAP Forum, 19-February-2000. URL: http://www.wapforum.org.

[15] "WAP User Agent Profile", WAP-174-UAProf, WAP Forum, URL: http://www.wapforum.org/

XHTML

For mobile devices, the future of WAP lies in its close alignment with widely accepted Internet standards. The WAP Forum and the World Wide Web Consortium (W3C) have successfully defined mobile Internet standards over the past several years. Most recently, the WAP Forum has moved to adopt the XHTML Basic ™standard from W3C as the basis for WAP 2.0. This transition to XHTML Basic will strengthen the mobile browser 's position within the Internet mainstream and allow for a far greater range of presentation design and formatting possibilities than previously possible. Consequently, consumers will enjoy a wider array of services, more intuitive user interfaces, and a generally more useful experience. At the same time, it will also enable carriers to exercise more control over the look and feel of the services they provide through their wireless portal.

According to the W3C specification, XHTML Basic defines a document type that is rich enough to be used for content authoring and precise document layout, yet which can be shared across different classes of devices – desktop, PDA, TV, and mobile handset. XHTML Basic is the mobile adaptation of XHTML 1.0, and includes everything in XHTML 1.0 except those capabilities,such as frames, that are not appropriate for devices with small screens XHTML itself, according to the W3C, is the first major change to HTML since HTML 4.0 was released in 1997. In fact, the latest version of HTML (v.4.1) forms the basis for XHTML: all tag definitions and the syntax are the same. XHTML simply adds modularity and enforces strict adherence to language rules. As a result, XHTML brings a clean structure to Web pages, which is especially important given the small screens and limited power of mobile devices. W3C is recommending XHTML for all future Web development for desktops as well as all other devices, including mobile handsets.

XHTML is the keystone in W3C's effort to create standards that provide richer Web content on an ever-increasing range of platforms. Using XHTML, content providers will find it easier to produce content for a broad set of platforms and with better assurance of how the content is rendered. In contrast, the initial WAP markup language (WML 1.x) does not ensure consistent layout across different devices. This is of particular concern to carriers that are managing mobile handsets from a number of providers and with different user interfaces from handset model to another.

With XHTML Basic, a document can be presented on the maximum number of Web clients, including a variety of mobile handsets featuring different display formats and presentation capabilities. And since XHTML Basic is sanctioned by both the W3C and the WAP Forum, it is an integral part of the overall Internet standards set, which will ensure its widespread acceptance now and continued viability over the long term. This section describes the key features of XHTML Basic for mobile browsers and its benefits for content developers, carriers and end users. It will also present examples demonstrating the display opportunities and presentation flexibility enabled by XHTML Basic.

Key Features and Capabilities

Support for XHTML Basic will bring a number of positive changes to the mobile browser. XHTML is a stricter progression of an established markup language as it is based on the same HTML tags used in every Web page today. This means that existing Web design and presentation tools can work with it. Similarly, a developer can view an XHTML Basic document in a standard desktop Web browser and the layout will be consistent. There are also significant differences to HTML.

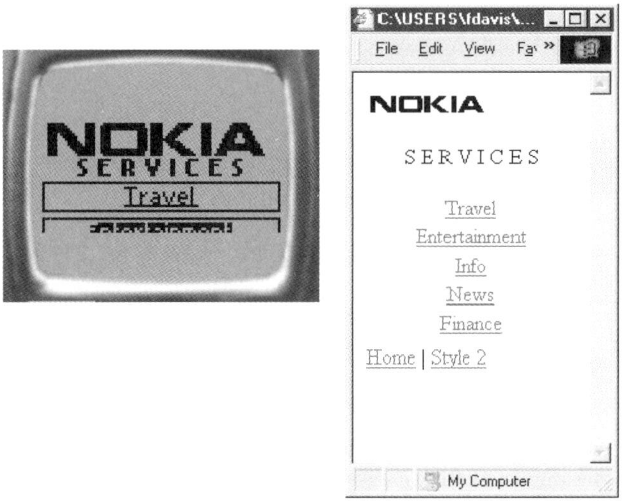

The same XHTML page, viewed using the Nokia WAP Browser (left) and Microsoft Internet Explorer (right)

Well-Formed XML

From a technical standpoint, XHTML Basic strictly enforces the rules for well-formed, valid XML (eXtensible Markup Language). XML is a W3C standard for specifying well-structured documents that can be processed by computers, e.g. a browser. A well-formed XML document conforms to all the rules for XML syntax. For example, every document must have an identifying XML header that is written properly, each tag (instruction) is matched by a closing tag and correctly nested and attribute values are handled correctly. By following all the XML syntax rules XHTML Basic enables a layout that is consistently displayed on different browsers. By comparison, HTML need not be well formed, but the resulting display might contain elements that do not appear as intended on some browsers.

HTML

```
<p> This is a paragraph.
<P> This is a another.
. . .
```

XHTML

```
<p> This is a paragraph. </p>
<p> This is a another. </p>
. . .
```

HTML tags are not always closed but XHTML tags must be closed.

Document.xhtml

```
<?xml version="1.0" ?>
<!DOCTYPE wml PUBLIC
"-//WAPFORUM//DTD WML 2.0//EN"
"http://www.wapforum.org/DTD/wml_2.0/xml">
<head>
```

An example DTD specifier (not actual approved DTD)

Valid XML and the DTD

Valid XML (as noted above, all XHTML documents must contain only valid XML code) simply means that the document uses the language correctly — that it includes only defined tags, and each tag is used in the correct context. For example, in XHTML a table row tag is only valid inside a table tag, but a table row would be senseless outside of a table. The rules for XHTML (or any valid XML language) are defined by its Document Type Definition (DTD). The DTD (or the address at which to access the DTD) appears in each XHTML document's header. The header forms the beginning section of a Web document and contains information that is not part of the actual page content. It also contains controls for the overall document.

According to the W3C, the DTD facilitates hassle-free production and presentation. The DTD identifies what each tag means and specifies how it should be treated. In short, it provides the rules used to validate the online document, while it ensures that the document can be displayed on any XHTML browser. This is particularly important in the mobile Internet, where devices offer a wide range of display formats and capabilities.

By strictly following XHTML rules, the memory cache in mobile browsers can be reduced, allowing for a smaller memory footprint and greater performance efficiency, which is essential in mobile devices. If you look at the size of standard desktop Web browsers, much of the complexity comes from handling the ambiguities of HTML. As the Web moves towards XHTML, applications designed for mobile devices will lead the way.

Cascading Style Sheets (CSS)

Central to XHTML Basic is its support for cascading style sheets (CSS). CSS describes how documents are presented on screen in the browser. The W3C has actively promoted the use of CSS on the Web with all browsers, desktop and mobile. Through the use of CSS, document creators can control the presentation of documents without sacrificing device independence or adding new markup language tags, as was done with WML 1.x. The use of well-known standard HTML tags will reduce content development costs by eliminating the need for developers to learn new tags, to store multiple versions of content, or to master different tools. In addition, servers will not experience the added overhead that results from the expensive transcoding required to prepare content for an array of different markup languages.

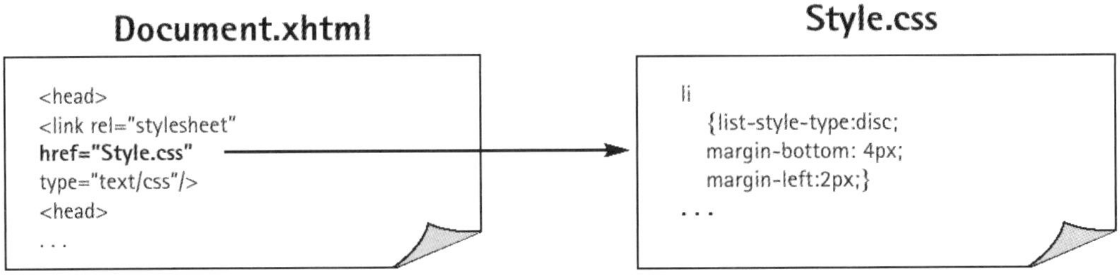

An XHTML document referencing a style sheet.

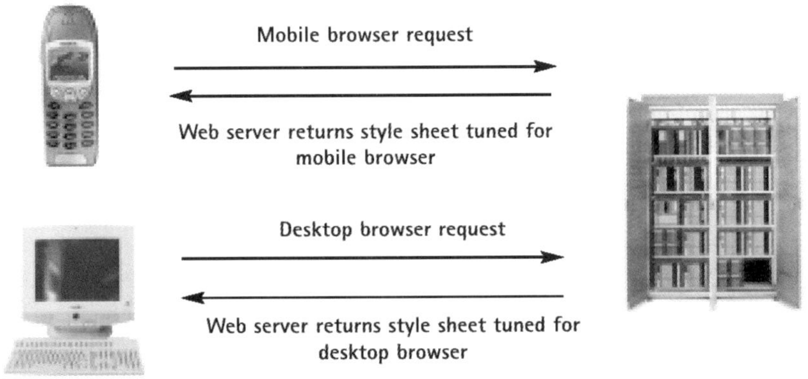

All pages on the web site can be the same for any browser; only the style sheet is tuned for different needs

Through CSS a document creator can specify the presentation of a Web application in one place, the style sheet. If the presentation needs to be changed at any time, the change is made once in the style sheet and the modification is dynamically reflected throughout all the pages on the site.

CSS separates the content of the document or application from the presentation. This allows developers to easily create browser-specific versions of the same content simply by creating the appropriate style sheet. For example, when a user requests a Web page, the content server identifies the requesting device and returns the content with a link to the appropriate style sheet. The style sheet is downloaded once and cached by the browser for use with subsequent pages, which speeds the rendering of all pages on the site.

The power of CSS lies in the precise control it offers document creators and the ease with which they can optimize content for presentation on any device. Every aspect of the document – positioning, fonts, text attributes, borders, margin alignment, flow – can be defined in the style sheet. A change to any aspect of the document needs to be made only once.

CSS also gives carriers greater control over the look and feel of the services they provide through their wireless portal. Through XHTML Basic, the operator can define a default style sheet for all devices it supports, which will ensure a basic look and feel consistent for all devices, even those from different manufacturers. For example, a carrier might specify that all links are underlined or all headings are bold and centered. Of course, content developers can choose to modify that style through their own use of CSS if different behavior is required, but the default style sheet determines the overall look of all applications for a particular browser.

CSS Browser Examples

The precise control provided by CSS makes it easy to reformat content for different mobile devices.

In the first figure you saw the same XHTML Basic pages displayed for different browsers. Through the power of XHTML Basic the presentation, style, and navigation are easily reformatted for each browser. The following figures show other examples of the same page with different formatting applied with different CSS style sheets.

Note how XHTML Basic's support for CSS allows padding and paragraph indentation to achieve precise positioning. Even sophisticated typographic treatments such as drop-caps can be used. Flexible alignment (center, left, right), borders of varying thickness and margins can also be used to create advanced effects such as floating paragraphs.

Identical files,but a different CSS style sheet changes appearance.

The different bullet styles are not defined in the XHTML page.

Transformation

As an XML-based standard, XHTML Basic allows for the automatic parsing and transcoding of content through the use of XSLT (eXtensible Stylesheet Language: Transformations). A part of XSL, the style sheet language for XML.A W3C standard, XSLT provides a language for transforming XML documents into other XML documents.

Using XSLT, a system can automatically transform the same XML content into multiple markup languages depending on the browser. Through such transformations, content can be created for one device and automatically transformed to appear on another device. For example, a news service could create its news feed once in XML and use XSLT to dynamically convert it to XHTML Basic and HTML for presentation on mobile handsets or other devices.

Better still, as XHTML becomes the standard for all Web browsers, a Web site could use XHTML for all its content. In this way, developers would create content once for display on different devices, eliminating the need to write different versions of the same content for different devices.

Comparison with Compact HTML

XHTML Basic is similar to Compact HTML (cHTML), the language used by DoCoMo's i-mode wireless service in Japan, with a few exceptions. The two biggest limitations of cHTML are that it is not a W3C standard and it does not support cascading style sheets. By comparison, XHTML Basic is a W3C and WAP Forum standard widely accepted throughout the industry. Furthermore, cHTML need not be well-formed, which means that it can produce inconsistent results from browser to browser, and it lacks the XML header and DTD specifier, which gives XHTML Basic much of its capabilities.

Mobile browsers that support only cHTML cannot take advantage of CSS. As a result, content providers have to expend considerably more effort to present documents on different devices, and carriers will not have the control CSS gives them.

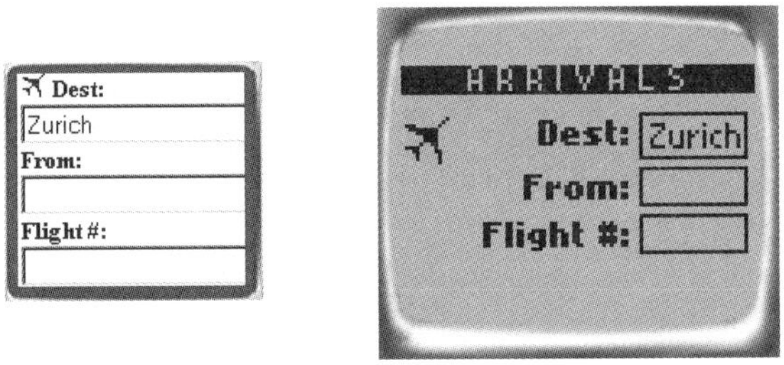

cHTML screen on a mobile device vs. layout with XHTML using the Nokia WAP Browser.

Benefits of the XHTML Basic Browser

XHTML Basic is leading the way to the next generation of WAP technology. It has been widely embraced by the industry, and its modular architecture makes it easy to adapt to any changes that result from the evolution of new technologies. In addition, the XHTML Basic browser delivers a number of specific benefits for all parties involved in mobile computing. These are described in the following table.

Group	Benefit
Device manufacturer	Ensures device complies with the latest WAP Forum and W3C standard, accepted industry-wide
	Results in a more attractive browsing experience because of enhanced layout control
	Makes the browser content easier to read due to the use of sophisticated typographic elements
	Reduces the memory footprint and increases performance due to the smaller size of XHTML Basic and its strict validation rules
Carrier	Control the look and feel of content delivered through its wireless portal to all supported devices

	Enables the carrier to define a default CSS style sheet for all supported devices, creating a consistent look across all manufacturers' browsers
	Standards-based browser makes it easier to attract content providers/ developers
	Automated, dynamic transformation capabilities ensure access to more content
Content developer/provider	Create content once
	Use existing tools
	Develop for an industry standard
	Precise control of content layout and display
	Style sheets allow for the easy application of design changes to the entire site in one place
Mobile device user	Browser that is more aesthetically pleasing
	Easier to use and navigate
	Enables access to more content

XHTML is the Future

The adoption of XHTML Basic within the mobile industry as the standard browser language brings wireless communications into alignment with the latest advances in Internet content development and presentation. Supported by the WAP Forum and the W3C, XHTML Basic establishes a global standard for all browsers. Even the i-mode service (which uses proprietary markup languages), appears to be moving toward XHTML compliance.

By adopting XHTML Basic as their browser standard, mobile device manufacturers, carriers, developers and end-users alike can get a head start on experiencing the many benefits its offers. Foremost among those benefits is compliance with the global standard for mobile Internet computing going forward.

SIP and VoIP

There are certain issues when applying SIP and other Internet application signaling protocols to the wireless environment. Currently 3GPP is specifying an IP Multimedia Subsystem (IMS) on "top" of 3G GPRS access networks to offer voice and multimedia services for 3G wireless terminals. IMS uses several of the protocols described previously, such as SIP, SDP and RTP. It can be assumed that a high-end wireless terminal will also include protocols like HTTP and RTSP. IMS architecture is shown in following figure.

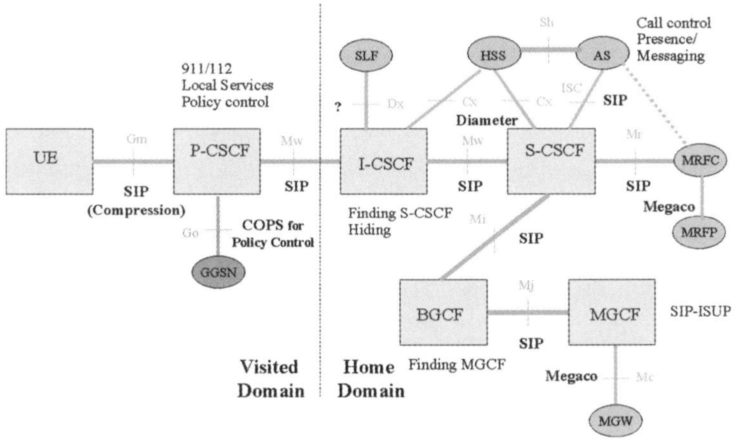

3GPP Release 5 IP Multimedia Subsystem Architecture.

The figure only shows the IMS components, the radio network or GPRS core is not present except for the Gateway GPRS Support Node (GGSN), which has a direct interface to IMS. One would be situated between User Equipment (UE) and Proxy-Call State Control Function (P-CSCF) providing basic IP connectivity and mobility management below it. UE is basically the mobile terminal. It talks with SIP to P-CSCF which is very similar to a SIP proxy server. If the user is roaming in a visited domain P-CSCF is located there. The serving CSCF (S-CSCF) is always located in the user's home domain, and it resembles SIP registrar and proxy servers, so a user registers there using SIP via P-CSCF. Interrogating CSCF (I-CSCF) is another SIP proxy responsible mainly for finding the correct S-CSCF for a given user. The Breakout Gateway Control Function (BGCF) and Media Gateway Control Function (MGCF) are used for telephony network interconnection. Media Resource Function Control/Processing (MRFC/MRFP) are related to multi-party conferencing, network announcements and media transcoding. The Application Server (AS) is a SIP element dealing with services. HSS (Home Subscriber Server) and SLF (Subscriber Locator Function) are related to user profile management and authentication.

Basically, IMS is just an instance of a SIP network. It has a number of proxies and a registrar, although they have strange acronyms. User Agent is situated in the terminal, UE. When two terminals establish a session they talk to each other via the CSCF elements in the same way as presented in the previous figure. RTP media flows do not go via CSCFs but go directly between the terminals over IP.

There are, however, some differences. One of the most stressing issues with the wireless environment is the scarcity of bandwidth. This means that each protocol should be as compact as possible when sent over the air. In practice, the Internet protocols do not meet this requirement as such. For example, when voice is carried in RTP/UDP/IP packets, over half of the size of the packets may be header information. In order to cope with this, IETF has defined Robust Header Compression (ROHC) for RTP/UDP/IP, reducing the size of the headers to a couple of bytes in most scenarios. In this way, VoIP tranport becomes fairly efficient compared to traditional cellular voice. 3GPP has adopted ROHC for this purpose.

The same applies for signaling. All application layer signaling protocols presented in this section are text based and rather large compared to traditional cellular signaling protocols. For example an SIP INVITE can be almost 1000 bytes long and it takes a while to send that amount of data over a narrow radio link. This is why signaling compression is needed. In this type of compression, dynamic dictionaries can be used to boost the compression efficiency. Thus a relatively good compression ration can be expected. The signaling compression will be placed on the application layer in between transport and the actual signaling protocol.

Another issue in the wireless environment is the number of roundtrips needed to establish a session. Currently the QoS interworking extensions for SIP do not solve the issue in an optimal way.

Conclusions

In the wireless environment, compression is needed to make SIP efficient enough for the air interface. The future will show what value mobile terminals will bring to the Internet applications. Location-based services are a possible candidate.

References

Session Initiation Protocol, RFC 2543
Real Time Transport Protocol, RFC 1889
Session Announcement Protocol, RFC 2974
Session Description Protocol, RFC 2327
Real Time Streaming Protocol, RFC 2326
Resource Reservation Setup Protocol, RFC 2205
E.164 Numbers and DNS, RFC 2916
3GPP Release 5 IP Multimedia Subsystem, 3rd Generation Partnership Project; Technical Specification Group Services and System Aspects; IP Multimedia (IM) Subsystem - Stage 2; (3G TS 23.228 version 5.0.0)

Instant Messaging and Presence with SIP

Instant Messaging and Presence (IM/P) is a great success on the Internet. In the mobile world people on the move have to keep in contact with their peers. IM/P will help them to maintain their essential social links when on the move. As we will see in this article, the Session Initiation Protocol (SIP) is a perfect fit in future communication services. First, we will introduce SIP by describing the protocol basics, highlighting the end user services and introducing the SIP service creation model. Second, the future of SIP will be outlined by introducing where SIP is defined. Finally, the conclusion will show that SIP has a great mobile future. Let's learn more about SIP.

SIP

SIP is going to be the standard signalling protocol/mechanism to support Voice over IP (VoIP) for 3rd generation networks. This choice ensures that future VoIP services will be truly Internet based. SIP is defined by the IETF Session Initiation Protocol workgroup in RFC2543 and is very different from traditional signaling methods used in telecommunication networks. It shares a lot with existing Internet protocols like HTTP [3] and SMTP [4] (protocol header formats). SIP was designed to establish, modify and terminate multimedia sessions or calls. It is text based (See Figure 1) which makes SIP relatively easy to implement, easy to debug, extensible and modular. The modularity helps to implement SIP clients in different device categories from small devices with limited capabilities to desktop computers with full features.

SIP uses Transmission Control Protocol (TCP) [11] or User Datagram Protocol (UDP) [12] as a transport mechanism. Actually SIP is independent of the lower layer transport protocol. SIP has its own reliability system and it can be run over any protocol that provides byte stream or datagram services. The SIP message is composed of headers and a payload (MIME type). The payload includes SDP (Session Description Protocol)[6], which can be used to describe multimedia sessions.

```
INVITE sip:j.k@n.com:5060 SIP/2.0
Via: SIP/2.0/UDP midas.research.nokia.com
From: "Bouret C" <sip:c.b@n.com>;tag=C39854
To: "Kinnunen J" <sip:j.k@n.com>
CSeq: 1 INVITE
Call-ID: 1844275253@midas.research.nokia.com
Content-Type: application/sdp
Content-Length: 143
Subject: Project Questions
Accept: application/sdp, text/plain
Contact: <sip:chris.bouret@midas.research.nokia.com:5060>

v=0
o=username 0 0 IN IP4 midas.research.nokia.com
c=IN IP4 midas.research.nokia.com
t=0 0
m=audio 1038 RTP/AVP 0
a=rtpmap:0 PCMU/8000
a=ptime:20
```

Headers

Payload:
Mime type
(SDP)

An SIP Message

SIP deliberately utilizes similar global URI (Uniform Resource Identifiers)[2] based addressing as Internet e-mail (sip:<user>@<domain>). SIP addresses a personal mobility using personal unique identity. A person may originate, receive and access services from any terminal and any location.

The Nokia 9210 Instant Messaging and Presence Client

SIP-Based Services

Voice is not the only service available on the mobile Internet. Mail, Web browsing, Instant Messaging and Presence will be the first services available and voice over IP will come as a complement to existing services, i.e. SIP allows initiation of any interactive session between any kind of network endpoint.

The mobile Internet offers a way to combine these services and therefore SIP offers a perfect evolutionary path by supporting:

- Instant Messaging (exchange of content between a set of participants in real time)
- Presence (subscription to and notification of changes in the communications state of a user).
- Voice can be introduced seamlessly when required Quality of Service (QoS) becomes available on the network. The voice path is established with some suitable protocols not part of SIP e.g. RTP [5].
- Click-to-Call from an electronic document (sip:chris.bouret@nokia.com). This means that the document includes a SIP link as a tag.
- Transport of MIME types (reference to Web/WAP pages, mail, pictures) in SIP signaling
- Interactive games with SIP signaling
- Signaling to any kind of SIP enabled elements in the network e.g. servers, databases, devices.

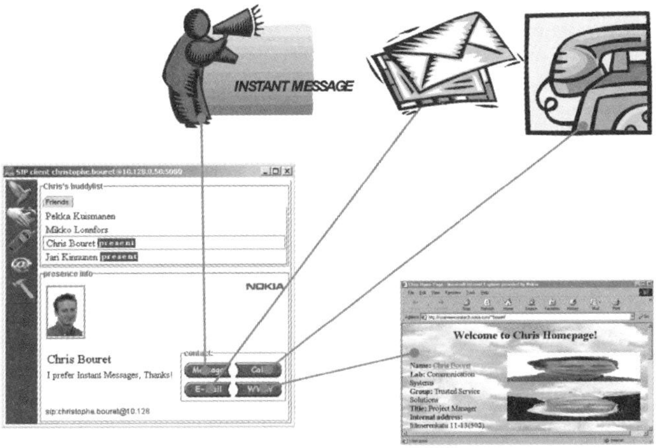

SIP Services

SIP Instant Messaging and A Presence Message Flow Example

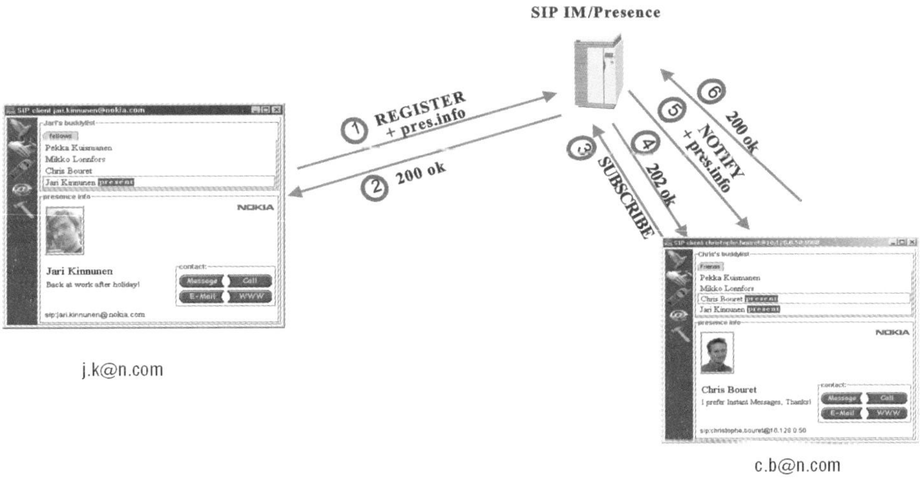

A Presence Call Flow Example

Presence flow description: The message flow describes SIP presence functionality. First, Jari sends a registration request to his Presence server (message 1 and 2). The register message includes his presence information coded in XML format. Chris, interested by Jari's presence information, subscribes by sending a message to Jari's presence server (3 and 4). After authorization from Jari (not shown here) Chris gets a NOTIFY message including Jari's presence information.

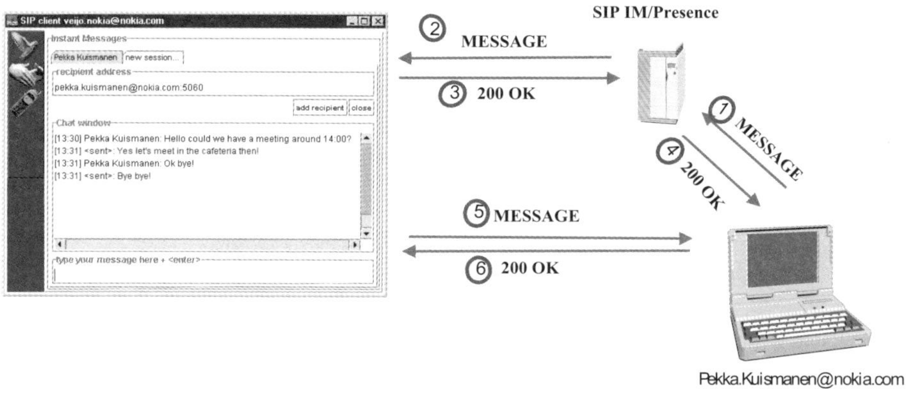

An Instant Messaging Call Flow Example

Simple Instant Messaging message flow:
The first time Pekka sends a message to Veijo (message 1) it goes to the IM/P server that forwards or redirects it to Veijo's actual address (message 2). Subsequent messages (message 5 and 6) can be sent directly from peer to peer but using proxies allow messages to go through firewalls.

Another alternative is to establish a standard SIP session between endpoints and to deliver messages in a separate negotiated media channel (RTP, TCP or other). The standardization work is still open on this issue.

SIP Service Creation

The SIP creation model is once again taking a page out of the Internet book. SIP services are taking advantage of the Internet structure:

- the core network is transporting and routing the signaling
- end user services are executed on application servers or directly on the terminal

The Internet architecture opens possibilities for third party service providers, such as the Application Service Provider (ASP), corporate or even end users, to create and control services. On the application server side, SIP service creation uses existing well known interfaces like the SIP Servlet [10], the SIP Common Gateway Interface (CGI) [7,8] and the Call Processing Language (CPL) [9] to ensure a large base of potential service creators. These standard interfaces also help to create services starting with a http request, for example click to call that starts a SIP service setting up the actual call. On the terminal, the service is integrated in the SIP client, making a secure environment such as Java very attractive. Thus, applet like service logic may be executed on the terminal.

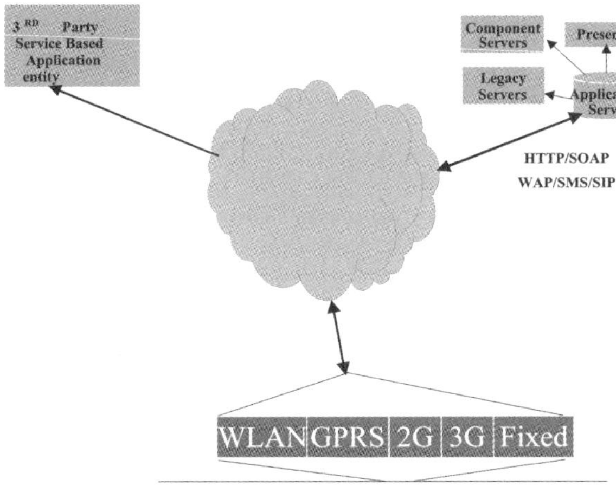

A Network Architecture Example

Current Status and Future of the Technology

SIP IM/P in IETF

SIP is currently being extended by the SIMPLE (SIP for Instant Messaging and Presence Leveraging) workgroup of IETF to provide SIP with presence information transferring capabilities. Presence is defined as subscription to and notification of changes in the communications state of a user. This communications state consists of a set of communication means, a communication address, and the status of that user. A presence protocol is a protocol for providing such a service over the Internet or any IP network.

Providing information about user presence is particularly well suited to SIP. Since SIP registrars and location servers already hold user presence information (through REGISTER messages coming from the UAs) and SIP networks already route INVITE messages from any user on the network to the proxy that holds the registration state, SIP can also provide presence services to the users. This means that SIP networks can be reused to establish global connectivity for presence subscriptions and notifications. In order to handle these kind of event notifications with SIP, it was extended with two new methods, the SUBSCRIBE and NOTIFY methods.The presence document supported by IETF SIP is defined by Common Presence and the Instant Messaging (CPIM) Data Format.

```
<?xml version="1.0"?>
<!DOCTYPE presence>
   <presence>
      <presentity uri="chris.bouret@10.128.0.50"/>
      <atom atomid="1810353">
       <address
uri="sip:chris.bouret@10.128.0.50;method=&quot;INVITE,MESSAGE&quot;">
          <status status="open"/>
          <note>I feel good!!</note>
       </address>
      <atom atomid="1810353http">
         <address uri="mailto:chris.bouret@nokia.com">
          <status status="open"/>
          <note>My mail!</note>
       </address>
     </atom>
      <atom atomid="1810353http">
         <address uri="http://www.nokia.com">
          <status status="open"/>
          <note>The web page!</note>
       </address>
     </atom>
   </presence>
```

A Presence Document Example

Since all Presence and Instant Messaging solutions defined by the IETF will support this format, it will be easy to translate form one transport protocol to another.

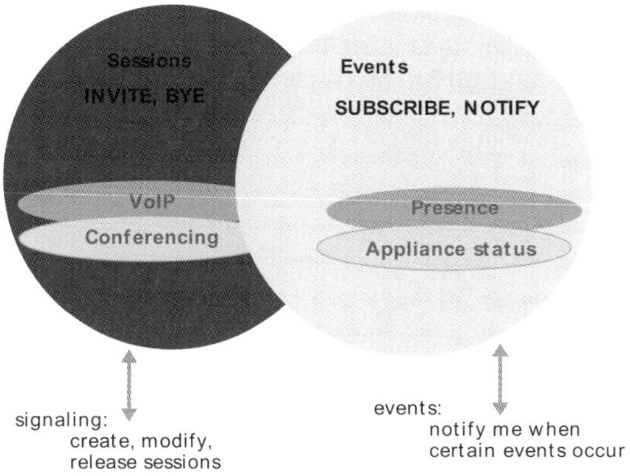

IETF based solution

SIP IM/P in 3GPP

Since in third generation networks the Call Control and the Service Creation Environment is based on the SIP protocol, it seems feasible to introduce SIP based presence services in 3G networks as well. The proposed solution is to place a Presence Server in the IP Multimedia Subsystem [13] of the Release 5 architecture [12] as part of the Application & Services cloud. The basic SIP solution will support the IETF SIMPLE definition as much as 3G architecture permits.

SIP and the Wireless Village Initiative

Founded by Ericsson, Motorola, and Nokia, Wireless Village, the Mobile Instant Messaging and Presence Services (IMPS) Initiative, was formed in April 2001 to define and promote a set of universal specifications for mobile instant messaging and presence services. The specifications will be used for exchanging messages and presence information between mobile devices, mobile services and Internet-based instant messaging services, all fully interoperable and leveraging existing web technologies. Through its supporters, the Wireless Village initiative aims to build a vibrant community of end users and global business partners where Internet and wireless domains converge. Wireless Village will support Short Message Service (SMS), Multimedia Messaging Service (MMS) and SIP solutions.

Conclusion

On the mobile internet Instant Messaging and Presence is a key feature. As we have seen in this article, SIP is a perfect fit to implement IM/P. SIP based services are integrated with existing Internet services and SIP service creation is easy and simple. Finally, the SIP solution is defined in different standardization bodies working in co-operation. We believe that the SIP Instant Messaging and Presence solution will prevail in the mobile Internet opening the door to rich multimedia communication services.

Authentication Methods and Technologies

Authentication, i.e. building trust between participating identities, can be done leveraging several different methods or combinations of these. Depending on the methods or combinations used and whether the identity is real, i.e. linked to physical person or device or whether it is a pseudonym, a different level of authenticity is required.

The most common high level way to define authentication methods is to divide them in to three different prime categories: 'What you know' authentication is based on the idea that a client knows something that verifies its authentication. Normal user id - password authentication is a good example of this category. 'What you have' authentication prooves an identity if some physical or virtual identification item is used. Examples of this category include a corporate ID card, a driving license or a normal key. 'What you are' authentication prooves an identity based on a unique physical characteristic of a person, such as a fingerprint.

If combinations of these methods are required at same time, the certainty of authenticity is increased. The disadvantage is that the usability of a system becomes worse, making people to try finding ways to bypass the security. This can be avoided with choices in system design. A term 'two-factor authentication' is used, if two methods are required to proove identity. In practice, this means a person has to hold some physical identity object like smart card and in addition know a password or PIN. The term multi-factor is used, if more than two methods are requires in one authentication transaction. This method increases security, but may decrease usability leading to neglectful usage of the security system, if special care is not given to usability design.

One challenge when designing the authentication infrastructure is to build reliable, functional and secure key distribution and management. As described above the authentication is based on shared secret or PKC private/public keys. The shared secret can be a password or symmetric key known by both communication parties. An important factor is how to deliver and upgrade secrets or keys in a cost-effective and reliable way. This can happen by physical means like posting passwords or smart cards containing the secret or over the network utilizing some authentication protocol.

The above described basic methods are implemented in computing systems using typically one or a combination of the following authentication technologies:

- Password authentication
- One-time passwords
- Hardware token (e.g. RSA SecurID, RFID)
- Symmetric key based solutions (e.g. Kerberos, GSM Authentication)
- PKC based solutions: Digital Certificates, Digital Signatures, PKI & Certificate Authorities, Public Key Encryption, WIM, WPKI
- Biometric authentication
- RADIUS, DIAMETER
- Single sign-On

These technologies are not equal in scope, but parts of different technologies are often needed to build a complete authentication system.

Password Authentication

In private and public computer networks, including the Internet, authentication is commonly done through the use of log in passwords. Loggin in is the procedure used to get access to an operating system or application, usually on remote computer. As log in almost always requires the user to have a user id and password. Often, the user id must conform to a limited length such as eight characters and the password must contain at least one digit and not match a natural language word. The user id can be freely known and is visible when entered at a keyboard or other input device. The password must be kept secret and is not displayed as it is entered.

Knowledge of the password is assumed to guarantee that the user has access priviledges. Each user registers initially or is registered by someone else, by using an assigned or self-declared password. On each subsequent lo gin, the user must know and use the previously declared password. Some Web sites require users to register in order to use the site; registered users can then enter the site by logging in.

In the simplest password authentication systems the user id and password are transferred openly over data networks. This enables eavesdropping crimes. In more advanced systems, challenge and response schemes are followed, and where secret information is encrypted with a symmetric or public/private keys utilizing either a software or hardware solution.

The weakness of a password authentication system, especially for transactions that are significant in value, such as the exchange of money, is that passwords can be stolen, revealed accidentally, or forgotten.

Onetime Passwords

Onetime passwords protect against eavesdropping, as the password is invalidated after one use. In off-line implementations, the most common way of using the onetime passwords is sending a list of passwords to the user beforehand. Every time the user accesses the protected resource, one password from the list is used. A more modern way to utilize onetime passwords is to send one to the user when needed. This may happen, for example, by sending a password in an SMS message to a mobile phone. The SecurID card, described in more detail under Hardware Token based authentication, also generates onetime passwords.

For onetime passwords it is important that a password and/or the password list is never sent over the same communication channel at same time as the user identification number or hardware security token.

Hardware Token Based Authentication

SecurID and RFID tokens represent the 'something you have' security method. They can be combined with the 'something you know' scheme, if a PIN is required to get final approval for access.

RSA SecurID

Essentially, SecurID tokens are onetime passwords for user authentication and can be used to authenticate the right to access restricted services. SecurID can be used together with a RADIUS system to increase the security level from a simple user-id password approach. The time-synchronized SecurID card has an LCD display that shows a string of numbers that changes every minute. When authenticating, the user is prompted for a user id and the Passcode, which is generated dynamically by the SecurID card when the user enters a PIN via the keypad on the card. The authentication request is then sent over the network to the RSA ACE/Server, which determines the validity of the Passcode. A validated Passcode will grant a user network access.

RFID

RFID (Radio Frequency ID) is a passive radio receiver and transmitter absorbing its energy from radio waves. Nowadays most implementations handle the basic challenge-response protocol with encryption support, which protects against the simplest man-in-the-middle attacks.

Symmetric Key Based Solutions

The most popular symmetric key authentication technology is one used in GSM authentication utilizing SIM-smart cards. Another popular example is Kerberos authentication. In addition to these, we will also briefly review smart cards used in PC login. The disadvantage of the symmetric key systems is that they cannot provide bulletproof non-repudiation.

Smart C ard Authentication

Smart cards can store either digital certificates or just log in information. Certificates are described in the Public Key Cryptography section.

The beauty of using smart cards is that the log 0in process is transparent to the user. As with an ATM, the user cannot access the system without a card. Once the card is entered, the user has to enter a PIN and the system takes over from there. Smart cards can also feature the ability to lock out a user after a set number of unsuccessful login attempts. Once this occurs, an admin must unlock the card before it can be used again.

SIM Authentication

The SIM (Subscribe Identity Module) defined in GSM Specifications is a form of smart card authentication. The basic idea is to encrypt all traffic between mobile station (i.e. terminal device) and base station. Opposite to that the operator network is considered as secure. To reach this target the client and the Authentication Center that are linked closely to the Home Location server share a common secret called Ki. To be more specific, this symmetric key, Ki, is stored in the client terminal SIM (Subscriber Identification Module). When a client wants to authenticate itself with a network, the network generates a random challenge called RAND in the GSM standard. The secret, Ki, is used to calculate the response, called SRES (Signed RESponse), to the challenge. If SRES matches with one calculated by the network, access is granted. The home operator, as opposed to the visiting operator, is always the one verifying the authenticity of the user. The visiting operator acta as a mediator, passing requests from the visiting terminal to home operator's Authentication Center and vice versa.

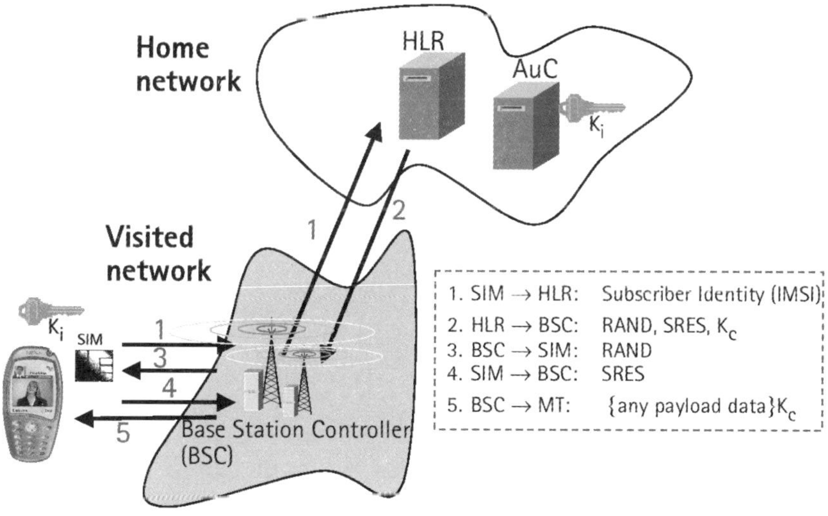

Simplified GSM Authentication Protocol.

After successful authentication, the terminal calculates a session key, Kc, from the random authentication challenge. The standard describes that from this point on all communication must be encrypted. However, in certain situations, encryption might be bypassed. In some countries, encryption is not even allowed. Generally, if encryption is not used, it is considered a lack in the quality of service defined by the standard. It should be good to noted that encryption is used only on the radio channel aa the operator network is onsidered a secure entity.

Kerberos Authentication

Kerberos is a network authentication protocol created in the early 1980s in Massachusetta Institute of Technology. Kerberos is, by nature, a distributed authentication service that allows a process (a client) running on behalf of a principal (a user) to prove its identity to a verifier (an application server, or just server) without sending data across the network that might allow an attacker or the verifier to subsequently impersonate the principal. Optionally, Kerberos provides integrity and confidentiality for data transmitted between the client and server.

The Kerberos authentication system uses a series of messages encrypted with shared i.e. symmetric keys to prove to a verifier (target service) that a client is running on behalf of a particular user. Initially, both the user and the service are required to have keys registered with the Kerberos Authentication Server (AS). The user key is derived from a password that he chooses; the service key is a randomly selected key (since no user is required to type in a password).

When a client wants to access a service (verifier, v) on the network, the client requesta authentication from the server (AS). The server then creates two copies of the session keys, also known as the shared key (Kc,v). The session keys are accompanied with other data relevant to the protocol like a time stamp. Time stamps are used to prevent a hacker from capturing the session and playing it back. After this, one of the shared keys (Kc,v) with related data is encrypted with a client's secret key (Kc) and the other is encrypted with a verifier's secret key (Kv). Next, both encrypted packets are sent back to the client, which forwards the session key to the verifier. The verifier can now decrypt the message using its private key (Kv). Only an authentic client and verifier can decrypt the shared session key (Kc,v) enabling further communication between the parties. The protocol also enables mutual authentication. A simplified description of the protocol is shown in the Figure below.

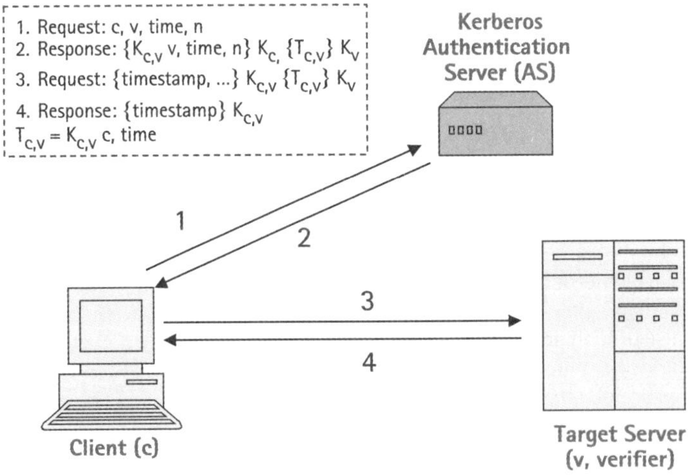

Simplified Kerberos Authentication Protocol.

One of the main benefits of using Kerberos is that it provides single sign-on capability using the SSPI (Security Service Application Programming Interface) standard. This is realized with the help of an additional server system part of the Kerberos server called a ticket-granting service. It is used to support subsequent authentication after the initial authentication without need to re-enter a principal's password. Cross-realm authentication approaches also exist, where the authentication of a principal registered with a different authentication server is enabled.

There are several enhancements to Kerberos. For example, in the Windows 2000 version of Kerberos, the initial user authentication is done using public key certificates instead of the standard shared secret keys.

The description of the Kerberos protocol in this section is simplified for clarity. Readers should consult RFC 1510 for a more thorough description of the Kerberos protocol.

Public Key Cryptography

Weaknesses in password authentication and symmetric key systems for transactions have led to a need to develop something more advanced. Mobile commerce, Internet business and many other transactions require a more stringent authentication process. The use of digital certificates issued and verified by a Certificate Authority (CA) as part of a public key infrastructure is to become the standard way to perform authentication on the Mobile Internet.

The distinct advantage of public-key-based remote authentication over mechanisms that mimic authentication in the local environment is that sensitive authenticating information, such as a password, is never sent over the network. In other authentication methods, if server Alice holds a copy of client Bob's password or thumbprint, Bob must authenticate himself by proving that he knows or has this information. This is typically accomplished by Bob conveying this information to Alice upon sign-on. During the process the information is vulnerable to third-party attacks.

Essentially, when PKC is used in a mutual authentication, the user's terminal device and the destination server's service present their certificates to each other. The certificate containing the name and public key of the holder are signed by a private key of the certificate authority contained in a root certificate. If the exchanged certificates are decryptable with the public key of CA, the authenticity and integrity of the certificate is prooven. Now, if the initiating party of the communication is able to sign the challenge sent by the other party with its private key and the first party is able to decrypt it with a public key from a valid certificate, the authenticity of the signing party is prooven. The same can also be done in the other direction. An alternative source for certificates, instead of getting them directly from other communicating party, is to fetch them from the central directory service.

The root certificates provided by a trusted authority are stored on the user's device. Root certificates are the basis for certificate verification and should be supplied only by a system administrator. The certificate authority can also invalidate a certificate by marking it invalid on the revocation server, from where the communication party should check the validity of a certificate.

Biometric Authentication

Fingerprint readers, voice recognition software, facial identification cameras and retina scanners are among the options used in biometric authentication. They represent the 'something you are' authentication principle.

Even with the promise of high accuracy and ease of use in biometric systems, they have not not been widely used. Current high cost, unreliability and lack of accuracy of these methods are major concerns. While biometric systems continue to develop, they still suffer from the effects of environmental factors, e.g. sweaty or wet fingers.

RADIUS Authentication

RADIUS (Remote Authentication Dial-In User Service) is a client/server protocol and software that enables remote access servers to communicate with a central server to authenticate dial-in users and authorize their access to the requested system or service. The primary functions of RADIUS are authentication, authorization, and accounting. RADIUS allows a company or network service provider to maintain user profiles in a central database that all remote access servers can share. It provides a single point of administration for multiple remote access servers. This eases setting up a unified policy for clients independent of access point to the network. Having a central service also means that it is easier to track usage for billing and to keep network statistics. RADIUS, is a de facto industry standard and is a proposed IETF standard.

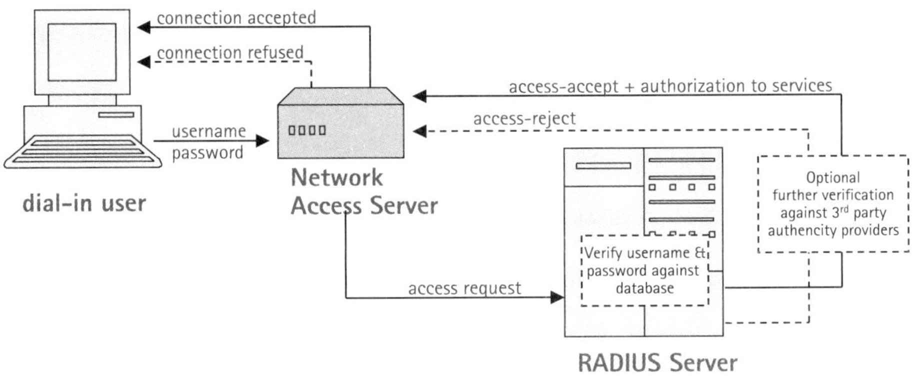

RADIUS Authentication and Authorization.

A remote Network Access Server (NAS) operates as a client of RADIUS. The client is responsible for passing user information to designated RADIUS servers, and acting on the response that is returned. RADIUS servers are responsible for receiving user connection requests, authenticating the user, and returning all configuration information necessary for the client to deliver the service to the user. A RADIUS server can act as a proxy client to other RADIUS servers or other kinds of authentication servers, such as ActivCard, ActivEngine or SecurID ACE/Server. The RADIUS server can support a variety of methods to authenticate a user.

Transactions between a client and a RADIUS server are authenticated through the use of a shared secret, which is never sent over the network. In addition, any user passwords are sent encrypted between the client and the RADIUS server, to eliminate the possibility that someone snooping on an unsecure network could determine a user's password.

When a client is configured to use RADIUS, any user of the client presents authentication information to the client. This might be with a customizable log in prompt, where the user is expected to enter their username and password. Alternatively, the user might use a link framing protocol such as the Point-to-Point Protocol (PPP), which has authentication packets that carry this information. Once the client has obtained such information, it may choose to authenticate using RADIUS. To do so, the client creates an access request containing the necessary authentication information.

If the correct authentication information, such as matching user-id and password pair, is provided to the RADIUS server, it responds with access-accept message indicating an accepted connection. The server also sends authorization information about the services the user can access. Instead of providing immediate access, the server may prompt further proof of authenticity, e.g. in the form of a SecurID challenge. If the user is not able to provide correct authentication information, the RADIUS server sends an access-reject response indicating that this user request is invalid.

The figure above shows the authentication and authorization process.

A more detailed description of RADIUS can be found from RFC 2865 - Remote Authentication Dial In User Service (RADIUS) and http://www.livingston.com/marketing/whitepapers/radius_paper.html.

Diameter

Diameter is being specified in the AAA (Authentication, Authorization and Accounting) working group of the IETF. Diameter can be considered as a next generation Radius, to support authentication, authorization and accounting in a number of situations. Diameter is primarily designed as a network-to-network interface (NNI), with Mobile IP extentions for key distribution purposes.

Diameter attempts to solve IP roaming issues between home and foreign networks, taking into consideration the authentication of a user, user access and accounting issues. Additionally, Diameter allows key distribution, where the AAA servers can act as a key distribution center that generates and distributes keys.

Currently, Diameter is being used in 3GPP for subscriber management. The specification work is still under way, but the Diameter Base protocol is expected to be ready before the end of 2001.

Single Sign-on

Single sign-on (SSO) is a mechanism whereby a single action of user authentication and authorization can permit a user to access all computers and systems where he has access permission, without the need to authenticate explicitly to all systems by for example entering multiple user-ids and passwords. Single sign-on reduces human error, a major component of systems failure, and is therefore highly desirable. User friendly, secure and reliable single sign-on can be seen as one of the primary catalysts to boost mobile commerce.

Many current computer systems or platforms provide a single sign-on feature. However, none of these satisfy the need for real cross-platform single sign-on that is needed when multiple service providers utilizing varying technical platforms are involved. This happens easily on the Mobile Internet, as it is not likely that one vendor will run the same computing platforms. To meet the requirement of cross-platform single sign-on, cross industry activity is needed.

There are four basic alternative ways to realize single-sign on:

- In the 'Standard identity and secret' scheme, the same protocol, identity and secret (e.g. password) are used for all systems. This model is unrealistic, as people want to be identified differently in different situations. To reach its ultimate target, this scheme must be followed some extent, as there is not enough memory and computing resouces in mobile terminals to build support for several different protocols.
- The second way to implement single sign-on, is to hide the differencies between multiple protocols and authentication information in a terminal device. This is called the 'Intelligence in terminal' scheme.
- The third solution, 'Intelligence in proxy', hides the differences of multiple protocols and authentication information on a trusted server in the network. The advantage is that the differencies are hidden from the terminal device, which allows single-sign on regardless of whether the terminal device has been specifically designed for it or not. The disadvantage is that the user must trust the server and the communications path between the terminal and the server. This is how Nokia mPlatform currently implements single sign-on functionality.

- The fourth way is to setup a centralized authentication server or a distributed cluster of authentication servers being able to communicate securely with each other. Once authentication is accomplished, the server generates a security token, which prooves the user's identity. The token may be called a ticket or a voucher and may include credentials that describe the rights of the user in the destination system. Examples include HLR + Authentication server in GSM, Home Subscirber Server (HSS) and Diameter in 3G and Kerberos or Microsoft Passport in the web domain. The server or the cluster can be called an 'Identity broker'. The advantage of this approach is that the broker does not become a bottleneck in the system as easily as in the proxy case, because the broker is consulted only in the authentication phase and is not needed in other communication.

None of the presented models work in their pure incarnation and a functioning system usually has elements from all of them.

A Single sign-on solution clearly has merits and risks. The two main merits are, (1.) Convenience: Users need only identify/authenticate themselves once to the "primary" domain and not for every platform or application. (2.) Security: Because there is only one sign-on, the risks inherent in remembering multiple login/password combinations are eliminated. Accidentally typing the combination user-id and password of one service, when authenticating to another service, gives the service provider an opportunity for fraud. Even worse risks follow from typical habits to use the same user-id and password for multiple services. The two main risks of SSO are, (1.) Security: If an intruder compromises a user's account or password, that intruder could have extensive and easier access to far more resources. (2.) Cost: single sign-on implementations can be expensive, both in the cost to purchase and in the manpower to deploy. Two-factor single sign-on, where access is granted based on the combination of what a user knows (a password or PIN) and what they possess (a biometric device, for example), is very secure.

Summary

Identification and authentication are fundamental elements on the Mobile Internet. They are needed for many important use cases, enabling the setting up of trust relationships between communicating parties. Globally functional interoperable solutions are required for mobile, web and proximity domains that support the users' preferred set of identities on a proprietary security level. The robustness of the complete system must be taken into account, as the terminal device and frequency spectrum are limited. Meeting all these requirements require applying the described authentication technologies. One size does not fit all. The complexity of the system needs to be hidden from end users with proper user interface implementation. Single-sign is an essential target in reaching an enjoyable user experience.

References:

- http://whatis.techtarget.com
- Security Engineering, Ross Anderson (ISBN 0-471-38922-6)
- RFC 2865 - Remote Authentication Dial In User Service (RADIUS), C. Rigney, S. Willens, A. Rubens, W. Simpson. June 2000. (Status: DRAFT STANDARD), http://www.ietf.org
- http://www.livingston.com/marketing/whitepapers/radius_paper.html
- http://web.mit.edu/is/help/kerberos/
- RFC 1510 - The Kerberos Network Authentication Service (V5). J. Kohl, C. Neuman. September 1993. (Status: PROPOSED STANDARD), http://www.ietf.org

- http://www.activcard.com
- http://www.rsa.com
- http://www.microsoft.com
- http://www.baltimore.com
- http://www.verisign.com
- http://www.projectliberty.org

Trademarks referred to:

- SecurID owned by RSA Security Inc
- ACE/Server owned by RSA Security Inc
- ActiveCard owned ActivCard, Inc
- ActivEngine owned ActivCard, Inc
- Windows 2000 owned by Microsoft

DRM

The timely convergence of numerous technologies for the distribution of digital media will dramatically impact the way in which copyrighted content will be created, distributed, sold, marketed and consumed in future years. Recent developments indicate that producers of content are under pressure and have a desire to a) profit from these new developments and b) reduce their vulnerability to the risks.

The risks are more obvious to content producers than are the potential benefits of new technologies. Copyright protection systems of the pre-digital age consisted of legal mechanisms to prosecute individuals and groups that ran large-scale illicit reproduction facilities for profit. These pirates of intellectual property required physical assets to reproduce the physical media of books, music or video and were thus exposed to traditional law enforcement techniques. The added complications imposed by the distribution of these contraband copies made them even more vulnerable to detection. From a consumer perspective, these illegitimate copies were less interesting because quality suffered and they were not always available as promptly as legitimate copies.

All that has changed in the digital world where flawless copies are infinitely reproducible and may be transmitted anywhere in the world in an instant. We have moved from a paradigm where a large number of individuals made a few copies to one where relatively few individuals can make many copies.

When cassette tapes were first introduced, record companies were concerned (records from the early 1980s carried the slogan "Home Taping Is Killing Music"). Eventually this lead to mandatory licensing fees (so-called 'levies') paid by cassette tape manufacturers to the rights holders.

Content producers are rightfully concerned with this new capacity to cheat them of a fair return on their intellectual property. It is for this reason they have shown considerable reluctance in taking advantage of the opportunities presented by digital commerce. Yet digital commerce offers the potential to increase earnings while cutting the high overhead costs of production, distribution and warehousing their goods while presenting new business opportunities. It is believed that if content producers were sufficiently confident in their ability to protect their assets in digital form, they would gladly take part in such a system.

While the protection of digital content shall continue to be enforced by legal and regulatory means, a deterrent is required to make the illicit copying and distribution of copyrighted content difficult and traceable. For this reason, the deployment of a trusted end-to-end solution for the management of digital rights is a necessary precursor to digital production, dissemination and consumption of copyrighted content. The management of rights to intellectual property must be considered a two-way street. Another precondition for the success of this market is consumer protection. The same technologies that protect content must be applied to protect consumer privacy and access to information usage historically protected in some regions by a concept known as 'fair use'.

The Evolution of Copyright Protection

The introduction of trusted systems to manage the digital rights of copyright holders change the landscape of copyright protection dramatically. Originally copyright law dealt with gross infringement- the industrial copying and distribution of media-through the legal system, both national and international. It had its own 'sweet spot' and thereby considered policing of individuals to be 'over protection'. The lessons of a compression format called 'MP3' and distribution technologies such as Napster have caused considerable alarm among the record labels. Other copyright-protected media will not remain safe for long.

Forgetting for a moment the ethics of the Napster dispute, it is clear that ignoring the infringing acts of the individual in the digital age has lead to 'under protection' in the market today. One of the major labels studied consumer reaction to Napster and found that, on average, after a consumer has downloaded less than a half dozen songs without paying, they will expect this to be fair and normal.

DRM system implementations are the tool by which protection may be pulled back into the 'sweet spot'. Properly designed and implemented they represent value propositions for all participants in the value chain of media distribution.

Since DRM systems are enforced by trusted systems that reflect the intentions of the original rights holder, new applications may be considered. This is no longer the application of blunt, legal instruments that simply enforce copyright in the face of widespread infringement. Rather, it is possible for the rights holder to package their content with a set of rules for the use of their intellectual property. This may be considered as a set of terms and conditions that represent the intentions of the rights holder in a far more detailed and trusted manner than simple copyright law ever could have before.

Terms and Conditions for Consumption

In order to consume the work, it must be played through speakers, rendered on a screen or printed. At this time the work is most vulnerable to copying therefore mechanisms must be in place (for very high value content in high trust systems) to support watermarking and fingerprinting capabilities. Note: while it is possible to require that a book only be printed to a trusted printer with watermarking capabilities, there is little that can be done when the text is rendered on a screen. While this is a potential security hole it is considered that-with good design-it will have minimal effect. After all, any content may be copied once it is rendered. Audio and images may be recorded in analog format and then have their watermarks stripped and books can be re-keyed for pennies.

The intention is not to make a sealed, perfect copy prevention system. Rather, that it be sufficiently inconvenient and the resulting work sufficiently degraded that the legal copy-appropriately priced-will prove more competitive on a free market.

Example rights that might be sold to an individual:

1. Play / View
2. Print
3. Pay per view

The addition of a trusted clock provides even finer control of the user consumption of a Work. Fundamental time intervals can be fixed (from start date to end date, or for a set number of time units), sliding (a fixed time interval that begins after a user action such as viewing an eBook for the first time), and metered (the rendering time is recorded in a secure manner and is later used for billing). A timed preview allows a work to be rendered for a particular time-based portion, for example a 30 second preview of song or a 10 minute preview of a movie.

In order to provide the above-mentioned features, the trusted system must support one or more of the following features:

1. Accurate and tamper-resistant clock
2. Support for international time zones
3. Fixed time interval
4. Sliding time interval
5. Metered time intervals
6. Preview

It is now possible to imagine a wide variety of consumption models that have never been available to a consumer simply because they have been unenforceable. How about renting a song collection for active Saturday night parties? Or perhaps listening to a particular type of music an hour before drifting off to sleep? In the past, this required the purchase of complete Works, typically, a CD with many unneeded tracks. One can either live without the music, pay the price, or download a song unprotected from the Internet.

DRM systems allow more choice for both the rights holder and the consumer in the way they consume media. This benefits both parties. The rights holder provides the consumer with a Work under certain terms and conditions for its usage and an associated price. The consumer decides the desired usage pattern and pays the relevant price. It is hoped that, with appropriate competition, the price will be in line with the perceived value of the goods for the consumer. But what if it is not?

Misuse of Copyright

What happens if content providers decide the best pricing model is whatever the market will bear? Unlike other commodities on an open market that gravitate to an acceptable price for value as determined by supply and demand, media is different. Only one studio owns the rights to a particular film. Only one label owns the rights to a particular song and only one publisher owns the rights to a particular piece of literature. A typical deal between a creative artist and the publisher is typically an exclusive one.

One might argue that the consumer always has a choice to purchase at the seller's price or not. Nobody is forcing you to boogie down to Staying Alive at your next party, after all.

In the hypothetical case, where a publisher charges the maximum attainable for a Work will have unfortunate consequences. What kind of a society would it be, if only the rich could enjoy the pleasures of art, literature and music that influence and shape our modern society?

What about the rights of the artist that cares more about freedom of expression than the dollars and cents it brings in?

Past Precedents of Abuse

If all this were simply hypothetical musing, then perhaps this could be considered simple paranoia or misguided projections. Unfortunately, there are a wide range of precedents from the entertainment industry that demonstrate that abuse takes place.

Recently, the Federal Trade Commission of the United States ruled against the RIAA which represents over 85% of the US market share. It ws found guilty of price fixing in the CD market with the result that consumers had paid an estimated $480 million more than they should have.

According to FTC complaints, the companies required retailers to advertise CDs at or above the Minimum Advertised Price (MAP) set by the distribution company in exchange for substantial cooperative advertising payments. The restrictions applied to all advertising, including television, radio, newspaper and signs and banners in the retailers' stores. The restrictions even applied to advertising funded entirely by the retailer. Under the policies, large music retailers would lose millions of dollars a year if they failed to follow the MAP restrictions.

The complaints detail how MAP policies were adopted to squelch discount music retailing. In the early 1990s, many new music retailers, including major consumer electronics stores, started to sell CDs at low prices to gain customers and market share. The more traditional music retailers also lowered their prices to compete. This retail "price war" led to lower CD prices for U.S. consumers as prices for popular CDs fell as low as $9.99. The record companies adopted the MAP policies in 1995-96 to extinguish this "price war," the Commission contends.

The Flip Side

While it is easy to demonise the record labels for exploiting artists and consumers alike, such a reaction is too simple and overlooks the dynamics of a complex environment. Record labels exist in a highly competitive and speculative market. The cost of promoting an act is huge, tours are lucky if they break even, recording studios are very expensive and the majority of artists never turn a profit for themselves or the label. A former Sony executive told the author that the labels were, in some ways, in the business of providing high risk loans without the benefit of a due diligence process. By adding in the additional losses due to consumer copying and large-scale piracy, one quickly understands why the operating profits of labels are so low.

Privacy and Fair Use

Even without considering the extreme case of the potential for price-fixing enabled by DRM systems, there are many more subtle areas where such a system may not be beneficial to consumers. Privacy is a good example. One of the features of DRM systems we considered in the previous section was the ability to provide pricing based on some metered usage of a Work. This requires that the system has knowledge of how long is spent consuming a specific Work that is uniquely identified and that information be passed back to the rights holder for billing purposes. Now, if that rights holder is in Europe, then they will be legally obligated to safeguard privacy in very specific ways. However, if the rights holder is in America no such consumer protections exist. This raises many questions of jurisdiction and sovereignty, for example, if I am a European living in Asia reading an American novel, what protections can I hope for? A precious few, one would expect, which puts the rights holder in a very powerful and unregulated position.

Fair use is a provision in American copyright law that is protected under the first constitutional amendment concerning the right of free speech. Essentially, it allows a consumer to copy sections of the work with the intention to comment upon, criticize or parody the Work. This is a fundamental concept of consumer protection, yet it is likely the first one a publisher will want to make unavailable in the DRM system as it opens a great opportunity for abuse.

The key then is to feed in these consumer requirements-as depicted in the next figuer in such a way that consumer needs are reflected in rights management policies. Furthermore, we need-as a global society-to consider how this might be achieved, while respecting the jurisdiction of nations and the citizens they govern.

What is Digital Rights Management?

The Original Role of Rights Protection

Prior to the availability of digital reproduction and distribution capabilities introduced by the wide spread adoption of computers and the Internet, there existed tools to limit the illegal reproduction and distribution of intellectual property. These tools can broadly be categorized as the social and business requirements that affect legislation on legal instruments, such as the laws that cover patents, copyrights and trademarks. In order to defend the principals set out in these legal instruments, legal tools of detection, enforcement and litigation are deployed.

The next sections discuss these concepts in more detail.

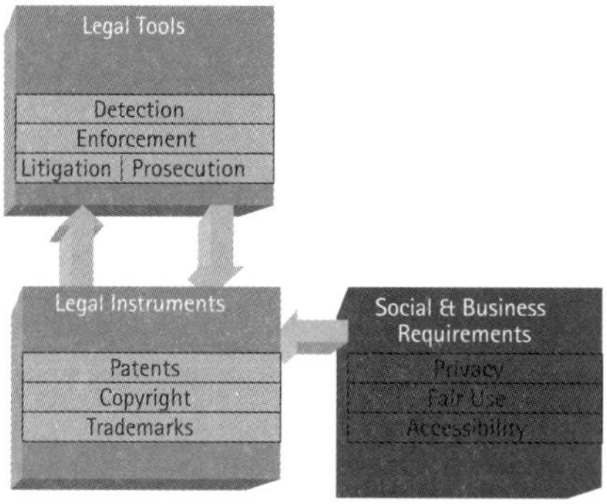

Social and Business Requirements

Social and business requirements are based on regional and national interests. For example, protection of privacy is a legal requirement in Europe, while it is considered a business requirement in America. Fair use is a concept that ensures that protection of intellectual property does not infringe on the first amendment provisions of free speech in the United States constitution. It allows individuals to copy and use parts of a work for the purposes of criticism, comment or satire. Accessibility to information for people with disabilities is a principal that should be enshrined in legislation of all nations even if it is currently hardly addressed.

These social and business requirements - the will of the people - are eventually codified in both national and international legal instruments.

Legal Instruments

The first national copyright legislation was introduced in London, England over one hundred years ago. The purpose was to promote authors to create and share their works for the benefit of society. Society, in return, would protect the intellectual production of the author by recognizing the author as the owner of that work. The Berne convention on copyright was the first international convention that extended the rights of an author.

Patent legislation originated in the United States with the vision of Thomas Jefferson. The intention was that an inventor should be encouraged to share the results of an innovation to the benefit of society. In return, the government would provide the inventor with a 20 year exclusive right to use or license the use of the innovation.

Similarly, trademarks are used to uniquely represent an individual, or more typically a corporate entity and the control of this intellectual property is provided exclusively to its owner.

These three categories of legal instruments have in common the rules by which intellectual property may be shared, and the attribution of ownership of ideas to those that created them.

In order to enforce these rules, legal tools have been used to protect these knowledge assets.

Legal Tools

The issue of rights management and protection has existed for as long as there has been legal protection of intellectual property. Up until recently, the mechanisms to circumvent legal instruments have been heavy and cumbersome. To profit from such circumvention has been even more difficult.

This is due to the fundamental nature of copying atoms (as opposed to 'bits'). To make and sell illegal copies of a music CD, it is necessary to have a CD replication facility, an illegal distribution network and a black market for vending such goods.

These are all activities that are detectable. The majority of nations enforce international intellectual property conventions. Their judiciary is active in the detection of illegal copying and commercial activities. National laws enforce international conventions and those that illegally copy, distribute and sell such goods are prosecuted and the guilty are suitably punished. Mechanisms are in place to require all nations to respect international conventions. For example, it is a requirement for participation in the World Trade Organisation (WTO) to have national legislation supporting international instruments with respect to intellectual property.

Up until very recently, these legal mechanisms have been sufficient to keep the infringement of intellectual property laws in check. This is due to the nature of copying the books, CDs and videotapes- the medium used to transport intellectual property. The physical difficulties imposed by copying and distributing atoms required a relatively large effort by numerous people to compromise relatively few intellectual property assets.

With the introduction of personal computers, and cost-free, worldwide distribution provided by the Internet has made the copying and distribution of bits simple and effortless. Intellectual property is now represented by bits, and as such, is weightless and infinitely replicable.

This enables the paradigm where many copies are made and distributed by relatively few people. The situation is complicated by the fact that individual consumers are capable of compromising intellectual property on a wide scale that was once the domain of commercial pirate organisations. To make matters worse, the process is practically impossible to detect as copying and distribution does not occur in the physical world.

Enter Digital Rights Management

Trusted systems offer a new opportunity for the management of intellectual property. Trusted systems were introduced with the first military computers that were used to calculate missile trajectories and later to model atomic bomb explosions over 50 years ago. These computers carried highly sensitive information and therefore required protection of their digital assets. Perimeters were set around the massive buildings that housed these computers, armed guards and intrusion detection systems were put in place. Rules were established on who was allowed into the secure facility, under what conditions, and what they were allowed to access. If someone tried to illegally access or copy data, they were dealt with swiftly and conclusively.

These measures are clearly not acceptable to protect intellectual property in the personal computers of the home and in business. The concept of trusted systems is now implemented in tamper resistant software and hardware systems. They enforce rules for how and who can access digital assets and provide intrusion detection and renewable security to circumvent illegal access.

In the late 1990s, trusted systems began to appear from several vendors, including Folio, IBM, Intertrust, and Xerox.

Trusted systems differentiate between different uses, such as making a digital copy, rendering a work on a screen, printing a work on a colour printer, or extracting a portion of a work for inclusion in a new work. When asked to perform an operation not licensed by a work's specific terms and conditions, a trusted system refuses to carry it out.

There is a wide range of techniques for implementing trusted systems. They include the use of encryption, public key infrastructures, policies on devices and servers.

Digital rights management services are essentially trusted systems that act persistently on behalf of the content provider to enforce the rules by which they have allowed a consumer to use the work.

These systems are trusted by the producers of intellectual property to protect their assets when they are digitally distributed. Of equal importance to ensure a market for the legal distribution of digital media, is trust from consumers. Consumers must be assured that their privacy is protected, that provisions for fair use are respected and that digital content will be available on fair and reasonable terms.

In fact, the framework for commercial vending of digital content has changed from what appeared to be the purchase and ownership of a physical good - for example a CD - to the licensing of the intellectual property.

The terms under which an asset is licensed are represented in publishing rights policies. These policies are determined by business and consumer requirements of protection under the umbrella of applicable legal instruments. They are expressed in a Rights Expression Language (REL) and enforced by the Digital Rights Management system as illustrated in following figure.

A Digital Rights Management System.

A Digital Rights Management system-or digital copyright protection system, as it is sometimes called-will coexist with societal norms, which include opportunities for legal recourse and insurance-based compensation for parties that fall victim to cracks in the system. The DRM system will exist within this framework of digital copyright protection. The framework includes:

1. legal rights which are the intellectual property rights of a particular work and include such familiar instruments as copyright, patents and trademarks;
2. requirements that represent the needs of individuals, corporations, governments and intra-governmental organizations. These requirements are often complex and conflicting. It is difficult to reconcile an individual's privacy needs with requirements of the state, for example;
3. rights management, which is the area covering policies, procedures and their applicable licenses as defined by legal rights and the requirements. This could also include the definition of a rights management syntax that would ideally be an open standard;
4. DRM System, which is the technically trusted component of the framework and the subject of this paper;
5. legal tools, which include litigation, arbitration and other legal means to protect legal rights.

Time Value Content

Works are created, produced, sold and distributed for the value derived, including profit, prestige and education, to name but a few motives. Works are 'stolen' by illegally copying and / or fraudulently misrepresenting ownership/authorship for similar motives. The effect of time on the value of different titles varies widely. Time must also be considered when we are protecting works since the protection must persist into the future. Attack methods, tools and techniques are expected to improve dramatically over time, while little will change with a given security implementation. For this reason, works that must be protected for the life of their copyright present greater challenges than works that lose value after a few days of their publishing such as yesterday's weather forecast.

The next figure shows the relationship between time and value and suggests six example categories. This matrix is shown for the sake of illustration. In the real world there is substantially more overlap among these categories.

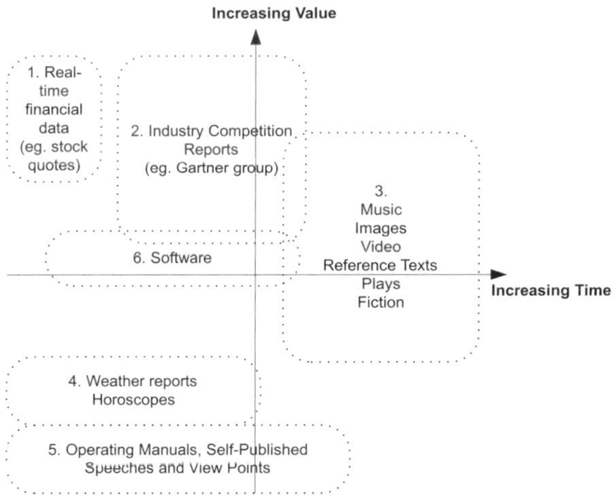

Time Value Content.

The sample categories include:

1. High value-short life: this includes stock quotes and other financial information that has very high value but a lifetime of only a few dozen minutes. The analysis of this information aggregated over time continues to hold value, but that value is a function of its presentation rather than the content and value life.
2. High value-medium life: examples include reports on highly dynamic, competitive industries that are used for business intelligence and strategy.
3. Medium value-long life: music, images, video, some reference works and all fictional works will retain their value for the lifetime of the copyright.
4. Low value-short life: some types of information are given away for virtually free. They also have a very short value life. Examples include weather reports and horoscopes that are available for free and are only valuable for a few days.
5. Low value-long life: this category of information covers authors that publish in order to share their viewpoints. Copying and wide dissemination is encouraged as long as the content is not changed and the author receives the appropriate credit for the work. Value of the works extends to the lifetime of the copyright. Examples include Amnesty International reports and political papers.
6. Medium value-medium life: a very particular example is software, which is made rapidly obsolete through advances in hardware and pressure from a highly competitive environment.

Trust Levels

The ability of the system to preserve integrity and protect content derives from the ability of the participating components to provide the trust services identified above with a known level of assurance. Because current technology cannot provide perfect security without incurring other undesirable characteristics (e.g. high cost, difficulty of use, inflexibility, etc.), we approach this problem by permitting components with different trust capabilities to interoperate, but enable the copyright holder to specify a minimum trust assurance for the handling of a particular protected title.

Four factors rank high in assessing and certifying the trust level provided by a component. First is the scope of a particular failure. If a single feasible failure has more serious consequences in one device than another, then the former ranks at a lower level of trust. Second is the skill level required to defeat the protecting mechanisms. The higher the level of skill necessary to wage a successful attack, the more trusted is a component. Third is the sophistication and availability of the tools necessary to wage an attack. The more widely available or accessible are the tools required to defeat a particular security mechanism, the lower the trust level of the component. Fourth and probably most important, the overall level of resources, including manpower, time, computer processing capacity, etc. factors into the trust level for a component. An encryption algorithm that succumbs to a Pentium-day exhaustive key search attack offers a lower level of trust to an algorithm that requires multiple Pentium-years to crack by key search.

By measuring the time value content and the vulnerability to attack, we can protect content sufficiently, while not over burdening the system with unrealistic requirements.

Implementing Trust Levels

The DRM system should provide trust levels that are necessary and sufficient to provide reasonable assurance to all parties-publishers, authors, insurance agents and their loss adjusters and law enforcement that adequate precautions have been taken to protect the intellectual rights of authors and publishers.

The trust rating requirements allow publishers to make a time value content estimation of their work and then set a corresponding trust requirement of the components allowed to access their work. In this way high value, long life works may be protected by highly trusted systems while lower-value, short life works need not be burdened by these requirements. The intention is to provide flexibility in the system in such a way that the needs of all parties are be met. In particular, it allows the publisher to select the right level of protection to maximize the value of a title, avoiding losing sales to excessive piracy at one extreme and to difficult protection mechanisms or offensive enforcement at the other end of the spectrum.

This concept is illustrated as the 'sweet spot' in the figure below.

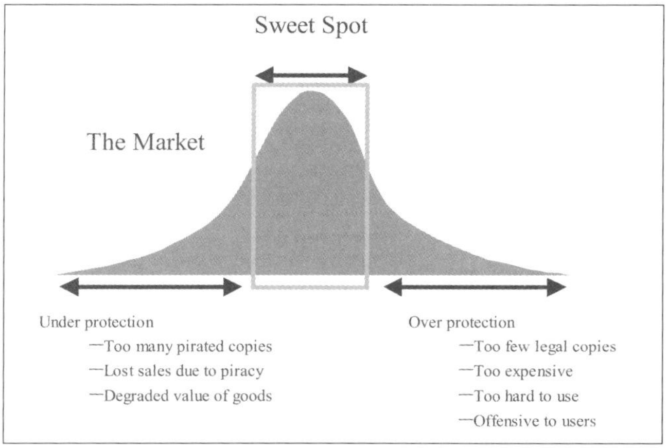

Combining the concept of the 'sweet spot' with 'time value content' leads to Mobile Digital Rights Management. The types of content, and the means to protect such content both vary greatly with respect to the old framework established in the personal computer and fixed line Internet domain.
The principle difference is the ability to acquire content any time and anywhere. The existence of easy to use, metered billing systems operated by mobile operators further facilitates commerce for digital media in the mobile world. Finally, the close integration of hardware, operating systems and players provides a trusted system for both the consumers and producers of content.

Mobile Digital Rights Management

A key aspect to DRM is that the usage of a piece of content on a consumers device is controlled by rules. These usage rules are expressed via a Rights Expression Language. However, expressing rights alone is not sufficient to implement a DRM system. Enforcement of the expressed rights is critical for a working DRM system. In Mobile DRM, there are two broad ways to implement this enforcement issue, (1) by Policy Protection and (2) by Encrypted Protection.

Policy Protected Environment

This environment is created of trusted terminals and trusted network elements. Content is not encrypted and thus it is terminal and network policies ensure that such content is only distributed in a trusted environment.

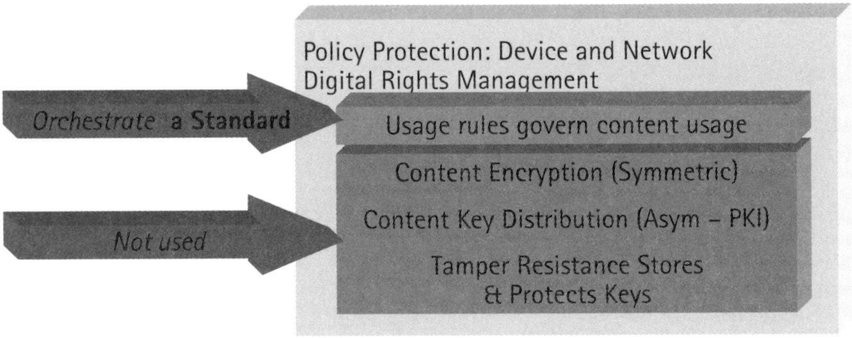

Usage rules, closed terminals and network policies are key to Policy Protection.

The Policy Protected Environment is based on the following terminal related principles:

- Mobile content and associated usage rules are distributed in plain-text and not encrypted.
- Mobile terminals are closed devices (i.e. not user programmable like the 9210 Communicator).
- Mobile terminals understand and enforce usage rules received with content.
- Mobile terminals will not forward content to other terminals. This could be enabled with explicit permissions as expressed in the usage rules.
- Content with no usage rules is treated as free content.
- Mobile terminal identification is possible so that the network is capable of determining if a particular terminal is one which respects usage rules.

The Policy Protected Environment is based on the following network related principles:

- The network content retailing service is capable of ascertaining the terminal identification to ensure that content with usage rules is only sent to supporting terminals. There may be non-supporting terminals that can mine the plain-text content, which would be a leakage.
- The content in this environment is normally statically packaged with the correct usage rules and uses standard and existing payment mechanisms.

This environment is appropriate for very low value content where the time-value is so short that to pirate the content is not worthwhile.

Encrypted Protection Environment

Where the Policy Protected environment creates a closed community of devices and network services, the Encrypted Protection environment is completely open. The content pieces are encrypted resulting in a Content Encryption Key (CEK). The encrypted content pieces can be freely distributed as no user can access the content without the CEK. Finally the control is based on the controlled and protected distribution of the CEK and associated usage rules.

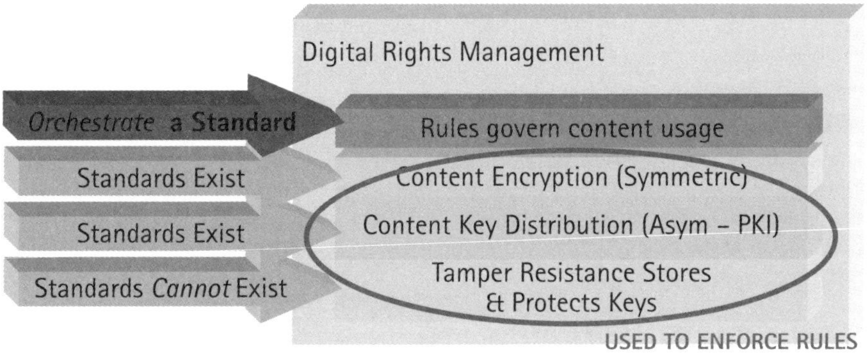

Usage rules, content encryption, secure CEK distribution with PKI, secure content consumption with PPE.

This environment is targeted at content with low to medium value and with a slightly greater time-value. Thus, it is important that the content is directly protected with encryption. As CEK distribution becomes the critical issue, it requires that terminals are capable of protecting the CEK. This is typically achieved using Public Key Infrastructure and a Protected Processing Environment (PPE) in the terminal. The PPE can be either software or hardware based. Hardware based PPE is of even higher security, and thus is appropriate for high value content.

Rules Govern Content usage

Usage rules are used to control the usage of content on consumer devices. They are created using Rights Expression Languages. These RELs can also be used to enforce contracts in all parts of the content value chain, but this section has concentrated on thecontrol of content consumption. A variety of rights expression languages exist which are mostly based on XML.

Usage rules apply to one or more pieces of content (although it can extend to any IP).

Usage rules contain:

- Conditions of content rendering, modification and distribution.
- Rules enforcement support using encryption/protection mechanisms.
- Information on rights holders.

Usage rules are created and used as follows:

- Usage rules are created and stored in a 'file'.
- The rules are associated with the content, often by binding them with the content.
- Eventually, the content (and rules) are received by a consuming terminal.

- The rules are rigorously evaluated by the terminal.
- The user is only given the possibility to use the content in accordance with the rules.

Content Encryption (Symmetric)

Symmetric encryption is used for 'bulk' encryption of content because it is much faster than asymmetric encryption. Old standards like DES (lasted more than 20 years, now retired) are being superseded by new standards such as Advanced Encryption Algorithm (AES). The typical Content Encryption Key (CEK) is an AES 128 bit key and frequently there is only one CEK for each unique content object. CEK's must then be distributed securely to clients.

Content Key Distribution (Asymmetric, PKI)

Well understood standards exist in this domain as Public Key Encryption has been used for a multitude of purposes such as transport security (e.g. TLS) and email security (e.g. PGP). Public key security is based on a 'difficult' mathematical problem with a trap door. These difficult mathematical problems cannot be solved in polynomial time (years left in universe etc.) The trap door is the private key, and must be kept secret. RSA and ElGammel are the only known algorithms that may be proven secure mathematically. But still, the private key must be protected by tamper resistance.

Tamper Resistance Stores and Protects Keys

Tamper resistance is used to build Protected Processing Environments. Such tamper resistance techniques cannot, by definition be standardized. Often they are not even patented, but rather are protected as Trade Secrets. It is difficult, costly and risks security to implement more than one tamper resistant environment in a device.

Converged World

A key objective is the service interoperability of mobile content protection systems and leading Internet DRM technologies. The convergence of Internet DRM and Mobile DRM is expected to take place as the capabilities of mobile devices increase. A key tool for this convergence is a standard for the expression of rights governing a particular content object. These are called rights expression languages.

In order to address different DRM requirements, two separate recommendations are made for the digital rights management of services. Both these systems utilize the same or interoperable rights expression languages.

- The Policy Protected DRM technology, for initial deployments where content with low value utilizes vouchers based on a standardised rights expression language.
- Encrypted Protection to be used in cases where a high level of protection is required, content encryption, targeting a device and the necessary key management, must be deployed. The same rights expression language would be used here also.

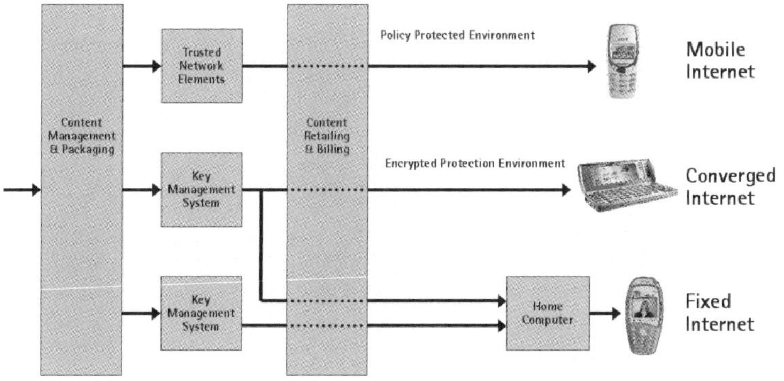

Unified View of Internet Media Distribution

A key element for enabling interoperability between various content delivery methods is a common language for expressing the governance information for a given content object.

In the figure above we depict three elemental means of content delivery (from top to bottom):

- Through a trusted network, from operator to a trusted terminal.
- Over-the-Air delivery without assuming a trusted network.
- Delivery from fixed line Internet, through a PC client to a device.

Note that all three delivery paths are enabled by creating the same content package. This allows economies of scale through the implementation of a standard rights expression language. A single plug-in is therefore sufficient for digital asset management (DAM) systems as well as content packaging tools.

Standardization

We have outlined a way to build a powerful mobile digital rights management system that combines the power of peer-to-peer distribution with easy and efficient mobile payment collection. Turning this opportunity into a mega-business will require a combination of standardization and technology innovation. The key stakeholders of the industry should standardize at least the following parts of the architecture to ensure interoperability and maximize growth:

- A Rights Expression Language (voucher): The expression of usage rights and their interpretation by mobile terminal middleware.
- Content encapsulation with rights metadata: The metadata included in the content objects that enables the user's payment service to talk with the retailer.
- A Rights Management Protocol: between mobile device, retailer and payment server: The protocol for exchanging parameters between different architecture components in order to complete a rights clearing transaction.

Summary

Media is the fundamental conduit for transferring knowledge and values in the knowledge society, therefore the value of intellectual property must be protected and those that generate it must be rewarded. This argues for the establishment of safeguards for artists and their publishers.

At the same time, and for the same reasons, consumers require assurances that media distribution be equitable and benefit all levels of society.

Consider the first figure, which depicts a framework where traditional legal intellectual property constructs interoperate with consumer and business requirements to produce rights management policies. These policies may then expressed as the terms and conditions to be enforced by the distributed DRM systems. Legal tools to police the leaks, breaks and misapplications of the DRM system would then concentrate on the exceptions where the DRM system is not in accordance with rights management policies, or those situations where the DRM system is compromised.

This suggests a means by which the interests of all parties may be best served, but it also requires a co-ordination, integration and articulation of business, consumer and legal requirements in an unprecedented way. Each of these groups will have differing requirements that will often be conflicting. The challenge is to reconcile these requirements to provide rights management policies acceptable to all the stakeholders.

Bibliography

1. The Bit and the Pendulum, Balancing the Interests of Stakeholders in Digital Publishing, Mark Stefik and Alex Silverman, 1997, Xerox Palo Alto Research Center, Palo Alto, California 94304
2. A Cyberspace Independence Declaration, John Perry Barlow, Cognitive Dissident, 1996.
3. Is a Stitch Online a Crime? P. J. Huffstutter, 1 August, 2000, L.A. Times, L.A. California
4. O'Reilly P2P Directory, http://www.oreillynet.com/pub/q/p2p_category
5. Peer-to-Peer Pressure, The Economist, 2 November, 2000 http://www.economist.com/displaystory.cfm?story_id=413334

PKI

One of the major reasons that electronic commerce has not really flourished so far is the lack of security, or rather lack of perceived security. Consumers are not willing to do business with companies they do not know and trust. On the Internet it is easy to impersonate a respectable business. Therefore, drastic methods for building trust are needed and PKI is the current "state of the art" mechanism for this.

PKI (Public Key Infrastructure) provides a secure way to communicate in a basically unsecure public network like the Internet. PKI also provides methods for increasing trust between the communicating parties through trusted third parties. This is achieved by utilizing PKC (Public Key Cryptography), digital signatures and certificates that bind public keys to a name. CA's (Certificate Authorities) work as TTP's (Trusted Third Parties) to assure the integrity of a certificate.

Benefits of PKI

A full PKI solution addresses all the major problems of security: Privacy, Authentication, Integrity and Non-repudiation, which are also known as the PAIN of security.

- Privacy means that the information is available only for authorized parties.
- Integrity means that unauthorized parties can not modify the information
- Authentication provides assurance of the identity (or name) of the origin of the information.
- Non-repudiation means that an entity can not later falsely deny a valid transaction.

Privacy is achieved by using cryptography. Integrity, authentication and non-repudiation are achieved by utilizing digital signatures and usually also certificates.

The Components of PKI

Public Key Cryptography

PKI bases heavily on PKC (Public Key Cryptography), which is also known as asymmetric cryptography. Asymmetric cryptography means that different keys are needed for opening and creating information. These keys always form a pair of a public and a private key. This very fundamental differentiation from the traditional symmetric cryptography provides the advantage of not needing a shared secret key. Following figures illustrate the difference between these two methods.

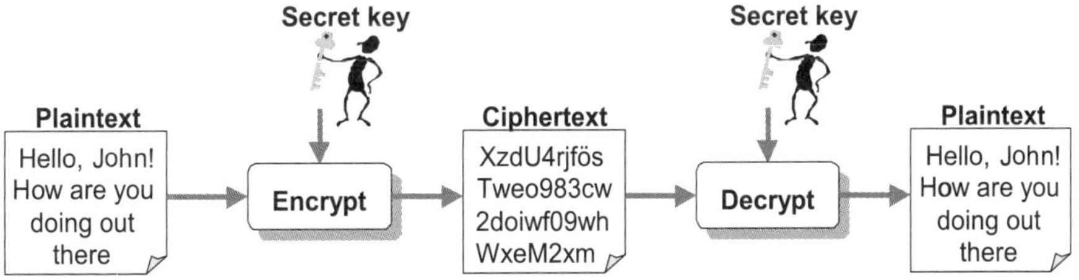

Symmetric encryption

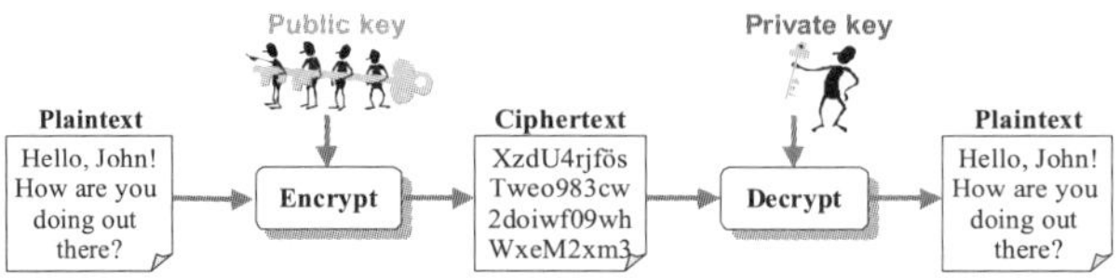

Asymmetric encryption

One benefit of asymmetric encryption over symmetric encryption that is not clearly presented in the figures, is that in order to achieve confidential one-to-one data transfers between many parties, less keys are needed in asymmetric encryption.

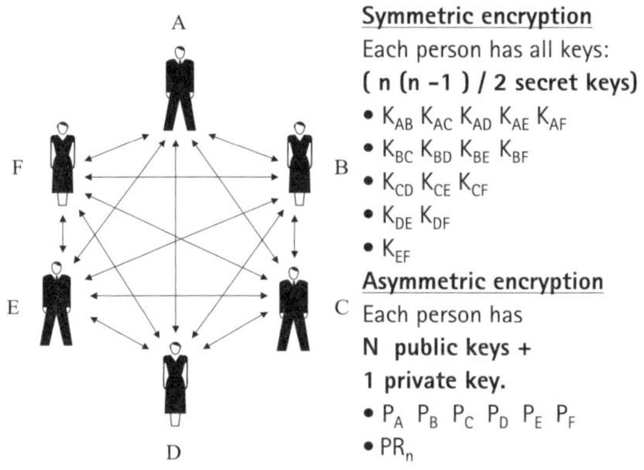

Number of keys needed in N to N private communication.

The downside of asymmetric encryption is the complexity of the algorithms. Most common algorithms (e.g. RSA and Diffie-Hellman) need very long key lengths (usually between 512 and 4096 bits) to be considered as secure and they are computationally very heavy, whereas symmetric algorithms (e.g. 3-DES, AES and RC5) only need key lengths of about 128 bits for a similar level of security and they are computationally relatively light. For this reason, a hybrid of these methods is usually used (at least with large messages).

In a hybrid mechanism the data itself is encrypted with a (pseudo)randomly generated secret key using a symmetric algorithm and the secret key is then encrypted with the recipients public key using an asymmetric algorithm.

Digital Signatures

Digital signatures provide data integrity, source authentication and non-repudiation of the transaction. Asymmetric encryption algorithms and the private key are used in creating the signatures. Signatures are verified with the corresponding public key by doing a reverse encryption.

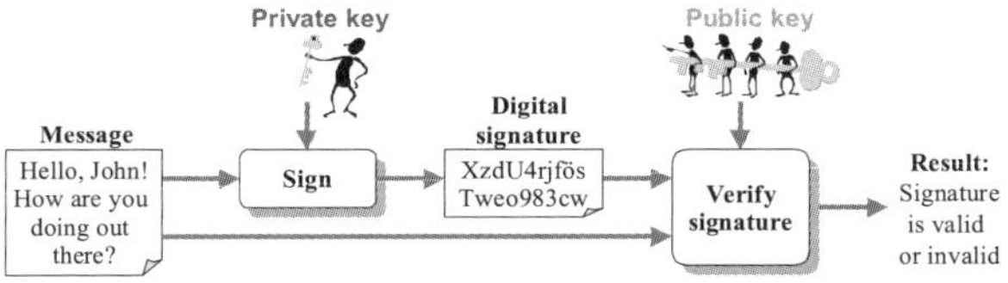

Using a digital signature.

Because it is often not feasible to sign the whole data, certain hash algorithms are usually used to create a fingerprint of the data. This fingerprint is then signed and sent to the recipient with the encrypted data. The recipient then decrypts the message and uses the same hash algorith to create a fingerprint of the received data, which is then verified with the sender's public key.

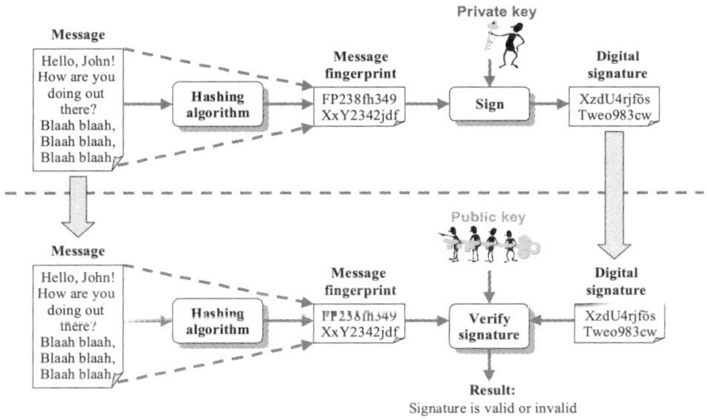

Digital signature with a hash algorithm.

Hash Algorithms

Hash algorithms are used to create a fixed size fingerprint of a message. The size of the fingerprint is usually 128 bits (e.g. MD5) or 160 bits (e.g. SHA-1). Hash algorithms are very efficient one way functions where Hash (X) = Hash (X') should be very difficult. This means that it is very hard to find different messages that provide the same fingerprint (hash value).

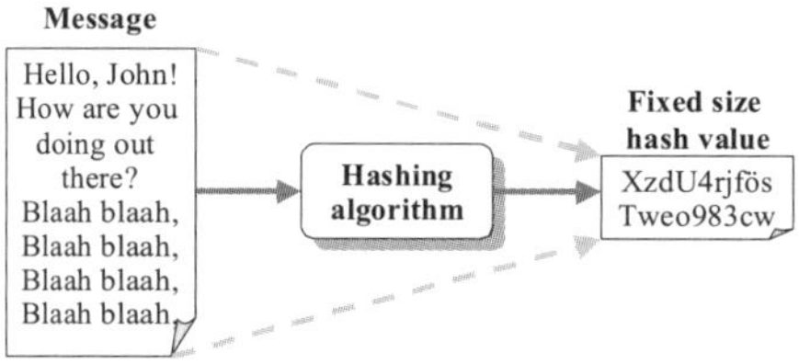

Using a hash algorithm

Certificates

Certificates bind a name to a public key and therefore provide authentication of the owner of the public key. Because it is usually not sufficient that entities provide such information by themselves, certification authorities are needed as trusted third parties.

Certificates (based on IETF standard X.509) contain at least:

- Public Key
- Name of the key owner
- Name of the certificate authority (issuer of the certificate)
- Serial Number of the certificate (unique within the CA)
- Validity period of the certificate
- CA's digital signature

These certificates can be published in a public directory, but entities can also distribute the certificates themselves. The idea of the certificate is that if an entity knows CA's public key and trusts the CA it can also trust the content signed by the CA. Now, if the same entity (relying party) gets a certificate signed by the trusted CA containing another entity's public key and a name, it can trust that the owner of the public key has the name that was in the certificate. If this other entity then sends a message signed with its private key to the relying party, the relying party can verify the name of the sender (authentication) by using the public key that was in the certificate.

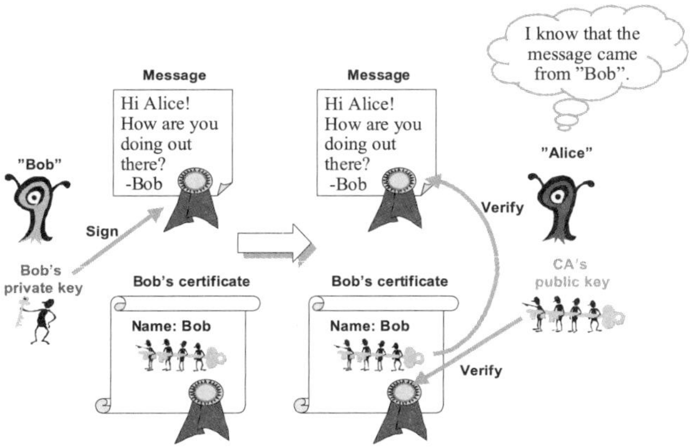

Using Certificates.

Certification Authorities

A certificate authority (CA) issues and verifies digital certificates. CAs also publish certificate revocation lists in order to block the fraudulent usage of certificates.

Revoking a Certificate

If for some reason a private key gets into wrong hands, or someone uses a certificate in a fraudulent way, there needs to be a mechanism to revoke the certificate. For this reason, CAs periodically issue a Certification Revocation List (CRL), which is much like the "black list" of credit cards. It's a responsibility of the relying party to do a revocation check before trusting a certificate.

Known Problems of PKI

There is really no standard way of implementing a PKI system. There are many different algorithms for key exchange and encryption (RSA, Diffie-Hellman, etc.) and there is also a variety of PKI vendors with their own implementations. Different certificate authorities also have different policies in issuing the certificates.

A name in a certificate does not really provide identity. Names are not universally unique and therefore a name in a certificate might not refer to the same entity the recipient associates with that name.

Private keys need to be safely stored. They can be stolen like any other data from a PC or a server and the passwords protecting the usage of the key can be cracked (especially because people tend to use too easy passwords for the sake of convenience).

What is Happening in PKI

EEMA PKI Challenge

A two-year project, which started in January 2001, the PKI Challenge aims to provide a solution for interoperability between PKI related products, and to develop specifications and the best practices in the world standards area.

The PKI Challenge is funded by the European Commission and organized by EEMA (European Forum for Electronic Business). The project is run by a 13-strong management consortium.

SPKI Development

SPKI (Simple Public Key Infrastructure) Working Group has developed a standard form for digital certificates whose main purpose is authorization rather than authentication. So, the key in a certificate is directly related to authorization and the name in a certificate is mainly used for logging purposes.

Privacy

The growth of the Internet and mobile communications has rapidly increased the amount of personal information that is collected about individuals. As consumers surf the Internet, use mobile services and engage in electronic transactions, they leave a vast amount of information behind. This information has an increasing commercial value and can be used in a way that is harmful to a consumer's privacy.

Moreover, much of the personal information that is requested from a user by a service, has potential uses for many purposes and by many companies beyond those originally indicated or reasonably expected by the user. Such use may occur without the consent or knowledge of the user. Many facts about a user, from his social security number to descriptions of his family, works and consumption patterns is being actively collected, exchanged and traded. Much of this personal data is openly for sale.

As a consequence, users have become increasingly reluctant to provide personally identifiable information online. The majority of Internet users have at some point of time refused to release information about themselves because of privacy fears. A smaller but growing minority of users systematically avoids services that require the submission of personal information.

Privacy is not to be confused with security. These concepts are not entirely unrelated, however. From a technological viewpoint, security is an important privacy enabler. Security methods can provide the means to verify the indentity of another actor. Security solutions can prevent unauthorized parties from getting or misusing private or proprietary information. However, good security does not guarantee privacy. In practice, even though the communication link between the user and the service may be secure, it does not mean that the consumer trusts his/her communication party on that party's handling of the delicate information that may have been transferred during the communication. Trust is an end-to-end relation.

Trust is a key element in privacy. Users must be able to trust both the devices and the services they are using. This means not only providing secure communication mechanisms, but also acceptable methods of handling the collection, use and release of information. A good way to achieve trust is by giving users control of their own information and letting them decide what, when and to whom their information can be released. Both security and privacy play a role in creating trust. Utilizing these methods will thus leverage existing businesses as well as enable new businesses that guarantee proper identity sharing.

Privacy is a human right, which relates to such fundamental values as dignity, freedom of association and freedom of speech. The definition of privacy is broad, but in MITA the focus is on what personal information is revealed about oneself and to whom. Fair Information Practises is a widely accepted set of guidelines on how privacy should be handled. They can be summarized through seven key principles as follows:

- Notice: An organization must inform individuals about the purposes for which it collects and uses information about them, how to contact the organization with any inquiries or complaints, the types of third parties to which it discloses the information, and the choices and means the organization offers individuals for limiting its use and disclosure.

- Choice (Opt out / Opt in): An organization must offer individuals the opportunity to choose whether their personal information is (a) to be disclosed to a third party, or (b) to be used for a purpose that is incompatible with the purpose(s) for which it was originally collected or subsequently authorized by the individual. Individuals must be provided with clear and conspicuous, readily available, and affordable mechanisms to exercise choice. Opt out basically means that a user must actively choose not to diclose information, and without doing so, the information may be used freely. Opt in means that a user must specificly choose to disclose information, and without doing so, the information cannot be used.

- Onward Transfer: To disclose information to a third party, organizations must apply the Notice and Choice Principles.

- Security: Organizations creating, maintaining, using or disseminating personal information must take reasonable precautions to protect it from loss, misuse and unauthorized access, disclosure, alteration and destruction.
- Data Integrity: Consistent with the Principles, personal information must be relevant for the purposes for which it is to be used. An organization may not process personal information in a way that is incompatible with the purposes for which it has been collected or subsequently authorized by the individual. To the extent necessary for those purposes, an organization should take reasonable steps to ensure that data is reliable for its intended use, accurate, complete and current.
- Access: Individuals must have access to personal information about them that an organization holds and be able to correct, amend or delete that information where it is inaccurate, except where the burden or expense of providing access would be disproportionate to the risks to the individual's privacy in the case in question, or where the rights of persons other than the individual would be violated.
- Enforcement: Effective privacy protection must include mechanisms for assuring compliance with the Principles, recourse for individuals who are affected by non-compliance with the Principles, and consequences for the organization when the Principles are not followed. Sanctions must be sufficiently rigorous to ensure compliance by organizations.

Personal information is something, which is related to a certain person, and can be connected to his identity. Examples are name, address, email address, credit card number, current location, habits, personal preferences related to food or hobbies etc. An important thing to notice is that although some information might not be connected to a person's identity as such, it is usually possible to make the connection by combining pieces of information.

On the mobile Internet, privacy concerns become greater, because even more personal information is involved than ever before. Since a mobile terminal is used and carried by a user all the time, it is possible to collect more accurate information about the user's habits, usage patterns and personal preferences. Once location based and presence services become available, it will be possible to determine, in connection with other data, increasingly accurate profiles of a person's activities.

Another difference of the Mobile Internet is that with mobile terminals real-time person-to-person communication is much more popular than accessing services. At the same time the mobile world and the Internet world are converging by making it possible for mobile users to access Internet services. From these capabilities, two privacy protection issues arise. In the service access case, protecting against uncontrolled collection of users' personal information is important. In person-to-person communication, the issue is to limit visibility to the other person, which closely relates to the concept of presence.

The basic problem with privacy today is that most users have little or no control of the collection and use of their personal information. An extreme solution would be to stop using any services or use only a very limited few. Yet, this diminishes the convenience of the Internet and that of mobile communication technologies. Therefore, more elegant solutions for the protection and control of privacy are needed; solutions that enable personalization without compromising a person's privacy. It is important that a user can easily control his or her privacy and have the option to stay anonymous when he or she so desires.

Privacy in Mobile Internet Services

Mobile Internet services can be divided into consumer to business and consumer to consumer services. A traditional example of the former is accessing a web service from a mobile terminal. A speech call would be an example of the latter. The categories have different privacy challenges, although the same privacy protection principles can be applied, as explained later.

Consumer to Business

Privacy is a major concern among users. Users feel that they have lost control of how their information is collected and used, and they would like to regain control of their own data. Users dislike automatic tracking of their online activities and are opposed to the release of their data to third parties without their consent. Users are reluctant to provide personally identifiable information online and the majority of Internet users have at some point refused to release information about themselves because of privacy fears. However, many users would be willing to provide companies information about themselves in exchange for some benefits, as long as they would be in control of how information about them is collected and used.

Users usually prefer opt-in type information collection to an opt-out process. Additionally, they are in favor of privacy policies that are clear to understand and that the companies are committed to implement. Many users are also in favor of stricter legislation to protect their privacy.

A user's privacy in relation to the outside world is often considered to cover his visibility, i.e. what he reveals to other parties. Another aspect, which is also a part of privacy, is awareness, analogous to what the user wants to see and know about other parties. It is likely that as the amount of available information and services increases, a user is not interested in all that information. Thus, the prevention of unwanted information overflow can also be seen as a part of privacy. Visibility and awareness are interconnected in the sense that if you choose to be very visible (e.g. you give out your email address), you may also affect your awareness mode (a lot of offerings and spam).

What makes privacy difficult is that it is present in most, if not all Mobile Internet services, and the setting often seems to be privacy against some other benefit. In order to get personalized service a user must reveal something to the service provider. For example, location based services are not possible without information regarding a user's location. Personalization is critical because Mobile Internet services are used quite often as people have terminals at hand, and because the terminals have small displays and use limited bandwidth connections, users require access to particular information. A mobile terminal capability profile can be considered personal information if it can somehow be connected to a person's identity. A regular user of some services probably wants to personalize them and must therefore provide personal information.

Two key issues are often forgotten: a small part of personal information is usually needed, and a person's actual identity is often not required. So, by providing only the necessary information, and applying anonymity towards the service provider, privacy can be maintained.

The problem for service provider is that users either do not use a service at all because there is a privacy threat, or users give false information when asked for it. A service provider can make the situation easier by clearly stating in their privacy policy how they handle personal information, and by collecting only necessary information and offering the possibility to use the service anonymously. After all, having a small amount of correct information to provide better service for customers is better than having huge amounts of false information.

The value of anonymity for a user is that he does not have to worry so much about what happens to his information because it cannot be connected to his identity. On the other hand, this creates a problem for service providers who would like to know their customers to enhance their service based on the customer base (like age or sex), provide some benefits for regular customers, or even provide targeted offerings or advertising. A solution is to use a nickname or pseudonymity provided by a trusted partner instead of full anonymity. The trusted partner between the user and the service provider is needed to map the real identity and the pseudonym of the user, and to assure to the service provider that there is a real identity behind the pseudonym. If the user decides not to use the service any longer, the pseudonym can simply be deleted, thus removing all connections between the service provider and the user's real identity.

An especially privacy sensitive area is mobile advertising. For a users the problem is that they do not want to receive information they are not interested in. Advertisers, on the other hand, do not want to waste money on targeted advertising when the target is not interested. Proper handling of privacy is an answer to both these problems: users are able to select what kind of advertising they want and can choose not to receive it whenever they want, and advertisers know which users are willing to receive information, and also what that information is. Users can also be rewarded for revealing part of their personal information and for accepting advertisements, and thus they are likely to have more positive attitude towards advertising.

When a user is using a service, the key privacy principle is chain of trust. This means that in order to enable the user to trust the next link in a value chain, the next link must either ensure that the following link can also be trusted, or provide protection through anonymity. The trust chain begins from a mobile terminal, goes through an operator and service platform, and through one or more service providers. The trust must go as far as real identity is needed. If a user, for example, wants to use a service he does not trust but a payment and thus identity is needed, there should be a trusted party in between offering anonymous payment for the service.

Consumer to Consumer

The most popular way to use mobile terminals has always been person to person communication. This can happen through rich call or messaging interaction modes, using services like speech call or chat. Privacy is an issue before the connection is made and during connection setup.

In general, people want to be reachable through their mobile terminals. The problem is that, in principle, everybody can reach you. So, from a privacy point of view there is a need to limit who can reach you. This can be done through limiting your visibility, meaning that your name is not listed in directory services, it is only visible to limited groups, or that you can only be contacted with a nickname. In addition to limiting visibility, it is possible to limit the methods of contact.

If visibility allows a connection to be made, the next privacy problem is what the opposing party can learn from you and what you can learn from him. A very old privacy problem of telephone connections, both wireless and fixed, relates to presenting the phone number of the calling party. Phone numbers can be used to link users with other collected information after matching the number with a name through directory services. Numbers can also be used for telemarketing purposes. Delivery of the calling party's number can usually be prevented, but this may cause a privacy problem for the called party: he does not know who is calling and must answer the call to find out, thus, compromising his privacy. The phone number is also present in more advanced wireless services like short messaging, or used as an identifier when accessing mobile Internet services.

On the Mobile Internet, the phone number is not the only information available in person to person communication. Therefore, for all the personal information involved there must be the possibility to choose whether the information is visible to the other party, both from the information owner's point of view and receiver's point of view.

Privacy Protection

Protection Methods

Privacy protection is a complex issue, which needs to be tackled on several levels, technological, societal and individual in order to be handled effectively.

There are four major mechanisms to protect privacy; comprehensive laws, sectoral laws, industry self-regulation and privacy protection technologies. Several of these can be used simultaneously to ensure a user's electronic privacy.

- Comprehensive laws: Comprehensive laws govern the collection, use and dissemination of personal information by both the public and private sectors. An oversight body ensures compliance with the law. This model is adopted by the EU.
- Sectoral laws: Sectoral laws are laws that govern privacy in specific sectors, such as the financial, medical and telecommunications sectors. Sectoral laws can be used alone or to complement comprehensive legislation by providing more detailed protection for certain

categories of information. The United States has so far avoided enacting comprehensive data protection laws in favor of sectoral laws and self-regulation.

- Self-regulation: Self-regulation means that companies and industry bodies establish codes of practice and carry out self-policing.
- Privacy Protection Technologies: Privacy can also be protected through the use of several technologies.

All four of these approaches have some drawbacks. The drawback of setting comprehensive laws is that it is very hard to define a law that provides a reasonable balance between the interests of all the affected parties. Using only sectoral laws to protect privacy would require new legislation every time a new technology or new ways of violating privacy is introduced. Thus, legislation frequently lags behind. Additionally, there is often no working framework or agency that oversees that the laws are followed. In self-regulation, the codes of practice are usually set with commercial interests in mind and do not always provide an adequate level of privacy protection for consumers. As in sectoral laws, the means of enforcement are weak or non-existent.

The drawback of privacy protection technologies is that there are no commonly adopted and well working standards. When it comes to proprietary solutions and commercially available services, privacy protection relies heavily on the trustworthiness of the service provider. Furthermore, the use of commercially available privacy protection technologies often require above average technical skills and the willingness to sacrifice convenience.

Disclosure of Personal Information

When the personal information of a user is given out to a service or application, the life span of the information can be divided into two major parts: before and after disclosure. It is important to understand that controlling the disclosure is much more easier than controlling what happens to information after that. The first part can be controlled with technology and supported by privacy policies, and the second part can be taken care mainly through policies. Laws can help with the enforcement of how policies are obeyed - if broken there can be legal consequences.

So, the threats are that in the first phase technology is bypassed or it is too difficult to use, and in the second phase there are no policies, a user does not understand them, or the agreements done at disclosure are not respected.

The basic principle for disclosing personal information is confirmed opt-in, which means that a user must give his consent for providing information to a requesting party. It should be noted that this applies to all methods to access personal information: whether it is a direct request from a service or applications, or an indirect way to get personal information, such as using cookies in a browser.

When personal information is to be given to another party, three possible scenarios to get user consent exist:

- Permission asked: The identity of the requesting party is authenticated and presented in an understandable form to a user who answers a specific question on whether personal information may be used or not. Additional information like the planned usage of the data could also be provided to help in the decision.
- Predefined permission: The user has configured beforehand which parties are allowed to use particular personal information. If making this configuration has been assisted by some party, such as an operator, the meaning of it must have been clearly indicated to the user. This especially applies to written or electronic subscriber agreements, where privacy preferences are set simultaneously with service subscribtions. As in the 'permission asked' case, the requesting party must be authenticated and additional parameters may effect an automated decision to

provide a particular piece of information. Changing preferences must be possible and easy enough for the user.

- Anonymous: The user may have agreeded to giving some personal information if he remains anonymous or has pseudonymity. In other words, his real identity is not known to the party requesting information. This could either happen through 'permission asked', where the user is assured that he remains anonymous, or through 'predefined permission', where anonymity is guaranteed by a trusted anonymizing service. However, it is important that a user gives consent in some form - anonymity alone is no permission to use personal information.

If the party, which is requesting personal information is inside a domain considered trusted by the user, such as operator premises or some trusted application environment, and the service or application can only be used by trusted parties, authentication is not mandatory. How a user is able to know which service to trust is a difficult issue, and normal privacy protection schemes like anonymity and asking for permission should always be applied.

General decision criteria for choosing one of these three possibilites to use is: anonymity should be used by default whenever possible, and when identity is needed the 'permission asked' or 'predefined permission' scenarios should be used, depending on how much the application is trusted (i.e. whether a permanent permission can be given).

Before disclosing personal information, the rules for its planned use should also be agreed on. This means that personal information is only given to be used for certain purpose, certain time etc. These preferences have either been configured by a user and are communicated to the other party with personal information, or the other party has presented a policy on how they handle personal information and the user has decided to accept it. Exchange of policies may happen as text documents, or in machine readable form (like P3P, for example).

In addition to carefully controlling the disclosure of personal information, privacy can be enhanced with traditional security measures like firewalls. The purpose is then to minimize attacks against privacy by hostile parties.

The Future of Privacy

Increased focus on customer information collection, and the emergence of information aggregation businesses, have led to a situation where massive amounts of personal profile data is collected and stored in numerous databases. The individual, whose personal data is in these databases, has very little control over how the information is used or traded.

Privacy legislation is developing, but is still not sufficient in most parts of the world. Self-regulation efforts have proved to be ineffective due to lack of enforcement.

Technology development helps to automate information collection in a way that is often unnoticeable and intrusive to users' privacy. New wireless technologies and ubiquitous connectivity pose additional privacy threats. Mobile devices are highly personal. They contain unique identifiers (e.g. phone number, Bluetooth or Ipv6 address) and they track location information, which enables even more detailed user profiling in a manner that is invisible to users. These environments will commonly involve communication with a party who is unknown or not trusted.

As technology brings new threats to privacy, it can also be used to protect privacy. Several standardization and industry forums, such as W3C, IETF, WAP Forum and LIF are defining better technological solutions for proper privacy handling. The danger is that the field of privacy protection technologies becomes too fragmented, and they are too difficult to use for a non-expert consumer. Therefore, the goal must be that a minimum number of technologies is used on the Mobile Internet, and that the user interface for controlling them is simple and intuitive.

Company investment in privacy protection has been low, as companies have undervalued the privacy issues. Internet services and vendors are, however, starting to experience the negative effects of users' concerns with privacy issues. More and more customers are questioning companies' information collection practices, and are refusing to purchase online because they fear how their information is being used. Similar reactions are likely to surface when it comes to the collection and use of location and other mobile data. Mobile Internet users need trust providers as most Internet services cannot be considered to be trusted. One natural choice for this are cellular operators.

Nokia is concerned about the emerging approach in online businesses, where the privacy consciousness is ignored and a consumer is not given any control over the use of his data. Merely adhering to general minimum privacy requirements may succeed if a business has a pronounced supremacy over its customers. However, this approach does not contribute to the increase of consumer trust and will not unlock business values captured in online identities.

As awareness of Internet and mobile privacy issues increases among users, the pressure to maintain or build general trust rises for companies. Companies need to take users' privacy concerns seriously. Whoever is able to provide working solutions that give control of the collection and use of information to the consumer is likely to gain the trust - and the transactions - of future Internet and mobile users.

Mobile Location Services

Mobile terminals have one distinctive advantage over fixed ones being by definition, mobile. Mobility, and hence location, is a critical attribute to be utilisied by all services. Mobile Internet services benefit from three major factors that boost the value of information to the end user: personality, time-criticality, and location-dependency. Enabling this third factor makes location a very valuable piece of information.

Figure below demonstrates how a service for finding a suitable bus route becomes much more usable when location information is utilized. This kind of service is typically needed when a user only has access to their mobile phone, and location is a critical attribute required for filtering suitable information for the display.

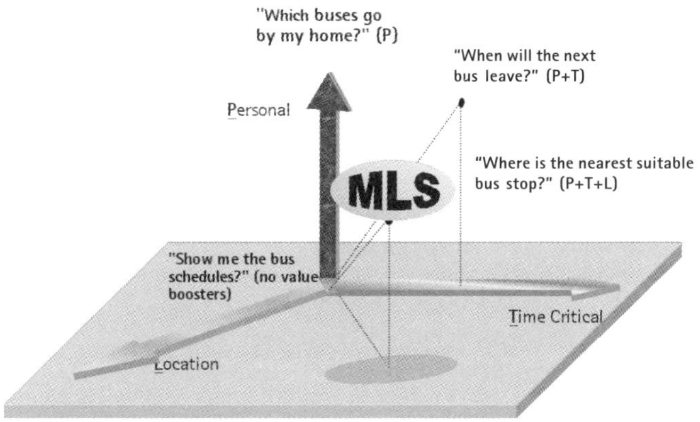

A new dimension to services brought by location information.

Wireless location technologies will become a crucial tool for providing the correct service, at the right time, for the right location to mobile customers. In the rapidly approaching age of Mobile Internet end users will not want to pay wireless Internet charges for irrelevant information. Service providers must be able to deliver pertinent and accessible information to their customers.

The ability to pinpoint the location of an individual also has obvious and vital value for emergency services. Pinpointing the location of people and valuable assets also opens the door to a new world of previously unimagined information services and mCommerce possibilities.

Mobile Location Services are value-added services that utilize the user's position information. From the user's perspective these services provide

- increased safety: positioning functionality increases user safety by being able to locate someone in distress.
- localized and up-to-date information: relevant up-to-date information in a particular location, delivered at the right moment for the right location.
- personalized information: information that is relevant to the user.
- increased efficiency: information can be more focused and of higher quality, when it is tailored to the user's location.

Location information is only a parameter for provisioning valuable applications and has no value in itself. Although location is a powerful instrument for enabling and enhancing the value of several applications, positioning should remain transparent to the end-user. On the other hand, because one's location is a very sensitive and private piece of information, the privacy of a mobile terminal user must be protected at the same time.

The purpose of location services is to make location information available for services, applications and users in a standard form, irrespective of what technology and infrastructure is used to determine the location of a mobile terminal. It is essential that location information is available for applications through appropriate software or protocol interfaces. When these interfaces are public, the development of applications becomes more attractive and their number will rise rapidly. Thus location services attain the critical mass they need to be highly popular and maintain this popularity. Another outcome of location information being easily available is that many existing and new applications will become location aware. This means that location can be used to enhance most applications without the application being a strictly location based application build around the location feature. It is quite likely that most of theses applications built with no other purpoise than to leverage the availability of location information will die away fairly quickly. Location awareness will then become an application feature on the same level as the graphical user interface.

Mobile Location Services is the general term covering both services that become possible when location information is available, and services offering location information. The figure below depicts the field of Mobile Location Services.

Location services are the services and infrastructure for determining and providing location information on a mobile device. The actual process of determining location is called positioning. Location refers to the physical location of a mobile terminal expressed in some coordinate system or in some other form known to the user of the information. A street address, a building name, or any other known system to identify physical locations can be used to express location.

When talking about services utilizing location, it is good to make a distinction between location based services and location dependent services, which both can fall under Mobile Location Services. Location based services are actual services that are visible to the user, e.g. find-a-friend, navigation, yellow pages, etc. Services that are dependent on a user's actual location, are called location dependent services. They are services provided either by a cellular operator or a 3rd party service provider that are available (pull type) or is activated (push type) when a user arrives in a certain area. Location dependent billing is a typical example of such a service.

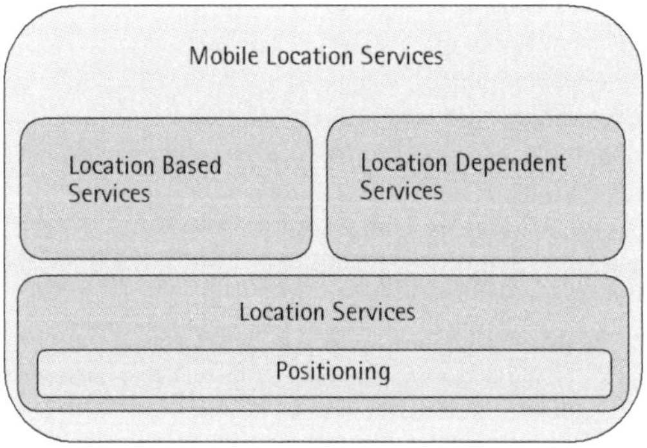

Mobile Location Services

Mobile Location Services will be integrated as the value-adding element in a whole range of services rather than being distinguished as a separate set of applications . For example, tourist information services which can let you know the price of fuel in a country do not deliver much value, but being directed to the cheapest petrol station near your location does so by combining the two services. Another example is a request to find the nearest hotel followed with further filtering criteria leading to an m-commerce transaction of booking a room. This in turn could be followed by direction information on the hotel's location, which could be displayed on a map if supported by the device. Location service can also be utilised within other services, such as a presence. An attractive set of location enabled 'killer' applications will be based on life management services supported by a dynamic portal that uses consumers' personalization details to change the type of information and delivery method according to the end-users changing time or context (home, office, morning, evening, etc).

Many existing information services can be enhanced with positioning technology. Using the Japanese example, operators do not charge extra for the location element but rather exploit the increased usage and hence revenue streams from customers who are attracted by the ease of use of location enabled applications. New applications are also being built by utilizing positioning technology. For example fleet management or group management applications can be developed on top of tracking services. Group management (locating your friends, family etc.) applications are expected to be very popular.

Mobile location services can be categorized in many ways. The next figure presents a popular way of categorization. The left column shows four location based service categories, and the top row presents three location service categories with differing levels of accuracy. The example services mentioned in figure already exist.

	Basic Service Accuracy of several hundred meters	**Enhanced Service** Accuracy round hundred meters	**Extended Service** Accuracy of tens of meters
Location based information	Yellow pages, public transport information	Find a restaurant, Find a Friend...	Turn by Turn Navigation, Find an ATM
Tracking services	Fleet Mgnt for long haul transporter	Fleet Mgnt for courrier	Fleet Mgnt for Taxis (safety), People tracking
Trigger services	Location dependent billing	Marketing push	Enhanced Mktg Push
Service assistance	Existence of a target round you	Emergency services	Roadside assistance

Mobile location services categorization.

Three factors will determine consumer trust in Mobile Location Services: privacy, security and ease of use. A clear prerequisite for this is that the customer has control over the use of their location information, not the service provider .

Of all the challenges facing Mobile Location Service providers, privacy is undoubtedly one of the greatest potential barriers to market acceptance (another is geocoded content availability). Whilst cellular operators typically treat position information as their own, it actually belongs to the consumer and its use is regulated by data protection laws. Even though a user has control over his location information, there has to be a trusted and open relationship between the user and any other party holding the information, as they both benefit from each other.

Because user privacy may be threatened, mobile advertising based on a user's location is a sensitive issue. However, introducing attractive benefits for the user for accepting advertisements, e.g. more favorable call/service charging schemes, could solve these issues.

Location Technology

This section discusses the positioning technologies that are available, and how applications are able to get location information.

Positioning Methods

In practice, there are two basic methods for automatic positioning of a mobile terminal as shown in the next figure, although other principles also exist. Triangulation refer to the measuring of signals by the mobile terminal from three sources. The propagation time of the signals together with the location of the signal sources enables determining the mobile terminals location. The Triangulation method can be built around cellular network basestations or satellites which have fixed positions.

In the second method the location of an access point, which provides the connection between a mobile terminal and a fixed network is used to approximate the location of the mobile terminal. In other words, the mobile terminal is located on the coverage area, or cell, of the access point. Two variations of this method exist: location information is delivered to an application through a service which converts the identification of the access point to a physical location, or the access point acts as a beacon sending its own location information. The first variation is called cell identification when used in cellular networks and the second one local positioning when used with short range radio technologies, e. g. BlueTooth. The Local positioning method can also be used within cellular networks, as has already been done by some GSM operators through a cell broadcast service. The disadvantage here is that it is then difficult for the operator to charge for location information.

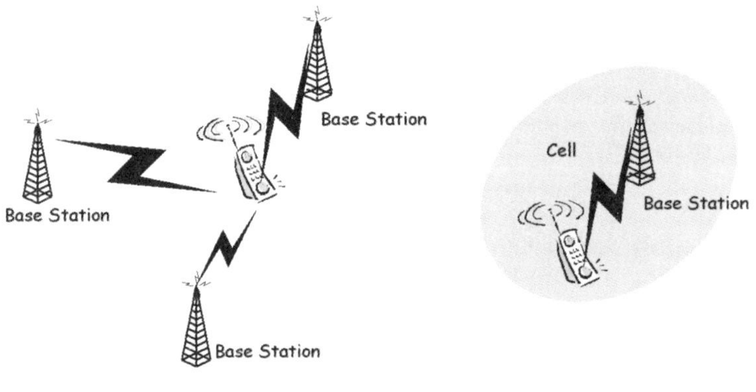

Positioning methods.

Triangulation based methods offer better accuracy, and their geographical coverage is usually adequate. Major disadvantage of triangulation technologies is technical complexity and high cost because new network elements or updates are needed. Access point location based methods are technically simpler, but their accuracy depends on cell sizes. Most of the methods available for positioning are implemented as a distributed functionality between terminal and network entities. Branding a method "terminal based" or "network based" determines the place of the actual location calculation and determination from measurement data. If measurements are also collected by another party in addition to the calculating party, or extra data for doing measurements or calculations is provided by the other party, a positioning method is called either terminal or network assisted. The additional data provided by the other party is often called assistance data.

Positioning technologies can basically be divided into handset-based, network-based and hybrid solutions each of which offer different levels of accuracy according to a three tiered service level structure of basic, intermediate and high level accuracy. Moreover, location accuracy and the degree of coverage dictate what types of application can be deployed. A successful positioning technology must meet the accuracy requirements set by the specific service, at the lowest possible cost and with good sensitivity and minimal impact on the network and mobile terminal equipment. An important consideration is that different positioning technologies are not necessary alternatives, but are complementary.

The three major emerging technologies for positioning are:

- Cell Identification with variations. These are all based on standard measurement and cell data used in cellular networks for mobility management.
- E-OTD for GSM and its variations AF-LT and OTDOA, for CDMA and WCDMA respectively. These are based on standard measurement and cell data in cellular networks enhanced with additional measurements.
- GPS (Global Positioning System), stand-alone or network assisted.

In addition to these positioning technologies, user self-locating and, for example, complementary local area technologies, such as Bluetooth Local Positioning may be used to improve coverage. Other short range wireless technologies can also be used to provide location determination. In multi-mode devices, they can be used to complement other positioning technologies, to enhance accuracy, coverage and/or acquisition time. For example, a Bluetooth 'Local Positioning' profile is being developed to provide meter-level accuracy, especially in places where GPS does not work well or at all, such as indoors.

Today, the availability of location information is a more important issue than its accuracy. End users feel that content and available services are more important than positioning methods and their accuracy. Many applications are already possible with the rough accuracy of cell identification methods, but that information should be easily available in a standard form for application developers and service providers. This is the challenge the MITA location service is fulfilling.

Location Interfaces

Location information is available for applications through software or protocol interfaces. When the application is a service running on a network server, a protocol interface between application server and the location server is needed. This interface has been specified by the the Location Interoperability Forum (LIF), and it is based on XML and runs on HTTP.

Another possibility for a service to get location information is the approach specified by the WAP Forum: a proxy entity exists between the user's browser and the service. The proxy either attaches location information to a HTTP request based on tables indicating the services that need location information, or the service gives an error response and requests location information, which is then provided by the proxy. The proxy itself gets information from a location server through an XML based interface specified by LIF, for example.

When the application is running on a mobile terminal the protocol interface introduced above can also be used, especially when a browsing session is active. The WAP forum has also specified a terminal based location interface which can be used if the terminal supports it. It is also expected that location interfaces will be available for other popular application execution environments.

Location interfaces are available in several environments. A common feature of these interfaces is a provision for the same primitives with a given content type. Therefore a location based application can be easily ported to different environments. The minimum level of service provided by the location interfaces can be summarized as the following primitives:

- Request
  - Normal - location response to be received immediately
  - Reporting - start or stop periodic or triggered location reporting
  - Emergency - privacy is overridden
- Delivery
  - Normal - response to normal request
  - Reporting - location report when triggered by time or other trigger
  - Emergency - either a response to a request initiated by the emergency service provider, or an emergency positioning request initiated by the user
- Interface capabilities query

Location contains information listed below. All of these are not mandatory, if the data depends on devices whose capabilities can be queried as mentioned above. Also, information other than that listed here is possible.

- Location
  - Coordinates in WGS84 form
  - Country & county & city & ZIP
  - Street address
- Error estimate
- Direction
- Velocity

After location information is released through location interfaces, it can also be used as geographical search criteria for services and in navigational services to help reach your destination. Therefore, when location information is used for comparing with other location information, it is essential that they are in the same format. For example, if a service is registered in an UDDI directory, it is easier and faster to look for services available near you if the location information in the directory can be easily compared against your location. There is no clear answer to the question whether location information should be stored in the directory as coordinates (WGS84) or as a street adress. Basic applications built on the idea of finding the nearest service are easier to implement with coordinate information, but this might not help in determinig if that service is easily reachable from your current location. Also, when a service providing navigational help is not available, it is better to have a location in street adress format because it is difficult for people to understand the coordinate form. As a general rule, when the location information or part of it is used or stored anywhere in a service value chain, it is better to use the same format as in the location information interface.

The Future of Mobile Location Services

There are three major drivers for Mobile Location Services. These can be identified as commercial, technological and regulatory.

Commercial drivers

Mobile Location Services provide a new revenue stream for operators, service providers, application developers, and content providers. Services will be a key weapon in the battle for customer loyalty. As saturation points are reached in the mobile markets, operators will be looking for ways to retain their customers, and issues such as cost, quality, and service differentiation will become very important. The Mobile Internet itself is one of the biggest driving forces behind Mobile Location Services. The possiblity to bundle location sensitive information into other applications, such as mobile commerce, brings the opportunity to develop appealing personalised services.

Technological drivers

The development of standards and cost efficient location technologies are key technology drivers. Cell-ID and enhanced Cell-ID based methods are available today, E-OTD will be available in wide scale in terminals and networks in 2002, and GPS will follow. In parallel with the development of better location based applications the accuracy of positioning methods will increase. Also, the deployment of 2.5 and 3 G networks, as well as the introduction of new terminal categories enable whole new service categories to be implemented and offered, on which location will be a key attribute. As terminals improve, their user interface will enable improved services, for example, applications on color displays with improved resolution and sound.

Regulatory Drivers

One should not forget regulations as one of the strongest drivers of Mobile Location Services. US FCC mandate for E-911 services as well as the equivalent European Union requirements for E-112 have an impact on the positioning technology adopted by mobile network operators. Also, regulatory issues have a major impact through setting requirements for privacy.

The ultimate target is to make utilizing location information part of everyday life among the users of the Mobile Internet. The use of location will be embedded in applications and services without users having to worry about it, but instead enabling highly personal experiences suited for the situation at hand.

4.
Nokia Solutions for Mobile World

Nokia All-IP Network Vision

The All-IP Architecture is a industry vision of future networks, with different access options seamlessly integrated with an IP packet network layer.

The Nokia All-IP Mobility System is Nokia's solution to address the needs of this future network. Specifically aimed at meeting the needs of wireless operators, it unifies the mobile network architecture and provides the best means to create and deliver services in a cost-effective and efficient combination.

Rapid upheavals and changes in the communications industry are influencing the ways we communicate today and in the future. Mobile communications, the Internet and digitisation are all changing when, what and how we communicate.

New business paradigms, in which mobility and the Internet are combined, will be necessary for mobile operators to succeed in this exciting and challenging 3G industry.

Operators who make the right choices will stand to reap huge rewards and market share. Key to this is the choice of technology and the right technology partner.

With 3G, the user will also benefit. New services, applications and features will become personalised, location aware and ubiquitously accessible anywhere, anyplace and anytime via a single terminal.

The Nokia All-IP Mobility System is the Nokia technology solution that addresses the business model of mobility and the Internet. Designed specifically to support the needs of mobile operators, it provides a scalable and phased approach to re-engineering existing wireless networks to the All-IP mobile network architecture of the future.

Future All-IP Network Vision

Today there is a blurring of the business boundaries between the traditional types of network operator, such as fixed, mobile and Internet. Similarly technology must also become integrated to support the future needs of operators.

These technology choices can be integrated into a single industry vision; the All-IP network.

All-IP enables seamless network integration of different access options, such as broadband, mobile Internet and existing wireless systems into a single IP packet layer.

IP allows all communication services to be carried over a single network infrastructure, enabling the integration of voice, data and multimedia services.

The All-IP networks will offer operators a number of important benefits including cost savings, scalability, flexibility, efficient network operations and most importantly new revenue opportunities.

Changing Traffic Patterns

Nokia predicts that by 2005 the number of mobile subscribers globally will triple and the traffic generated by each subscriber will double, leading to a sixfold increase in overall network traffic.

As profits from voice services decrease, many operators will shift to the new service opportunities available in 3G to generate revenue. These operators will find it costly to maintain both circuit-switched and packet-switched networks.

In areas with high mobile penetration, like Western Europe, the balance between packet- and circuit-switching will quickly become packet dominated.

To support this growth in data traffic, many operators are building GPRS packet networks that will more efficiently support data traffic. This is an important stage in the evolution to 3G and All-IP.

In the future, wireless networks based on the 3rd Generation standards will better support packet services and accelerate the explosion of data traffic.

The Nokia All-IP network solution supports both circuit and packet traffic in a single core network. 3G applications and services will have widely variable requirements for data speeds and bandwidths. The combination of high traffic levels and constantly varying demand can only be efficiently handled by evolving to All-IP.

Nokia All-IP Mobility System

The Nokia All-IP Mobility System is the wireless component of the All-IP network. The system is designed to meet the needs of wireless operators in areas such as service flexibility, service deployment, evolution and backwards compatibility.
Based on 3G Standards (3GPP) it is compliant to Rel4/ Rel5 standards.

Whereas 2G networks were primarily optimised to support wireless voice (and some supplementary services), the All-IP Mobility System is optimised to support real-time multimedia packet services.

With All-IP, network resources are used more efficiently and capacity can be deployed as necessary to meet demand. Standard network elements such as routers can be used throughout the network, which:

- simplifies network design
- shares capacity more efficiently
- significantly lowers investments, because off-the-shelf components can be used.

The different functions in the All-IP Mobility System are divided into different network layers, i.e. transport, mobility/connection/control and application and network subsystems, such as Access, Core and Services. These are all seamlessly integrated with open standardised protocols.

The benefit of this approach is that subsystems or even layers can be evolved separately or may co-exist with other layers and subsystems without affecting the overall network.

Several examples of this include:

- the high-speed backbone in the Gateway Layer can be upgraded (i.e. ATM to MPLS) with minimum impact on the Services and Control Layers
- the radio and core network can be upgraded without affecting the services and applications.
- control elements, gateway elements or service elements can be upgraded independently of each other based on users, traffic or types of service.

IP Multimedia Capabilities

With All-IP, voice, data and multimedia services can be carried over a single network. This yields two main benefits for operators: new and better ways to develop and offer applications and services, thereby earning more revenue; and cost savings from a single core network using common IP-technology.

The All-IP Mobility System is optimised to support real-time multimedia services. The adoption of Session Initiation Protocol (SIP) signalling is a key ingredient in providing this new functionality.

SIP is a common signalling protocol joining the Internet and mobile telephony. It provides integrated multimedia capabilities for IP enabled terminals.

For a mobile phone user, this means for example, integrated voice, video and Internet browsing in a single call. Numerous applications can be implemented with SIP, which combine traditional telephony with messaging and multimedia.

For instance, instead of a blinking telephone number on the display, the caller can introduce himself in the beginning of the communications session with a personalised logo or business information. The subject of the call can then be immediately displayed to the recipient as text, voice or picture depending on the terminal capabilities.

With IP multimedia, it will also be possible to create richer profiles that include additional features, such as sending tailored information to predetermined callers or integrate email/media-on-demand so that the voice call, for example, can be converted to a text message and forwarded to a web mailbox if the recipient is unavailable.

The Nokia All-IP Mobility Core can be connected to any major radio and data network standards, to bring a secure, complete, future-proof, end-to-end network for 3G Mobile Multimedia and a multi-service environment to operators.

IPv6 Enabled and IPv4 Backwards Compatibility

The versatility of the Internet Protocol makes it easy for operators to provide a wealth of customised services, and it is clearly the best way to transport everything from voice to broadband data. However, some challenges exist within the Internet Protocol itself.

The limited size and structure of the current Internet address space, lack of end-to-end security and sub-optimal performance are seen as critical weaknesses in Internet Protocol version 4 (IPv4), the most widely used IP protocol today.

IPv6 avoids these problems and represents the next generation of IP protocol.

The All-IP Mobile System will be IPv6 enabled from day one. IPv6 is seen as a key requirement to implement IP real-time multimedia services with inbuilt Quality of Service (QoS) indicators.

The most important benefit of IPv6 is the huge address base, meaning that all IP terminals will have their own globally unique IP addresses.

Security is also a major feature of IPv6, with IP Security (IPSec) providing the ability to encrypt/authenticate all traffic at the IP level, i.e. all applications running over the All-IP Mobility System will be secure.

Mobility is also catered for, with IP Mobility (Mobile IPv6) being a standardised part of the protocol. In Mobile IPv6, each mobile element (mobile node) is identified. This means that users are always connected.

Initially, IPv4 and IPv6 will co-exist with IPv6 gaining ground as its capabilities are realised. To support IPv4, dual protocol stacks, tunnelling and translation mechanisms will provide backward compatibility in the All-IP Mobility System.

Radio Access Network

All mobile radio standards have distinct network architectures and transport mechanisms.

In the future, many different access technologies, such as GSM/EDGE, WCDMA and WLAN, can be operated as one uniform radio access network using IP. Again, IP is the protocol of choice enabling closer integration of both radio access and core technologies. More importantly, end-to-end IP services will become possible from the wireless IP terminal to the Internet via the All-IP Mobility System.

All-IP Radio Access will drive down the operator's production cost per bit, improving operator profit capability. The All-IP Radio Access architecture provides this lower cost by combining multiple radio technologies as a single resource pool and by offering improved scalability without the bottlenecks associated with hierarchical network architectures.

Managing end-to-end Quality of Service (QoS) is a key requirement in future mobile networks. The radio interface is typically the "thinnest pipe", so controlling the QoS here is critical.

Nokia Common Radio Resource Management (CRRM) provides optimum use of radio resources, QoS control, and network automation. CRRM can add up to 100% to capacity for interactive services by combining multiple radio access technologies in the optimum way.

The resources of each radio access technology are used in the best way and common resources, such as transport, are efficiently shared by all the access media.

The Nokia All-IP access solution will introduce a new RAN architecture that is fully optimised to support IP-based traffic.

Multi-Services Environment

All-IP networks provide a more efficient platform for end-user applications. Compared to circuit-switching, All-IP provides more bandwidth, more efficient network usage, more freedom in service creation and more flexible billing.

An All-IP core network enables the implementation of services that are not possible in separate circuit-switched and packet core networks. For example, rich media calls are not easily implemented in circuit-switched networks, where there is a restriction on the capacity of a single circuit.

Operators and service providers that can provide reliable access to exciting and useful mobile services will develop brand loyalty and secure their profits. Those that cannot will be more vulnerable to tightening competition.

Nokia realises that making the Internet accessible through mobile devices opens a huge new business opportunity. However, since much of this information is freely available, the value provided by the operator will be based on pinpointing what is relevant.

Nokia sees four service groups having major relevance in 3G:

- information and entertainment content
- communication
- productivity-enhancing services
- business solutions.

When deciding to implement a 3G mobile network, operators need to consider these services as they build the infrastructure.

They need to discover creative ways of packaging these services and using existing services to prompt the usage of new ones.

Excelling in the following four key criteria will provide maximum value for users in the world of 3G:

- services that are personalised
- services that provide timely information
- transactions that are simple to complete
- services that are relevant to location.

Coverage, capacity and pricing are used as key differentiators today, but they will not be sufficiently competitive advantages to gain a profitable market share in 3G.

The success of 3G operators will be based more on their ability to provide a suite of personalised services to users, and thereby decrease the chance that the users will move to another operator for better services.

The All-IP Mobility System is designed to more economically and flexibly support existing services in GSM, GPRS and Release 1999 networks, by creating a multi-service environment.

Furthermore, it provides a solid foundation upon which future services can be created and deployed easily. Discrete applications and disparate services can be integrated into seamless service packages.

Coherent Evolution Strategy to All-IP

Although the All-IP Mobility System represents a fundamental change in mobile networks and their capabilities, it does not represent a discontinuity in the existing Nokia technology portfolio.

Nokia GSM, GPRS and Release 1999 networks can be evolved in stages to support the All-IP Mobility System.

There are three key phases:

1. evolution from Release 1999 to All-IP Mobility System
2. All-IP Mobility System based on Rel4/Rel5
3. All-IP Mobility System beyond Rel4/Rel5.

Step One: Evolution from Release 1999 to All-IP Mobility System

The process of evolving towards the All-IP Mobility System in the core network has already begun with General Packet Radio Service (GPRS).

Nokia has already delivered and integrated almost 50 of these networks world-wide and the experience of being first in volume GPRS deliveries and full commercial network roll out provides valuable knowledge of how to create the real mass market for 3G.

These packet-switched, always-on network services are the first stepping stones. They provide the basis for the mobile packet core network in Release 1999.

The Nokia Mobility Core network is complemented by a unique radio access network capable of supporting several radio interface technologies, ranging from GSM to EDGE, WCDMA and WLAN.

Release 1999 is focused on seamlessly integrating these different radio access technologies into a single core network.

GSM radio access networks will gradually develop into IP-based networks, offering end-to-end packet-switched transmission of 3G services.

This evolutionary process has already begun, with operators across the world deploying GPRS, and will be completed with the deployment of WCDMA technology. The Nokia solution for radio access offers smooth evolution, as EDGE and WCDMA networks can be co-sited with existing base stations, reducing rollout costs and accelerating the shift to an All-IP radio access network.

Step Two: All-IP Mobility System Based on Rel4/Rel5 Standard

By upgrading the GSM/GPRS network, the core network is able to continuously support circuit-switching for voice and mission-critical data while providing the benefits of packet-switching for other data services.

As 3G rich call and multimedia services become more widely available, operators can meet demands for quality, capacity and profitability by implementing an All-IP Mobility Core.

The Rel4 standard specifies a new element to support GSM circuit-switched terminals. However, the All-IP Mobility System bypasses the need to invest in this by supporting the function in the All-IP Mobility Core.

The 3GPP Nokia Rel5 All-IP Mobility Core will provide:

- a new IP multimedia core network subsystem
- roaming charging
- service creation based on IP mobility management interfaces.

New platforms are available to support IP multimedia, or alternatively, Nokia Release 1999 and Rel4 elements can be upgraded. With many types of radio access technology (GSM, EDGE, WCDMA, WLAN, etc.) and multi-band terminals becoming available, managing these resources becomes more complex. Fortunately the situation is simplified with a single radio resource controller. Nokia Common Radio Resource Management (CRRM) integrates GSM/EDGE and WCDMA into a seamless radio access network.

Step Three: All-IP Mobility System Beyond Rel4/Rel5

The All-IP network presents a future in which there is no need for circuit-switched network elements; all circuit-switched traffic will be re-invented as voice over IP or other real-time packet services. The All-IP RAN and All-IP Mobility Core ensure end-to-end transport and control of IP real-time/non-real-time services.

Backwards Compatibility with Existing Networks

The major factor determining the rate of transformation to an All-IP mobile network without legacy support will depend on how rapidly GSM, GPRS and early 3G terminals evolve into IP multimedia terminals.

As this is uncertain, the phased evolution of the All-IP Mobility System ensures that the flexibility remains with the operator. Nokia will ensure that the All-IP Mobility System remains backward compatible with GSM, GPRS and early 3G terminals.

Intelligent Edge and Technology Neutrality in the Backbone

The focus of the Nokia All-IP Mobility System is to add value to the operator's business by providing innovative technology. The Nokia focus is not to provide high-speed backbone routers, but instead, "intelligence" in the network through mobility, mobile services and radio access technologies.

In this respect, the Nokia All-IP Mobility System is essentially neutral to the packet backbone, enabling it to seamlessly integrate with the operator's high-speed backbone. It supports backbone packet technologies such as IP, ATM, MPLS and SDH.

Alternatively, if the operator does not have a high-speed packet backbone, Nokia is ready to provide a complete All-IP Mobility System, including the backbone, as a single supplier solution.

Our global partnering with Cisco ensures that packet backbone products provided in any Nokia All-IP solution will be compliant with industry standards and state-of-the-art backbone technology.

Conclusion

Nokia is able to provide a complete technology solution based on All-IP Mobility. This includes the network, rollout, services and support to optimise the network and introduce new revenue opportunities.

Leadership in Standardisation

Nokia is a global leader in mobility and IP and a driver within the standardisation forums for 3G networks.

With an active role in the standardisation of key system technologies, such as IPv6 and 3GPP, QoS and global leadership in mobility and IP, the Nokia All-IP Mobility System is a complete technology solution, a strategy and a vision for the operator.

Standardising the functions and architecture of 3GPP Rel5 (on which the All-IP Mobility System is based) is well underway. Nokia has been able to make significant contributions to the relevant forums and standardisation bodies with the aim of leading the industry towards a common IP-RAN architecture and unified mobile IP packet network architecture.

Nokia is playing a key role in 3GPP, 3GPP2, 3G.IP, MWIF and IETF, all of which are involved in some aspect of the standardisation of the All-IP concept and how it will develop. It is the intention of Nokia to stay at the forefront of the mobile and Internet industries by actively participating in these forums to ensure Nokia 3G products are compliant and future-proof.

Reducing Costs by Leadership in Radio Access and IP

The All-IP Mobility System can bring significant cost savings to a network. Firstly, ubiquitous access in the core network eliminates the overhead of operating two networks; one for data and one for voice. Additionally, the All-IP Mobility System provides unsurpassed flexibility in allocating bandwidth and QoS for services. This translates into a cost-effective network.

Secondly, All-IP Radio Access offers savings by optimising the use of resources, particularly in wireless networks where different radio frequencies must be managed.

Finally, the unified network management system provides an end-to-end perspective of the network enabling easy optimisations and fine-tuning of the network.

Cost Effectiveness with a Single IP Network

The Nokia All-IP Mobility System provides a cost-effective network on which a multi-service environment can be built.

The All-IP Mobility System has a simplified network design. All real-time/non-real-time data and signalling are carried by a common transport network. This is invariably less costly, both in investment and operation, than supporting two networks.

If the All-IP Mobility System is properly dimensioned with QoS mechanisms, the total capacity needed is smaller than in two separate networks. Each network requires a certain amount of extra capacity for traffic bursts. One large network requires less capacity than two separate networks.

New Markets and Revenue Opportunities

The business model for All-IP networks will be a convergence of Internet, media and wireless communications. Mobile operators are in a key position to exploit the new business opportunities brought about by the shift towards this convergence.

An operator's key assets in this new business model are: the Mobile Internet; control of the end-to-end quality of service for operator hosted services; the mobile network-specific enablers, such as location and presence information; the mobile network infrastructure and naturally, the subscribers themselves.

The Nokia All-IP Mobility system represents the best technology solution in addressing the operators business for the future and provides a smooth evolution path from existing wireless networks.

Nokia's Vision of Providing End-to-End Quality of Service in 3G

This paper summarises the Nokia approach to Quality of Service (QoS). The QoS concept is based on the goals of

- Defining tailored service quality for end users.
- Supporting the implementation of different operator business models.
- Using resources efficiently.

The QoS concept is discussed using the evolution of General Packet Radio Service (GPRS) and Universal Mobile Telephone System (UMTS) networks, including the multi-access All-IP network, which is an evolution of the R99 version of the 3GPP standard. Nokia supports QoS both for IPv4 and IPv6.

General

This white paper summarises Nokia's view regarding the provision of end-to-end Quality of Service over IP-based wireless networks. This first chapter introduces some fundamental issues related to QoS. UMTS releases are used as a framework and example for QoS.

The next chapter reviews the major forthcoming IP-based application service types and their demands on the network. QoS is clearly an end-to-end issue because it affects the terminal as well as the network solutions. In what follows, the Nokia approach towards QoS during periods of transition is described from the perspective of access evolution, IP network technology and terminal development.

Applying isolated QoS mechanisms is not enough. Instead, QoS approaches in different subsystems must be managed in a co-ordinated fashion. The last chapter examines the current strategic position of Nokia towards the concept of IP-based mobile network management, which is followed by a summary.

Third Generation Mobile Network Releases in 3GPP

The Third Generation Partnership Project (3GPP) is the ETSI standardisation body for 3rd generation mobile networks and defines standards of interoperability in different 3rd generation mobile network interfaces. The stable phases of evolving standards are organised into releases. The releases most relevant to the near future are listed in the table below. The releases will be referred to throughout the report by abbreviations, such as R99.

Nokia's approach is to introduce All-IP RAN into its R5-based networks. This will provide universal IP transport, both in the logical mobility core functions and in All-IP RAN. It makes cost savings possible through the use of lighter protocol stacks and common infrastructure in different parts of the 3G architecture. It is easier to highlight a single consequence of the previous, consistent management of resources if the network uses a single transport platform. Thus the direction of evolution tends towards the increasing use of IP as the transport solution. Nokia will continue to support interoperation with existing ATM-based systems nonetheless.

The All-IP architecture option of R5 enables operators to deploy IP technology to deliver full 3G services and to take a step towards greater convergence in IP and mobile technology. At the same time, various solutions are being developed in the IP network industry to provide the support needed for demanding services such as real-time voice and video. With all this in mind, a single packet switched network based on IP can provide efficient support for all services in the wireless domain. Support for version six of the Internet Protocol (IPv6) provides a natural route towards a wireless Internet with multiple access methods.

An explanation of the need for Quality of Service follows. We discuss the role of IP, especially in the context of All-IP networks, and we present a conceptual model of QoS.

The three first UMTS releases

Abbreviation	Noteworthy points
R99	This was the first release of the 3G standard. It contained the WCDMA radio interface, packet-switched mobility core, and circuit-switched RAN.
R4	This included the GERAN radio interface and added a media gateway and MSC Servers into the R99 architecture.
R5	Proposed the IPv6-based IP multimedia subsystem and made multi-radio All-IP RAN possible.

What is Quality of Service?

According to the QoS Forum, Quality of Service is a mechanism that provides a level of assurance that ensures that a service or application can be delivered to the end-user in a satisfactory time frame. In other words, QoS refers to the ability of a system to provide an appropriate transport service to deliver various types of communications traffic to different users. It can sometimes be difficult to define the exact technical parameters required to ensure such delivery, because the perception of "service quality" may differ from one user to another. However, delivering a service using specific technologies often clarifies which parameters may be set in a meaningful way, since the underlying protocols in the network often have definable variables. For a packet switched network, these parameters are typically given as packet delays, jitter and loss rates.

The requirements for transport services vary according to the types of application streams and end users. At the same time as application requirements are fulfilled, it is preferable if the system is scalable and uses network resources efficiently. Furthermore, the system should be flexible, allowing for the implementation of different business and service models by the network service providers.

From the fixed network point of view, the easiest way to provide a high service quality is to ensure that sufficient bandwidth is available at all times to all users and all services. This leads to the inefficient use of network resources, overdimensioning and excessive costs. Additional bandwidth can be added, but it will usually be used up quickly. The real challenge is to manage the existing network bandwidth in such a way that appropriate service levels can be delivered to all users all of the time. This results in an efficient use of resources.

From the wireless network point of view, making provision for a high service quality also affects the end-to-end performance. The air interface is likely to be one of the major bandwidth-limiting factors in a wireless data network. Radio is also a hostile medium for access, with problems arising from error-prone and time-variant link characteristics.

In summary then, Quality of Service does not create additional bandwidth. Instead, it manages the existing bandwidth in such a way as to provide for the efficient transport of traffic through the network. It allows for predictable delays, delay variation (jitter) and packet loss rates.

Quality of Service and All-IP networks

Quality of Service has become a topical issue with the transition towards the use of Internet Protocol (IP)-based transport platforms. An open IP-based network provides a cost-efficient, unified approach to transport. In addition, the IP environment facilitates the easy creation of a variety of new services, including different forms of multimedia applications.

Let us take recent developments in the 3G mobile network as an example. One of the most significant advances in 3G mobile communication technology is the support for a richer variety of services and a higher level of service personalisation. This is supported by the UMTS QoS framework in the 3GPP standard.

The main difference and potential benefit from the QoS provisioning point of view – as opposed to earlier mobile network solutions – is the unifying effect of QoS transport mechanisms. This applies from the multiradio IP base station onwards, all the way to the edge of the mobile network, under a single framework of IP QoS.

A Conceptual Model of Quality of Service

To understand the complete picture of end-to-end QoS, one needs to examine the issue at a more abstract level. The essential components are the application service domain, the bearer service (network transport) domain, and the management domain, which spans the other two domains.

The separation of Application service and Network transport service domains is an important aspect of this model, as is the principle of UMTS bearer services. The role of Application service is to provide QoS control from the perspective of the needs of the applications. The Network transport service domain is mainly there to provide the transport service.

On an abstract level, the role of each domain for QoS can be described in terms of the following aspects:

- Network service QoS – this covers the Radio access network, the Mobility Core, and IP transport network, including the backbone network, service capabilities and QoS control.
- Endpoint QoS – this covers the terminal, end user, and application service QoS awareness and QoS control.
- Network QoS control – this is a control function that administers the resources in the network in order to achieve a service behavior consistent with the needs of a specific service QoS.
- QoS management – this provides functions in the management plane (network & OSS) to provide a statistical feedback loop for network QoS control e.g. through service quality management and various functions in network, service and customer care layers.

Within a network transport service domain, QoS functionality can be further divided into the User Plane and the Control Plane. The User Plane includes Telecom/Datacom processing and forwarding services for user traffic, while the Control Plane provides Telecom/Datacom Control & signaling services.

Finally, the transport network services and protocols also span the User Equipment (UE). This provides an interesting double-role for mobile terminals, since they offer both application services and network services. Clearly, intelligence in the terminals plays an important role in the optimised mapping of service requirements to transport QoS.

In the chapters to follow, all of these aspects will be reviewed in more detail. First, different IP application types and services are discussed.

Application Types and Services

Some of the most important classes of existing and emerging end user services are:

- Rich call
- Streaming
- Browsing
- Messaging
- M-commerce transactions

In an IP network, these services use the server-client model. Applications such as streaming and rich call are real-time by nature, involving periodic transmissions of information in the network. Transaction-type applications such as m-commerce and Internet browsing produce less regular traffic patterns, but nevertheless pose delay limits for transmission, thanks to their interactive nature.

Perhaps the most important classification of applications – from the Quality of Service point of view – is one based on end-to-end delay requirements. For instance: Interactive real-time applications, such as voice or video conferencing, have the most stringent end-to-end delay requirements. For interactive transaction traffic types such as browsing and m-commerce, longer delays are acceptable. Streaming can allow for transmission delays of up to seconds in length. Instant messaging benefits from short delays, but in e-mail delivery delay is not an important issue.

The delay requirements of applications are reflected in the protocol stacks that can be used for transmission. Strict delay requirements often prevent the use of retransmission, which is characteristic of the Transfer Control Protocol (TCP). For example, Voice over IP is an application that needs rapid transmission, but can tolerate an amount of PDU loss. The User Datagram Protocol (UDP) is used instead of TCP to get better delay performance. An overview of QoS requirements and transport protocols is shown in the next table.

From the point of view of the applications, the whole transmission path from the terminals to the network must provide consistent QoS support.

Rough delay and loss classification for some application types and transport protocol. Legends: C-critical, I-important, L-less important.

	VoIP	Streaming	M-commerce	E-mail
Delay requirement	C	L	I	L
Loss requirement	L	L	C	C
Protocol	UDP	TCP/UDP	TCP	TCP

Technology Aspects of the Radio Access System

3G Quality of Service Framework

Within the Radio Access and Core Networks specified by 3GPP, QoS provision and resource reservation are based on the activation of a Packet Data Protocol (PDP) context. Every PDP context is associated with a set of QoS attributes to be negotiated in conjunction with the context activation procedure. The set of QoS attributes is referred to as a QoS Profile. Depending on traffic class, it includes the parameters that are essential for the application to describe the QoS needs of its media streams, including traffic class, target transfer delay, reliability, guaranteed bit rate and priority.

The UMTS framework supports multiple PDP contexts for the same IP address with different QoS requirements. This enables a multimedia session to describe the QoS needs of individual media streams. One UMTS bearer is equivalent to onc PDP context.

Case Study: Quality of Service Negotiation in a UMTS Network with All-IP RAN

The complete QoS-activation chain between the User Equipment (UE) and the Gateway GPRS Support Node (GGSN) is illustrated and described below:

1. Based on a request from an application, the UE maps the QoS needs to the relevant PDP context QoS parameters. The UE requests PDP-context activation from the 3G Serving GPRS Support Node (SGSN).
2. The SGSN first checks whether the subscriber is entitled to use such a QoS profile. If so, and if the SGSN has adequate resources available, it will signal the RAN Access Server (RNAS) in All-IP RAN to establish a Radio Access Bearer, based on the QoS request.
3. The RAN performs internal Admission Control and resource reservation. If successful, the SGSN will send an acknowledgement back to the UE, and a Radio Access Bearer will be set up. The SGSN now makes a PDP-context activation request to the GGSN.
4. If GGSN has adequate resources to provide the service with the requested QoS, a transport connection (e.g. a tunnel) is then set up between the SGSN and GGSN, and a PDP-context with the requested QoS capabilities is established between the UE and the GGSN.

Quality of Service Functions in 3GPP

The QoS functions allow the network to provide different levels of QoS services for the purpose of establishing, modifying, and maintaining UMTS bearer services. Not all functions are subject to standardisation, and are often only defined at a conceptual level, leaving freedom for implementation. 3GPP Release 5 considers the following UMTS functionality as the minimum in the Control plane:

- Service manager
- Translation function
- Admission/Capability control
- Subscription control

and in the User plane:

- Mapping function
- Classification function
- Resource manager
- Traffic conditioner

Service Manager

The service manager is responsible for providing all U-plane QoS management functions with the relevant attributes. This function resides locally in the UE.

Translation Function

The translation function sits in the UE and the gateway (GW). Its main task is to convert the QoS attributes of the external network into a form that can be understood by the UMTS network and vice versa.

Admission/Capability control

The purpose of admission control is to calculate what network resources are required to provide the requested QoS, to determine if such resources are available, and reserve them.

Admission control is performed by the SGSN server and also locally at other network elements, such as GGSN, RAN Access Server (RNAS) and Cell Resource Server (CRS). The Admission controller in SGSN has the responsibility for accepting or rejecting PDP context activation and the requested QoS parameter values. GGSN, RNAS, and CRS check locally whether they are able to support the new context and QoS values.

Subscription Control

This function is located in the SGSN. It checks the administrative rights of the UMTS bearer service user to use the requested service with the specified QoS attributes.

Traffic conditioner

The conditioner ensures that user data traffic conforms to the QoS attributes of the relevant UMTS bearer service. It exists in UE, GW and All-IP RAN.

Mapping Function

The mapping function marks each data unit with the specific QoS indication related to the bearer service transferring the data unit.

Resource Manager

The resource manager is in charge of determining whether the QoS requirements of a PDP context can be satisfied and allocates adequate resources to the incoming packets. The Resource manager takes into account the QoS associated with various packets and distributes resources among all bearer services.

In addition, to enable end-to-end QoS between the UE and remote end-points outside the UMTS network, 3GPP R5 introduces the concept of IP bearer services. These have the following functions:

IP Bearer Service (BS) Manager

The IP BS Manager uses standard IP mechanisms to manage the IP bearer service. The UMTS translation/mapping function provides the interworking between the mechanisms and parameters used within the UMTS bearer service and those used within the IP bearer service. It also interacts with the IP BS Manager.

IP Policy Control

IP Policy Control is a logical policy decision element, which uses standard IP mechanisms to implement policy in the IP bearer layer. IP Policy Control makes decisions with regard to network-based IP policies using policy rules, and communicates these decisions to the IP BS Manager in the GW.

Wireless LAN

Nokia supports the interworking of WLAN with the 3G network using the Operator WLAN (OWLAN) solution. The use of a single infrastructure for both networks is an advantage for authentication and the quality of charging & billing services. Nokia plans to add QoS to the OWLAN solution, as embraced in WLAN standards.

The Evolution of Radio Access in Wide Areas

Both GPRS, E-GPRS and 3G networks support IPv6 for user plane traffic between the UMTS terminal and the GGSN. This allows the technology to develop towards IPv6-based networks while maintaining support for QoS.

R99-based Quality of Service Solutions

Within the launch of the Nokia 3G network, the managed QoS support is built with the following key components:

- 3GPP-compliant radio interfaces and networks.
- The use of WCDMA air interface resources is optimised by mapping the application traffic onto appropriate radio bearers based on cell load monitoring and optimising coverage and capacity. Moreover, radio transmission uses prioritised scheduling based on application type.
- Differentiated Services (DiffServ) will be offered in the mobility core for IPv4 initially and for IPv6 in the next phase. DiffServ also forms the basis for QoS interworking with external networks.
- Downlink IP flows are mapped into PDP contexts based on Traffic Flow Templates (TFTs) at the GGSN. GGSN supports both IPv4 and IPv6.

In opening a PDP context, the terminal requests the UMTS traffic class, such as conversational, interactive, streaming or background. Advanced, per-PDP context QoS control is possible with the right bearer allocation, retention priorities, and traffic handling priorities. 3GPP R99 standard allows 250 different QoS profiles to be defined in HLR, each of which can be used for defining a PDP context QoS behaviour. For downlink traffic, GGSN maps the incoming traffic into PDP contexts based on Traffic Flow Templates (TFTs). The IP QoS mechanisms in SGSN and GGSN have been designed to reflect the UMTS traffic types.

Differentiated Services (DiffServ) provide a scalable way of implementing both service differentiation and the efficient use of resources at the IP network level. DiffServ is supported in 3GPP IP interfaces for both IPv4 and IPv6. Consistent QoS behaviour across different technologies can be obtained with advanced management techniques.

The Nokia NetAct™ management framework manages ADSL broadband access networks, as well as cellular components. This provides the operator with a consistent view of the entire system.

Evolution Towards All-IP Multiradio Access

For the All-IP evolution of the 3G network, some of the emerging key QoS technologies are:

- To provide for spectral efficiency in the All-IP air interface, Nokia is contributing to the Internet Engineering Task Force (IETF) on robust header compression.
- Common Radio Resource Management (CRRM) for optimising the use of resources during handovers between multiple radio technologies.
- The efficient use of resources in All-IP RAN through traffic engineering, using IETF standards as embraced by 3GPP.
- A functional split for improved packet performance and scalability.

In addition, QoS in the All-IP network will be supported by the monitoring of load levels both in the air interface and in the transport network. Network QoS support will be optimised using advanced rule-based management techniques.

The evolution of CRRM functionality allows for QoS control of macro, micro and picocellular handovers for WCDMA, E-GPRS and legacy interfaces, as well as wireless LAN (W-LAN) access points. This functionality allows for simultaneous implementation of QoS and the efficient use of resources.

IP Transport Network Aspects

One of the challenges in the transition towards the use of Internet Protocol (IP)-based transport platforms is the provision of sufficiently high-quality transport services for traffic that is very sensitive to delay. Furthermore, low bandwidth links are often used in cellular networks, so overprovisioning is clearly not a cost-effective solution.

From the perspective of QoS optimisation, the IP transport network may be divided into two sub-networks:

IP Access Network

The purpose of the IP Access Network is to connect base station sites to the transport backbone. It mostly consists of bandwidth-limited links (microwave or copper), and imposes the biggest cost in transport.

Backbone Network

The so-called backbone network usually means a general purpose network that covers, in a mobile network, the network between core sites (the long distance network) and the network between core sites and RAN (often a "regional network"). It may also cover a significant part of the transport network within RAN, depending on the implementation (a fibre-based network).

In practice, the selection of suitable QoS solutions for a specific operator and a specific segment of its transport network will depend largely on the traffic profile, the existing infrastructure, the cost and availability of transmission bandwidth, the need for scalability and so on.

There are several IP QoS -related mechanisms available for IP transport networks. These were developed by the IETF forum. The QoS management solution uses mechanisms such as packet prioritisation and traffic engineering protocols as building blocks in constructing operator solutions. A short summary of IETF protocols follows:

IntServ

IntServ stands for Integrated Services. It is an IETF-specified IP QoS mechanism using resource reservation in the routers. IntServ requires RSVP signaling between routers. In the per-flow QoS signaling variant, IntServ has poor scaling properties and is widely considered unsuitable for large-scale backbone usage. In general, adding IntServ support makes network elements more complicated and more expensive.

RSVP

RSVP stands for Resource Reservation Protocol, the IETF QoS signaling protocol used by IntServ to reserve resources. Aggregated RSVP is currently being standardised within the IETF.

DiffServ

DiffServ stands for Differentiated Services. DiffServ is an IETF-specified QoS method for packet prioritisation and does not require per-flow signaling. The method is scalable and provides relative per-hop prioritisation. It can be used together with Multi-Protocol Label Switching (MPLS).

MPLS

MPLS stands for Multi-Protocol Label Switching. MPLS, again a protocol from the IETF, can be used to create routes (called label-switched paths) for traffic aggregates. It can be used for fast traffic restoration and Traffic Engineering.

Nokia currently views the Differentiated Service model (DiffServ) as the most suitable framework for differentiating services while using the transport resources efficiently inside the IP access network. In addition to implementing DiffServ, various traffic engineering and bandwidth management schemes could be used to improve the QoS performance of IP networks.

DiffServ is simple and very well scalable. DiffServ routers do not have to store any information about the state of flows and it allows flexible, operator-specific PHBs to be used. However, with DiffServ alone, strict end-to-end QoS provisioning is more difficult, because DiffServ only offers relative per-hop-behavior. In addition, Traffic Engineering would be restricted to the existing routing protocols. MPLS, on the other hand, enables explicit routing. With MPLS, traffic aggregation capabilities can be used to improve the level of network usage and to avoid congestion. Extensions to current routing protocols offer another possible way to control routing.

DiffServ

DiffServ can be used for both IPv4 and IPv6. Per-packet classification and per-hop-behaviour (PHB) are the basic building blocks in the DiffServ concept. The DiffServ code point (DSCP) bits in the IP header identify the class of traffic. Based on that, traffic policing and relevant QoS control mechanisms can be applied to each packet and at each DiffServ-capable router. The basic types of PHBs defined by IETF are the Expedited Forwarding (EF) PHB group and the Assured Forwarding (AF) PHB group.

According to RFC 2598, which defines the EF PHB, it can be used to build a low loss, low latency, low jitter, assured bandwidth, end-to-end service through DiffServ domains. This is useful for applications such as delay-critical real-time streams.

The Assured Forwarding (AF) PHB group is defined by RFC 2597. It enables a provider DiffServ domain to offer different levels of forwarding assurance for the IP packets received from a customer DiffServ domain. Four AF classes are defined, where each AF class is allocated a certain amount of forwarding resources (buffer space and bandwidth) in each DS node. Within each AF class, IP packets can be marked with one of three possible drop precedence values. In case of congestion, the drop precedence of a packet determines the relative importance of the packet within the AF class.

Nokia has performed research on DiffServ for many years and has developed implementations of the scheme that cater for the delivery in different UMTS traffic classes. In addition, Nokia Research Center has developed a new PHB type, called dynamic Per-Hop Behaviour, which is a Service Level Agreement (SLA) to be implemented directly at the edge router.

In what follows, the operation of DiffServ is illustrated. An example of the results obtainable with DiffServ is provided. Finally, we discuss the implementation of network-wide QoS with DiffServ.

How DiffServ Works

DiffServ edge routers play an important role in preventing congestion in the transport network. The edge router controls the bit rate and burstiness of traffic as it enters the DiffServ domain. In the case of excessive bandwidth for incoming traffic from an IP host, the edge router may smooth burstiness by delaying packets (shaping), by remarking the packets to a lower DiffServ class or, as an ultimate sanction, by discarding packets altogether.

Class-based queuing mechanisms combined with edge policing create a powerful yet simple tool for using the existing bandwidth intelligently. DiffServ is also very well in line with the Internet's stateless packet forwarding paradigm and avoids the per user signaling needed in more traditional QoS schemes.

DiffServ Delay in IP-based RAN

Nokia has simulated DiffServ routers in a RAN environment. The results indicate that DiffServ doubles the bandwidth usage efficiency compared to a traditional best-effort router, allowing the use of reasonably thin links from the base station.

DiffServ does not create bandwidth. Instead, it uses the existing transmission more effectively than a traditional best-effort Internet. The model can be refined by introducing more traffic classes.

Implementation of Network-Wide Quality of Service with DiffServ

Because DiffServ is based on prioritisation and statistical multiplexing, routers must be configured consistently to obtain QoS on a network level. Nokia addresses this by applying policy-based management. The Nokia NetAct framework also includes tools for monitoring service quality.

What is Traffic Engineering?

The IETF defines traffic engineering as follows: Internet traffic engineering is that aspect of Internet network engineering dealing with the issue of performance evaluation and performance optimisation within operational IP networks. Traffic Engineering encompasses the application of technology and scientific principles to the measurement, characterisation, modeling and control of Internet traffic.

In what follows, traffic engineering is viewed from the perspective of resource management. Traditional IP routing protocols, such as OSPF, RIPv2 or IS-IS, cannot take into account the available physical link bandwidth and/or the statistical link loading level in the next-hop routing decision. As a result, traffic engineering may be needed to force the IP packets to perform load sharing. Multi-Protocol Label Switching (MPLS) is an emerging standard that addresses these issues to improve Traffic Engineering in router-based IP backbones.

Transmission network provisioning is the key to successful DiffServ deployment. DiffServ shares the available bandwidth efficiently between the different traffic flows, giving priority to the most urgent packets. The traffic profile and the physical transport network topology are the most significant parameters for dimensioning the packet flow. Inside the RAN, the traffic passing into the transport network is well known because of the limited capacity of cellular radio access. The packet-forwarding path between each base station and the RAN Gateway is also stable as long as there are no link or router failures.

Conceptually, dimensioning is similar to PVC set-up in ATM or TDM networks, but behaviour aggregates (BA) make the task easier. Capacity is not dimensioned for the packet flow from each individual host or application, but for a group of flows with similar requirements. SNMP and COPS protocols can set-up and change the provisioning remotely from the NMS.

MPLS and Traffic Engineering

MPLS is a protocol and set of traffic engineering techniques that can be used as a tool in building QoS support in networks. As the name implies, MPLS can be used to transport multiple protocols, not only IP. MPLS provides a way to define routing for flows or aggregate flows, while allowing for the use of lighter protocol stacks than IP transported over ATM/SDH/WDM, for example. MPLS with routing allows the automatic configuration of Label-Switched Paths (LSPs), avoiding the equivalent of setting up PVCs in ATM networks.

Technically, MPLS is a tunneling mechanism that implements LSPs on the basis of each packet's IP header, assigning a label to each item. At the edge of the MPLS network, an incoming flow is assigned a forwarding class based on the IP address prefix. The forwarding class is mapped into an appropriate LSP and an appropriate tag (label) is marked on the MPLS header. Inside the MPLS network, the MPLS switches direct the packets to the correct output based on label routing tables. For the output link, the label in the MPLS header is replaced with the correct one for the output link. Since labels are not global, LSP management is flexible.

The tunneling nature of MPLS means it can be used when normal IP routing is not available, in the case of header compression, for example.

The LSPs are defined by the consistent configuration of label routing tables in the MPLS switches. Several LSP setup methods are possible, and the building blocks for these are dedicated label distribution protocols and routing protocols such as Open Shortest Path First (OSPF).

Backbone in General

Nokia 2G and 3G GSNs support interworking with existing and evolving backbone traffic-engineering technologies, such as ATM and Multi-Protocol Label Switching (MPLS). Furthermore, the GGSN and call control elements support both IPv4 and IPv6 and can therefore fit into the multiple operator scenarios currently envisaged in the evolving backbone environment.

Separate white papers on IPv6 give detailed descriptions of the step-by-step evolution from IPv4 to IPv6 and the use of different interworking scenarios.

Within the Nokia GGSN, the QoS profiles of the PDP contexts are mapped to a corresponding DiffServ code point (DSCP). This then selects the desired per-hop-behaviour (PHB). Similar mapping is carried out in the IP BTS for the uplink traffic.

In native IP backbone environments, QoS between the GGSN and IP BTS network elements is provided by using DiffServ-capable routers between the two.

At the moment, ATM is widely deployed in operator backbones. Although most of the new services are based on IP technology, there are still many benefits to be found in using ATM in the backbone network. The key issues are the availability of reliable "hard" QoS schemes and the superior traffic management capabilities available when compared with native IP. When interworking with the ATM backbone, the use of dedicated PVCs for different classes of traffic provides a practical solution (for example, CBR is used for PVCs carrying real-time traffic). There is no direct interworking between IP and ATM.

The third potential backbone technology is MPLS. It is designed to combine the benefits of IP and ATM, and in many cases to act as glue between these two network technologies. MPLS provides both soft QoS (DiffServ-type) and hard QoS (using resource reservation) service capabilities. IP QoS mechanisms make it easy to implement QoS interworking with the mobility layer. DiffServ can be used for priority-based forwarding and MPLS for Traffic Engineering.

Traffic engineering is an important requirement for implementing manageable Quality of Service in the backbone. Standardised technologies and interoperability are further key requirements for QoS support in the backbone. QoS is managed with DiffServ-compliant interworking technologies.

Endpoint Quality of Service Aspects

In UMTS 3G, the terminal is responsible for opening a PDP context with proper QoS attributes. To this end, interfaces for application QoS control are supplied in terminals. Furthermore, the QoS requirements of different traffic types are supported with a combination of support from intelligent scheduling and proper assignment of traffic types to Radio Access Bearers (RABs).

For applications where delays are not critical, reliable protocols can be used for transport, based on retransmissions. On the other hand, applications such as voice telephony have strict delay limits, because protocol data unit (PDU) losses cannot always be hidden from applications.

QoS over radio interfaces in challenging situations is an example of just such a delay-critical circumstance and can be further improved by using adaptive coding methods. For example, the R99 standard Adaptive Multi-Rate (AMR) coder/decoder can adjust its output bit rate according to channel properties. Nokia is currently active in 3GPP standardisation for providing higher audio quality with the wideband AMR codec.

Quality of Service Management

Concept of Policy-Based Quality of Service Administration

The need for rule and policy-based network management arises from the implementation of services on the network level, as well as from the need to use network resources efficiently. The automated configuration of rules and set-ups in network elements make the task of configuration easier and allow for a higher level of flexibility in configuring the overall network. The need for easier configuration increases with evolving RAN architectures. Rule and policy-based management allows consistent QoS to be configured for multiple service types and user classes. The concept of rules means that the network can react to varying situations affecting QoS. Consistency here means that all the components in the system – for example radio bearers, All-IP RAN transport parameters, and the mobility core – are configured for certain QoS services by the policy-based management system.

For example, policy-based management is used to map UMTS traffic classes into DiffServ per-hop behaviours (PHBs) in gateway-type UMTS elements (SGSN, GGSN). Furthermore, the PHB mapping is complemented by the configuration of DiffServ routers so that adequate macroscopic Per-Domain Behaviours (PDBs), such as specific delay and packet loss requirements, are implemented. IETF is working on specifying a Policy Information Base (PIB) for DiffServ. This would allow policy-based management of the different functions of DiffServ routers, such as classification, policing, dropping, queuing and scheduling mechanisms and so forth. The policy-based management system for a UMTS network configures both DiffServ and UMTS parameters together, resulting in consistency between services.

Rules make the user plane transport network more responsive to network conditions. This feedback can be obtained by using information from the network management data processing and correlation system, as well as from more intelligent entities such as bandwidth brokers. Greater flexibility and a wider range of measures can be obtained by formulating new rules "on demand", based on feedback from the network.

Finally, feedback can be used by people to introduce large-scale measures relating to QoS.

Evolution Towards Policy-Based Management

Rule and policy-based management simplifies and automates the process of configuring bearers for different types of traffic in a mobile network. For R99 and R4 releases of UMTS networks, QoS policy configuration can be achieved using the Nokia NetAct management framework.

In R5, 3GPP plans to specify a policy control function between GGSN and P-CSCF. This will coordinate call setup signaling and the allocation of bearers.

Service Quality Management

3G services will bring new challenges for operators in terms of managing their networks effectively. QoS management at the UMTS QoS bearer provides service class-based management from the access network GGSN or external bearer.

Service quality management supports service level processes by providing the operator with an up-to-date view of service quality, based on QoS KPIs collected from the network.

One class of source for QoS KPIs are service-specific agents that can be used for monitoring performance indicators for WAP services, for example.

Using active measurement to verify the service quality implies testing the actual communication service, rather than simply collecting data from network elements.

Network probes (or stethoscopes) can be distributed around the network to perform regular testing of the services. These perform simulated transaction requests at scheduled intervals. The stethoscopes can be installed on some server platform at the edge of the network. By installing the stethoscopes at the edge of the IP network, the compound effects of network, server, and application delays on the service can be measured, providing an end-user perception of the QoS.

Summary

Nokia supports QoS in packet-based mobile networks throughout their transition from solutions for GPRS and IPv4 up to IPv6-based All-IP networks. The QoS support provides end user service differentiation while addressing the efficient use of resources.

QoS is based on the consistent management of radio and packet-based resources. To this end, the management systems evolve towards policy-based management and make increasing use of feedback from the network.

In Nokia's view, IPv6 provides a future-proof unifying platform for the evolution of multiple access networks. This includes support for QoS and security.

Virtual Private Networking

The business world today is increasingly dependent on communications. Companies requires and provides more real time information in order to stay competitive. This has impacted information security programs as connections to remote employees, business partners, vendors and others have new avenues for business communications. These communications provide immediate access to a wide array of systems and business information. New communications tools are required to help protect information as it moves beyond traditional boundaries.

There has been much discussion in the press and in industry about Virtual Private Network (VPN) products and services. Like many other new technologies, the VPN market has been suffering through the standards development phases to establish the guidelines for interoperability. Nokia has established the following criteria for today's VPN products.

VPN Requirements

VPN requirements are

- Scalability
- Performance
- Reliability
- Usability
- Ease of management
- Interoperability
- Protocol support
- Seamless Integration
- Authentication
- Accounting, auditing, and logging

VPN Defined

There are many VPN products and services on the market today and a wide variety of implementations. This has created some confusion about what the term VPN actually means. This paper will attempt to define and provide clarity on VPN solutions and what their uses are in business today. To properly define a VPN, it will be necessary to understand a few terms and definitions. The following terms and definitions should help provide some understanding of what a VPN really is.

- "Virtual" - being something in essence or effect though not formally recognized or admitted.
- "Private" - intended for or restricted to the use of a particular person, group, or class.
- "Network" - a system of computers, terminals, and databases connected by communications lines.

- "Public network" - a network generally accessible by many other networks where the connections to and from the network are generally not known and the entities on the network are free to join.
- "External network" - a network that is not publicly accessible but may have connections to other networks that are authorized.
- "Internal/private network" - a network that is primarily used by some entity and is protected from the public due to physical barriers.

Definition of a Virtual Private Network (VPN)

A network utilizing encryption technology to privatize data for transmission to a trusted party.

1. It is virtual because it can use a public or a private network to transmit data
2. It is private because it uses encryption to protect the data
3. It is a network because devices and systems communicate on a common path

Functions of a VPN

The core function a VPN provides is privatizing data between two points. The primary function is to protect the data during transmission over a given network path. This privatizing of data allows the use of public networks for transmission as though they were private (virtually private). The privatizing of data through the use of a VPN can also be beneficial where information and services must be protected from disclosure, modification, unauthorized use and interruption while traversing network segments.

Functions Not Performed by a VPN

A VPN does not protect or privatize the data while it is at point A or once it arrives at point B. It can also leave information exposed while unencrypted over a given network path (internal networks before the encryption device or external networks after the decryption device). This point is made because a VPN only addresses one aspect of information protection but doesn't address other issues with information that is on a disk, displayed on a screen, or printed to the local printer. Businesses must understand that there is no solution that does everything and that information security is a continuous process involving many policies, procedures, and controls. A VPN is a worthwhile component if deployed correctly within the enterprise. Security is only as good as the weakest link. Every business must assess, design, implement and maintain their security program and understand the risks to information assets.

VPN Business Uses

Common uses of VPN technology include:

- Connecting business units at different physical locations via a public network
- Connecting departments who transfer sensitive data over a shared corporate backbone
- Connecting to business partners to enhance the purchasing and sales cycles
- Providing cost effective remote access for mobile employees via a public network
- Connecting confidential systems to allow secured communications
- Organizations can utilize VPN technology throughout the enterprise. When VPN technology is

deployed correctly in the organization it can enable business processes that would otherwise be cost prohibitive. VPN technology can rightfully be viewed as an enabling technology that enhances the ability of the organization to achieve its goals and objectives by increasing opportunities and revenue.

- Security means more than keeping bad things from happening to you.
- Nokia's partners' security experts can turn security technologies, policies, and procedures into business enablers that will make it possible for your enterprise to maximize competitive advantage, achieve increased operating efficiency, and enhance revenues.

Return on Investment - Security as a Business Enabler

- Cut costs for profitability
  - Especially during economic slow down
  - Maintain business operations (and cut costs)
- Communication costs are large
  - What technology is used today?
  - How much does it cost?
  - What is the return?
- Communication technology benefits
  - Share information...
    - ...with whomever you want
    - ...whenever you want
    - ...securely
    - ...how you want to conduct your business

Remote Access VPNs can aid in maximizing current and future investment in mobile technologies and work forces :

- Replaces international/long-distance calls with local calls
- Charges vary by region
- Not based on office locations, but the location of remote users
- Other cost elements
- Monthly ISP charges for remote users
- VPN client deployment
- Dial-in server maintenance vs. VPN management
- Compare costs between a Remote Access VPN and the current total monthly recurring dial-up access costs

Site to Site VPN's can provide very significant reductions in cost whilst improving a business's flexibility :

- Compare Internet access and leased line costs
- Line costs add up quickly
- Payback periods vary
- Scalability ~ purchase additional leased lines
- Full mesh at no extra cost
- Compare costs for a traditional private WAN costs? For full mesh?

What the Industry Says

SalomonSmithBarney, July 14, 2000

- VPN saves approx. 60% vs. traditional secure networks
- The ability to connect from almost anywhere in the world

Forrester Research

Per 1000 users:	**Traditional RAS Costs**	**VPN Costs**
	$1.50M	$.59M• *Communications*

Industry Researchers, 1999

Typical cost reductions

- 50% to 80% - mobile user/telecommuter access
- 40% to 50% - long-distance leased line/frame relay replacement
- even higher - international connectivity
- Most VPN projects have very short payback periods (some are as low as several months)

The Management Challenge

There are three kinds of entities to every potential VPN deployment :

- Sites
- Partners
- People

How should these be managed without dramatic increases in costs and required time? What issues should be considered in the business case?

Ease of Control and Deployment

- Deploys Policy-based Tools
- Reflects business needs vs. network devices
- Emphasizes relationships and information flows
- Easily adapts to a growing environment
- Simplifies complex fixed/dial-based deployments
- Manages multi-node clusters as one logical entity
- Automates administration with a powerful scheduler
- Streamlines remote configuration
- Enables staged upgrades

Integration with an Existing Infrastructure

- Interoperates with IPSec devices
- Provides high availability for a mission critical site
- Ex. hub in hub & spoke

- Manages complex topology
- Sets the architecture template
- Adds new hosts
- Defines new hosts and protected resources
- Creates a policy update
- Distributes a policy update in one step
- Creates diverse VPN deployment
- All identities are known
- Aware of protected resources
- Advised of required configuration criteria

Protection of Existing Investments

- Supports SecurID and RADIUS expenditure
- Simple configuration from the VPN Policy Manager
- Does not weaken security controls
- Provides a clear migration path to future and merging technologies
- Includes an internal certificate authority
- Implements key PKI components
- Simplifies deployment and management using digital certificates
- Integrates with external PKI technologies
- Builds upon an existing infrastructure

Accommodation of Future Growth

- Corporate growth impacts the VPN architecture
- New relationships need to be created
- Requires flexibility
- VPN Policy Management provides a strong policy based VPN model
- Focuses on quick and accurate deployment
- Easily reflects various new security associations
- Adheres to industry standards

Examples of VPN Deployment

A Global Enterprise

- Current locations

  Headquarters: New York , USA

  Branch offices: Vancouver, Canada; Texas, USA; London, England; Singapore

  Satellite sales offices: California, USA; Brazil; Australia

- Deployed hub and spoke topology
- New York and London offices are hubs
- Additional offices connect through hub to access spokes
- Leased lines currently link each site

Phased VPN to Connect Remote Sites

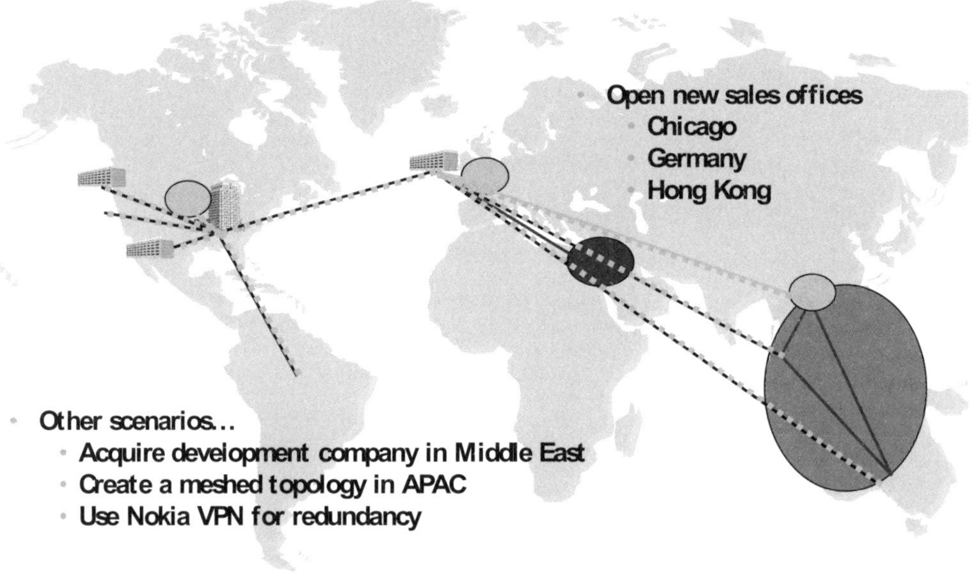

Business-to-Business VPN Application

Financial partners facilitate an electronic bill-pay system

- Links billers, consumers & banks
- Central data center processes most transactions
- Problem:

  Provide secure, open & interoperable infrastructure for electronic bill pay system

- Verifies available funds as customers initiate payments

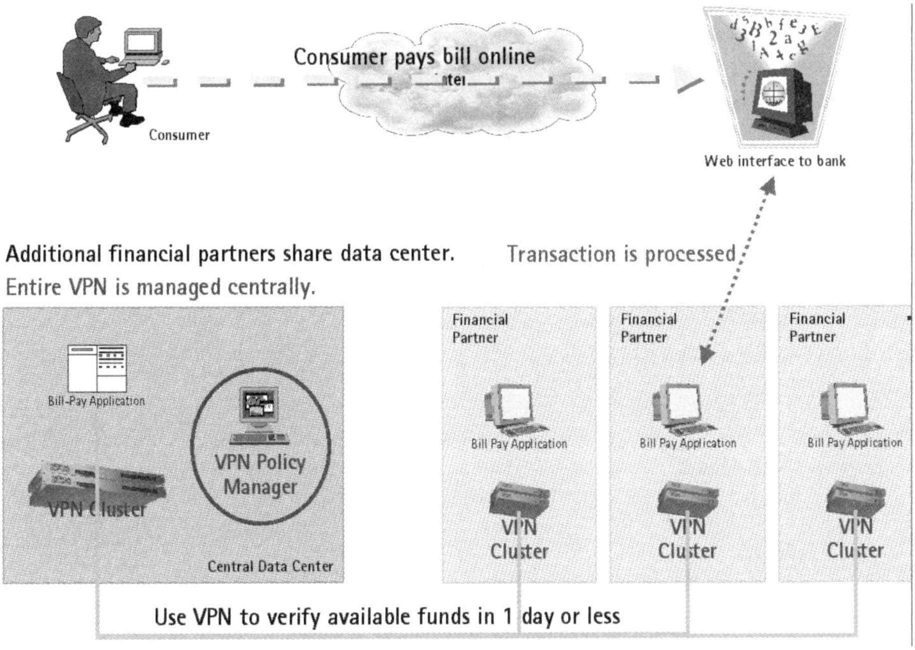

VPN Application Benefits

- Increases customer satisfaction
- Simplifies management
- Saves money
- Protects revenue

Requirements of a Well Designed VPN

Scalability

Scalability is more than just throughput. Throughput is how much data a given box can handle at any one time. The problems is how do we grow from there. Once we reach that limit what do we do to scale. Scalability must include ways to stack boxes together to work in tandem, similar to the concept of stackable hubs. If we need more connections we just add another hub to the stack. Of course there is always a limit to how many things we can stack, but the point is that we are not limited by the performance of a single unit. Scalability allows a solution to grow with the business and eliminates fork lift upgrades and usually provides load balancing and redundancy.

Performance

Performance is the raw throughput of a device. The VPN must be able to handle a large amount of traffic if it is going to work in the enterprise. Depending on the given scenario, a VPN should be able to process close to the input line speed that it connects to or to the line speed of the slowest link. If you are connecting a VPN device to a 100 Mbps Ethernet segment but the incoming traffic is routed over a T1 then you would only need a 1.544 Mbps throughput. If you were using the VPN to encrypt traffic across an internal ATM backbone then you would want the encrypted throughput to be close to 100 Mbps.

Reliability

For a VPN to actually make it in the business world it must be reliable. Reliability means that it should be available at all times, like a telephone. Reliability must include redundancy to allow automatic recovery with limited interruptions of service. As VPN solutions are adopted in the business world and begin to play a key role in the organization, reliability will be a necessary factor. Organizations are looking to use VPN technology as an enabling strategy to provide access to enterprise resources to increase opportunities and revenues while lowering the costs of providing connectivity.

Usability

The VPN needs to be very easy to use and understand. For a VPN solution to be successful the end users of the VPN should be able to use their services without realizing they are doing so. The VPN should be transparent to the user when tunnels are established and torn down. There shouldn't be any unnecessary requirements for the users to perform allowing for secured access to resources required to get their job done. Increased security and efficiency allow the workforce to meet the organizations goals and objectives. A good policy distribution system should allow the VPN to determine when to encrypt and when to send clear text information. The only thing the user needs to notice is the authentication to provide access to the system.

Ease of Management

For a VPN product to be deployed by any medium or large company, it must have a well-implemented management solution. In security products a solid management solution is key to ease the burden of administration, management and reporting. The management platform must have a simple way to design a security policy, an easy way to distribute that policy and an easy way to simultaneously manage a large number of devices. It should include the capability for separation of duties, distributed management consoles and fault tolerance.

Interoperability

To make complete use of a VPN product it should be interoperable with other VPN products. For a complete enterprise solution the VPN will need to allow you to communicate with business partners and others who may have deployed a different solution. To communicate securely you should ensure that your selection of VPN equipment is interoperable according to industry standards and protocols. This factor will enhance the ability of the organization to enable new forms of electronic commerce in the future and continue to capitalize on the cost advantages of a VPN as well as protect investment.

Protocol Support

If a VPN is a true enterprise solution it will provide a wide range of networking protocols. These protocols ensure that you can communicate with other business partners and various forms of dial devices that may be necessary in the organization. The following protocols should be supported:

- IPSEC
- IKE
- PPTP
- L2TP
- RADIUS

Seamless Integration

For a VPN solution to fit into an organization it must integrate into the network and other systems as a complementary service. A VPN merely privatizes data to protect information as it travels between two points. The VPN device should integrate into the existing infrastructure and provide value. This additional value is only recognized fully if the solution will work with the existing remote access servers, internet connections, certificate authorities and authentication mechanisms. A solution loses value if an administrator must maintain a completely separate authentication database to provide access.

Authentication

As we start distributing data through the use of VPN technology we must realize the need for authentication. This authentication will be required at several levels. First the VPN software must authenticate with other VPN devices that it is communicating with. This requires that device A and device B have properly exchanged information (keys, certificates, etc.) before they begin the encryption /decryption or authentication/ verification of communications. Second it may be necessary (as in the case of remote laptop users) to follow up with user authentication to ensure the person in possession of the device is who they say they are. This can be in the form of a username/password, tokens, certificates, smart cards or a variety of other user access controls. It provides an extra measure of protection to mitigate the risks of lost or stolen devices. To help with centralizing user administration, the VPN system should support third party authentication mechanisms.

Accounting, Auditing, and Logging

For any security solution to be complete, it must be able to log, account, and create audit trails. The system should log events to help administrators identify problem areas and understand when key events have taken place. Security management will want logging enabled to review breaches in the security policy, potential intrusions and other security related events. Others in the organization such as internal auditing and accounting may wish to view data on users, network usage, and other activities. A good logging function will provide the necessary information with the ability to notify selected individuals when is encountered a given event.

Understanding the Requirements of Implementing a VPN

Organizational Security Policy

One of the underlying requirements of implementing a VPN are the overall enterprise security policies and procedures. A VPN is merely a solution to solve a specific problem. A security policy should be identifying your need for a VPN based on the formalized risk assessment process.

Understanding Business Requirements

A VPN should be viewed as an enabling device to allow business communications to be achieved over public infrastructures or by other insecure means. A VPN solution will help businesses provide solutions for remote employees, business partners, and other electronic commerce initiatives.

Understanding Protected Property

Before deploying any solution businesses need to understand the problems and ensure that the solution they are deploying actually addresses those problems. A VPN only protects the information in transit from point A to point B and additional measures may need to be taken to safeguard the information once it reaches point B.

Understanding Legal Issues

With any encryption solution there are legal considerations. Encryption import and export laws must be understood when deploying international solutions and this must be taken into account during the design phase of the solution.

Understanding Integration Issues

Infrastructure requirements are a big consideration when looking at deploying a VPN solution. Organizations need to understand VPN technologies as well as other technologies to properly integrate the solution into the network. A solid understanding of information security, system security, routing, and other issues is required to deploy a successful enterprise wide VPN solution.

Conclusion

Organizations are becoming more aware of the opportunities available to meet strategic goals and objectives by leveraging security and technology to deliver services to employees, clients and business partners. The use of VPN technology has matured to a level where it will provide organizations with a solid solution to enable new avenues to drive revenue and reduce operating costs.

Mobile Terminals of the Future

Matching Technology to Consumer Demand

The big challenge and opportunity in developing mobile terminals for the mobile world is to match the seemingly limitless opportunities offered by the technological advancement to the speed of consumer acceptance and consequently to actual market demand.

In practice this means that we need to solve the problem of fulfilling the demand for small and inexpensive mobile terminals, which also provide the full range of features and services we believe the consumer will want to buy. Although these two aspects seem to be mutually exclusive, there are technologies that enable us to develop the terminals a consumer wants. Such technologies include the selection of displays, input methods and the supporting terminal service platform. To succeed, terminal vendors need to have a good understanding of customers and market segments.

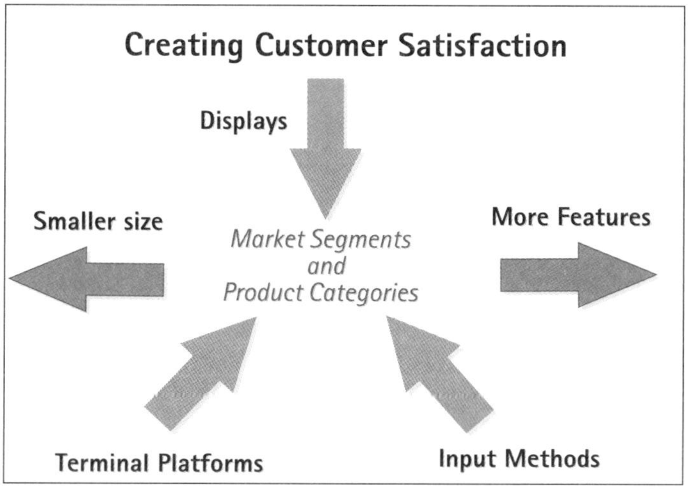

In order to create the winning value proposition it is necessary to focus on the most attractive application driven product categories, which optimize a specific application while providing a full set of mobile applications to ensure the usability of the terminal as the preferred personal companion.

To give an example, let us take two products optimized for different applications – an imaging phone and a Communicator. The key application to be optimized in an imaging phone is imaging. This includes capturing, handling, sending and storing images. The key application for this product category is multimedia messaging (MMS), which will utilize a large colour display and the terminal will facilitate taking pictures with an integral or plug-in digital camera. The terminal will be coupled with a service where images can be sent over the air and stored on a server.

In addition to the imaging application, the imaging phone will support all the necessary mobile applications (i.e. voice/SMS, PIM, games, etc.) that the already consumers are used to. These are necessary to secure the utility of the terminal for its owner as a preferred personal device.

The Communicator on the other hand is a product targeted for business use and provides the ultimate in efficiency for the mobile professional. Its key applications are those providing seamless integration into corporate IT solutions, as well as, a capability to support a wide range of third party business applications. Input devices can be for example a QWERTY keyboard or a stylus.

The Communicator also supports imaging as described above, for the imaging phone, but the technologies chosen for the Communicator product clearly optimize the applications for corporate use.

Path to the Future Terminals

The evolution of terminal categories must be coupled with a co-ordinated development of related services which together comprise the total offering for the consumer. The final purchasing decision is made by weighing the prioritized use of the terminal (i.e. sharing the moment with a friend with imaging or increasing professional efficiency) and design against the price of the terminal and its closest alternatives.

The three main steps for terminal evolution are:

1. From a cellular phone to a functionally and operationally optimized personal terminal – functional satisfaction supported by user delight and supported by industrial design
2. From "fixed functionality" to a "platform of continuous excitement" and emotional attachment – end-users can vary and upgrade the applications and synchronized their personal information with other databases.
3. From a specific communication device to a "central hub of personal communications, interactive services and life management" – all functionality can be executed via the terminal or with accessories.

Winning Terminal Categories

It is clear that the "everything for everybody" approach will not receive industry acceptance, as all stakeholders in the industry need structure and simplicity when deciding on the direction of their choices. As far as mobile terminals are concerned, it is important to decide on the applications and terminal categories which will have the broadest appeal for consumers and industry players in order to create a healthy business environment for industry growth.

At this stage of industry development, there are five winning terminal categories, which fill the requirements described above.

The Winning Terminal Categories

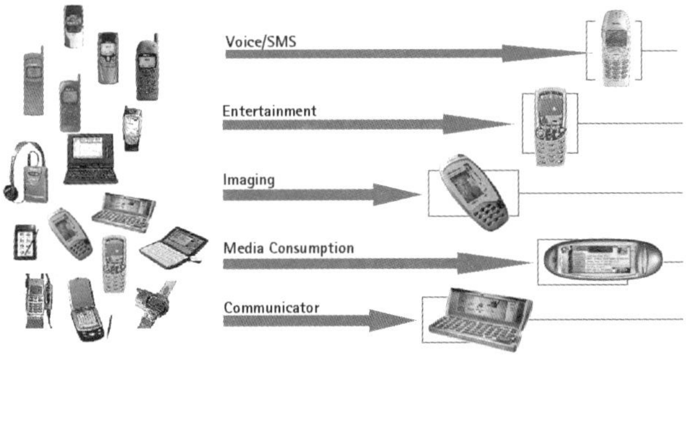

Voice/SMS Product Category

The Voice/SMS centric product category has set the benchmark for mobile personal communications by supporting the individual lifestyle of end users through multiform design, as well as, wide personalisation options. In this category, terminals emphasize voice centric usage and messaging applications, such as SMS, for personal use.

Key features:

- Excellent, personalized voice product
- Good messaging capabilities and efficient data solutions.
- Meeting the needs of individual lifestyles through design and personal relevance

Entertainment Product Category

The Entertainment category offers new experiences that extend the traditional scope of entertainment in the mobile context. Mobile entertainment terminals are primarily devices, which combine mobile connectivity with the most desired entertainment applications.

Key features:

- Messaging, music and games
- Personalisation
- New ways of social interaction
- Cost-efficiency
- Timeliness

Imaging Product Category

The Imaging category includes terminals with a strong emotional attachment, devices for sharing experiences through rich communication. Focusing on person-to-person(s) multimedia communications, with self-created content, the category represents a natural evolution of today's communications paradigm and takes us one step closer to "virtual presence".

Key features:

- High quality phone with excellent messaging and imaging capabilities
- Exciting and easy-to-use interface for instant sharing of emotions and experiences
- Compact, transformable concept, attractive design

Media Consumption Product Category

The Media Consumption category offers the preferred wireless platform for personalized access to Internet services, with PIM (personal information management) functionality and easy-to-use, integrated phone features. It provides anywhere, anytime access to services and seamless integration of both audio and data functionality. The category expands the terminal platform and mass appeal by supporting third-party applications and software development. In short, it provides full-scale wireless deployment of the Internet.

Key features:

- New category with global scope
- Ultimate Mobile Internet Companion
- Migration to comprehensive support for existing web content
- Distribution channel for third-party digital services and content

Communicator Product Category

The Communicator is a wireless business terminal with seamless integration into corporate IT systems andto content and applications provided by ISPs and portal operators. It includes e-mail, a networking office application suite and access to corporate applications, with full input and output capabilities. It is the most powerful platform with an optimum size/ergonomics ratio and facilitates real-time information consumption and decision-making, shortens customer response times and improves logistics, with optimum use of time, while on the move.

Key features:

- Complete business communication: voice, messaging, image communication, fax
- Seamless integration to corporate IT, telecom and Internet, with transparent mixing between terminal- and server-based applications and data (WWW, thin clients)
- Powerful tools for viewing, editing and creating business-related data
- Excellent development platform for third-party applications and services, in both terminal and server platforms

Behind the Corner

The product categories presented will clearly be the ones to provide the breakthrough for new communications methods in the mobile world. Their strength lies in their reliance on the fundamental communications methods in human history – talking, writing and messaging. Since the categories are application driven, it is very possible that they will change in nature with time or that they will be replaced by other product categories.

The introduction of stable new technologies and changes in the consumer behavior will be the driving forces behind these changes. They will also determine the speed of the change.

Nokia Mobile Software Product and Service Offering

Introduction and Summary of Key Service & Technology Drivers

1. Making the mobile services tornado happen begins with a question: What do consumers need and want, really? In our view and experience, consumers want three things: sheer fun, sharing of experiences/emotions/ information, and practical utility, in a personal(ised) fashion - and fast!

2. The key applications for service delivery are messaging and e-mail, games and personalisation, and web content download and consumption.

3. The key technologies behind the applications and services are MMS (Multimedia Messaging Service) for messaging and e-mail, Java for downloading applications, and XHTML for content search and access.

4. Intelligent middleware is needed to ensure that the Internet-based content is tailored to end user preferences and to terminal capabilities. We offer functionality that is categorised as Common Service Enablers. In this category, we offer Portal Services, Content Services and Commerce Services. The Intelligent Mobile Middleware consists of functionality that is needed for the service provider to be able to run this service delivery vehicle as cost efficiently as possible. The "Operator Service and Subscriber Management" category includes all the necessary functionality for orchestrating the Mobile Services business. It is elementary that we have the necessary interfaces in our offering for the operator's support systems, e.g. CRM, OSS, customer care and billing

5. The common transport layer will evolve from the current WAP/WSP to TCP/IP-based WAP NG.

6. When designing and delivering the mobile terminal services environment we are committed to:

- Support next-generation WAP/XHTML mobile services
- Support legacy WML content
- Future-proof software and service roadmaps
- Maintain consistent user experience across handsets
- Maintain flexibility to meet market demands
- Solutions are developed for fast time-to-market (mass-market)
- Everything is billable - faster time-to-profit
- Based on open standards, API's, and protocols for
  - Cost effective new services development and deployment
  - Easy third party product integration
  - Early influence specifications
  - Competition between suppliers

7. An example of a recent action taken:
"There are tremendous benefits yet to be realized from the wireless Internet," said Anssi Vanjoki, Executive Vice President, Nokia Mobile Phones. "Nokia is convinced that GSMA is accelerating the process in a way that is beneficial to the entire industry. To make that vision a reality, mobile phone and service platform manufacturers like Nokia must play a central role in creating the software and the services that will attract users and increase revenue. We are committed to building those solutions ourselves and in conjunction with other mobile phone companies. In addition, Nokia is working with other manufacturers to ensure the interoperability between mobile devices and service platforms."

Nokia has taken the first concrete steps toward the creation of this new software platform by also announcing the immediate availability of the beta version of the Nokia Mobile Browser 3.0. This browser natively supports both WML (Wireless Markup Language) 1.2.1, including push and the telephony features of WTAI (Wireless Telephony Application Interface) as well as XHTML (Extensible HyperText Markup Language) with Cascading Style Sheets. The browser is available in source code form for licensing by other mobile phone manufacturers interested in adding the latest functionality to their product lines.

The Nokia Mobile Browser 3.0 has also been designed to work seamlessly with Nokia Smart Messaging, which is already the world's most popular mechanism for downloading and interacting with content such as ring tones on a mobile phone. In an effort to spread the success of Smart Messaging, Nokia has also published the complete technical specification so that other companies can adopt this proven technology.

Nokia expects that new areas of collaborative development could include advanced browser implementations, standard multimedia messaging clients, and tighter integration between the browser applications such as content downloading and messaging.

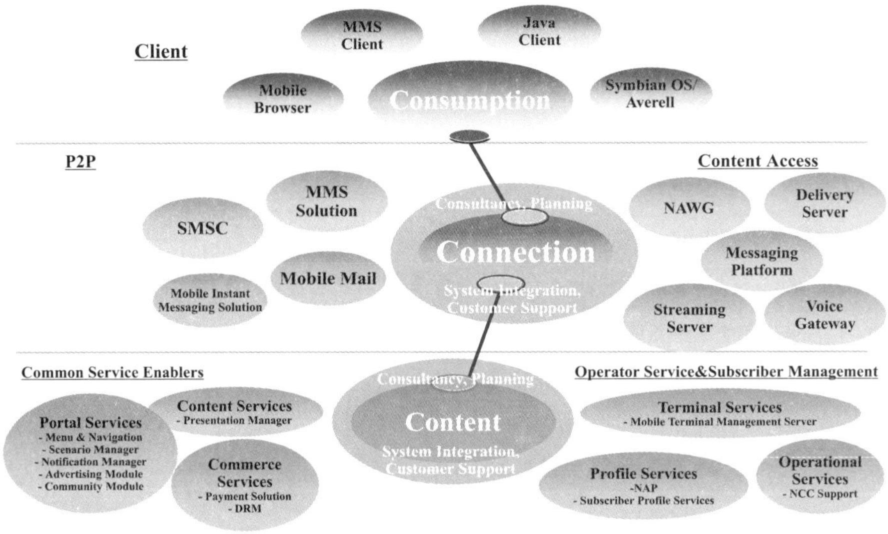

Nokia Mobile Software end-to-end offering, including mPlatform elements

Consumption - Client

Open Application Development Environment, including

- Consistency and openly shared interfaces and development tools to stimulate the innovation in service development
- Core development environment: Java, MMS, XHTML, Symbian OS

Open terminal-network and client-server interfaces: Open standards based connectivity middleware enables interoperability between terminal and infrastructure manufacturers (MMS, SMS, SyncML etc.)

Open service platform architecture: Modular intelligent service delivery platform with open interfaces and multivendor interoperability support

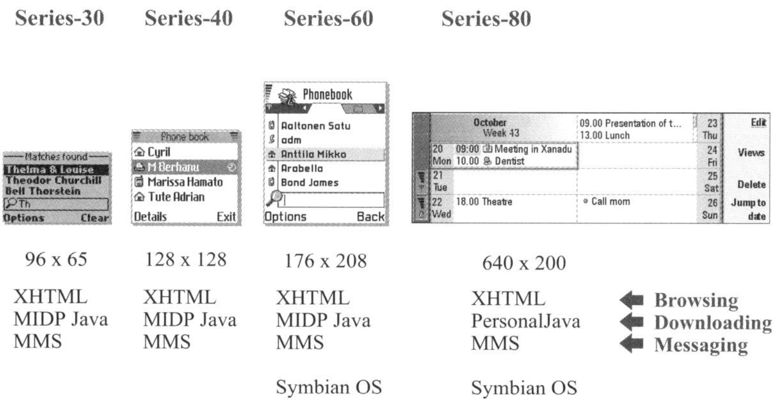

Nokia Mobile Client environment for mass-market services

Connection

Person-to-person

Nokia Artuse Short Message Service Center (SMSC)

Nokia Artuse Short Message Service Center (SMSC) is a scalable, high capacity solution for person-to-person short messages and value added services.

The point-to-point Short Message Service (SMS) provides means for sending messages to and from GSM mobile stations (MS). The provision utilizes a Short Message Service Center (SMSC), which acts as a store and forward center for short messages. Thus the GSM network must support the transfer of short messages between the SMSCs and mobile phones.

The Nokia Artuse SMS Center gives a network the ability to deliver short messages of up to 160 characters, to and from mobile phones. The Nokia Artuse SMS Center provides capacity, scalability, load sharing and efficient routing and also meets the demand for value added services. Using a commercial UNIX-based computer platform, Nokia Artuse SMS Center provides open interfaces to existing data systems, letting you provide a wide range of new, innovative services to your GSM mobile users.

Nokia Artuse Multimedia Messaging Service Center (MMSC)

Nokia Artuse Multimedia Messaging Service Center (MMS Center) is a service platform targeted at GPRS/3G operators who are looking into offering new interactive creative multimedia messaging services. With MMS, it is possible to combine text with richer content types such as photographs, images, audio and voice clips, and eventually also video clips, and send these combinations in the same simple manner as SMS today. Operators will be able to attract customers with new value-added services, and to differentiate and enhance their competitiveness by offering Multimedia Messaging.

Nokia provides complete MMS solutions based on its terminal and infrastructure expertise. The key element in the MMS network architecture is the MMS Center. The MMS Center enables multimedia messages to be sent with various content types from terminal to terminal and to and from applications, such as E-mail, with instant delivery. The MMS solution also provides permanent message storage with multiple access methods and offers services specially designed for legacy phone users to enable multimedia messaging between multimedia and non-multimedia terminals from day one.

The MMS Center will be able to handle several message types from pure text to video clips or any given combination of media types. The support for different media types and formats is being specified in standardization bodies and the aim is to enable interoperability especially between different devices from different vendors.

The very nature of the MMS Center is similar to the SMS Center - it is a store and forward platform. Upon receiving a multimedia message, the system will send an MMS notification to the phone of each arriving message. The result from an end-user point of view is dependent on the phone implementation - the subscriber might be prompted to receive the message (downloading takes place when the subscriber wants within a limited time period) or the whole process can happen automatically, as in SMS. After successful message delivery, the message will be deleted from the database. If the message cannot be delivered successfully, it is stored for later delivery.

The usage of bearer independent WAP will enable smooth transition from GPRS, where the service will be first introduced, to third generation networks (e.g. WCDMA).

The MMS Center has an External Application Interface (EAIF) for application development allowing service providers to introduce services. Using this interface, the Nokia Multimedia Messaging Solution boasts an enhanced rich set of MM services. These incorporate a Personal and Public Multimedia Message Album, Legacy Phone support, Call Completion Services, Multimedia Creation and email delivery.

The standardization work is a continuing process at both 3GPP and the WAP-Forum. Nokia's MMS solution is compliant with existing MMS standards . Nokia is also closely following, and contributing to, the standardization. Through the active participation, Nokia aims to maintain compliancy with the standards to come.

Nokia Mobile Instant Messaging Solution

The mobile instant messaging services will allow end users to update their own presence status from time-to-time and also find out the status of their friends, wherever and whenever. Users can communicate according to their friend's preferences indicated in their status information.

Mobile operators are very keen to increase customer loyalty and develop new revenue streams from exciting and innovative services.

The instant messaging solution from Nokia will allow the mobile operators to

- Offer a branded instant messaging service
- Offer subscribers a mobile instant messaging service with SMS, WAP and WEB interfaces
- Own the presence information of their subscribers
- Introduce exciting services based upon instant messaging, presence and location
- Retain 100% of earned revenue

- Position various services and avoid service cannibalisation
- Provide connectivity to fixed internet instant messaging services (AOL, ICQ, MSN, Yahoo)

The terms 'instant messaging' and 'chat' are sometimes used synonymously but they are very different services targeted for different usage situations. Chat services are intended for occasionally meeting new people with certain interests and sharing viewpoints on some themes, whereas instant messaging services are intended for keeping regular touch with known set of friends and other social contacts.

Instant messaging services are integrated with presence services that are a means of finding, retrieving, and subscribing to changes in the presence information of other users.

Presence and Instant messaging services are separate and can work independently of each other. However, by using the Presence Service, a user has a better idea as to whether a recipient is listening for instant messages. Therefore, the two services are often used in tandem.

Users can send instant messages via SMS, WAP and WEB over their mobile phones and PCs.

Contact lists for person to person and group messaging with available friends: Users can compose several contact lists and add their family members, business associates, colleagues and friends to these lists. Users can then exchange one to one or group messages with their available contacts. Alternatively, an email can be sent to users who are unavailable (or offline) for messaging. A contact list is also referred to as contact list or group list in this document. Online is referred to as available or on in this document.

Set own and view other's presence: Users can see the availability of their contacts as soon as they log in. Based upon the availability information, they can decide to call them, send a text message or send an email. Presence information is also referred to as status or availability information in this document.

Notification services: Users can find out when their friends are online on the Internet or available for messaging via mobile phones. As soon as a user logs into web instant messaging service, friends who have added this user in their contact list get a notification about the current status of the user. Mobile operators can use this service to allow users to view their online PC friends.

Cross-IM with other instant messaging services: Users can add contacts from Internet instant messaging services like Yahoo, AOL, ICQ and MSN to their contact list. Using this service, users can retrieve presence information of their Internet friends and exchange messages with them.

Content Access

NAMP

The Nokia Artuse™ Messaging Platform enables the mobile phone user to easily access useful information and services with short messages. The Messaging Platform allows an operator to create value-added messaging applications for all mobile users quickly and easily.

The Nokia Artuse Messaging Platform is a comprehensive and robust messaging platform that will enable operators to offer messaging value-added services. Nokia Artuse Messaging Platform links the Internet or company Intranets and the mobile network. As a result, information such as headline news, exchange rates, sports results, or stock quotes can be accessed in seconds. The Messaging Platform also makes it possible to download ringing tones, graphical logos and icons. Services can be accessed by any mobile phone with short messaging or Smart Messaging capability.

NAWG (WAP Gateway)

NAWG is a core element of the Nokia's WAP solution. The current release (release NAWG 3.0; GSM/TDMA/ CDMA) is compliant to WAP June 2000 Conformance Release specifications, and functions as a WAP gateway for mobile networks, including support for the WAP stack, WML content fetch from the Internet using the HTTP protocol, WAP content encoding/decoding, cookie support and Push Proxy over SMS functionality. The supported bearers are UDP/IP based bearers (CSD, HSCSD, and GPRS), SMS, and USSD.

To ensure data confidentiality, it supports Wireless Transport Layer Security (WTLS) and Secure Sockets Layer (SSL) security protocol layers. The billing interface, network management system interface, load balancing and high availability support, and subscriber identification are also features of the NAWG.

Nokia mPlatform Delivery Server

The Delivery Server can be used to extend the functionality of a mobile portal or Service with a reliable and well-controlled content download Service. Typically this could mean, e.g. a Java MIDlet game download as part of a operator's mobile portal. However, in addition to Java based games, other content types will be appearing in commercial download Services provided by operators.
Nokia mPlatform Delivery Server:

- Offers control for content download and charging
- Offers flexible download pricing and charging model options
- Selects content based on the terminal type
- Supports various content types
- Allows the evolution of business models
- Is based on open technologies

Mobile content downloading starts with content discovery. Multiple methods to discover mobile content are supported, such as mobile browsing and SMS initiated. After discovery, the actual download transaction takes place including, e.g. access control, content delivery and billing. The end user requires an URL for the download. This URL always points to a resource, which is available only through the Delivery Server, regardless of its actual location. Content storage is used to offer storing capacity and the mPlatform Delivery Server is linked with a charging system and offers multiple methods for billing.

Nokia mPlatform Streaming Server

The Nokia mPlatform Streaming Solution is a part of Nokia mPlatform Solution, which enhances existing mobile services by adding video and audio content. Streaming allows content to be viewed or listened to before it has been completely downloaded from the content provider's server. Nokia mPlatform Streaming consists of state-of-the-art streaming video and audio servers, as well as media caching servers. Applications can utilise the streaming capability to include live or on-demand video or audio clips in the presentation to deliver rich media presentation. The most important applications for streaming will be entertainment, as well as other services such as news, advertising, tourism and security, to name a few.

Nokia mPlatform Voice Gateway

The Nokia mPlatform Voice Gateway is a part of the Nokia mPlatform Solution. The Voice Gateway provides voice access to applications using mPlatform. With the Voice Gateway, operators and service providers are able to create more versatile services for a large user group with a wide range of devices. The Voice Gateway is based on techniques such as Automatic Speech Recognition (ASR) and Text-to-Speech (TTS), where speech is translated from VoiceXML.

Content

Common Service Enablers

Nokia Portal and Content Services

The Nokia offering for enhancing browsing functions is The Nokia mPlatform Solution. The mPlatform core holds four mobile servers relating to browser functions: Application Server, Data Collector, Notification Server and Data Server. These servers cooperate with each other allowing effective exchange of information. Some of the key features of these servers are discussed below.

Application Server

The Application Server within the mPlatform core is the intelligence driver of the browsing functions. It provides a web interface for portal management. Portals allow service customisation and menu tree hierarchy structuring according to operator or subscriber preferences. When using amobile terminal to browse content, the portal will make navigation more convenient through pre-organised settings. On the other hand, operators are able to categorise applications enabling the subscribers to start with bigger entities and then dive into the application level. Introduction of new services is as easy as clicking add/ remove service from a Graphical User Interface.

Smart Navigation

Smart Navigation assists subscribers to find services. The basic principle is that after using application A, the user is prompted to use application B. Suggestions come from the operator, who is able to build simple logical chains of actions. Another way to suggest new services iby using engines that utilise session data to determine application proposals. Third, subscribers can manage how the menu tree behaves.

Application Interworking

Integrated applications have access to middleware knowledge and can utilise the personal data when targeting services to asubscriber. Integrated applications can also read and write to/from the session cache. This allows the applications to share information about session transactions and reduces the need to rewrite information to different applications. Instead, middleware will suggest the information stored in session memory by default and the subscriber can accept it or change it.

Presentation Manager

Terminal markets are more or less fragmented, which causes different display properties in terminals. There are small and big displays with black and white or colour screens. When the same content is displayed in different terminals, the results vary greatly. What looks good in one terminal may look terrible on another. Middleware browsing functions should therefore include intelligence to do some optimisation before delivering the content to the terminal. This optimisation can be done in the network or by the terminal. The network approach includes functionality that retrieves terminal type information from personal data and attaches asuitable style sheet to the content. The style sheet has information about font sizes, colours and backgrounds. The content is scaled to match the display properties of the terminal.

Notification Server

The Notification Server is used to communicate subscriber transactions. It sends notifications to subscribers regarding browsing transactions supporting SMS and E-mail notification methods.

The Application reads the active user profile that includes information about the preferred notification method. The application contacts the Notification Server and requests to send a message. The Notification Server delivers the message to the end user. Advanced API is used to connect the Notification Server and an application.

Data Collector
The Data Collector is used to periodically collect log files and ensure that the collection is done only once. After this, it organises data into various tables in a database and makes them available for auditing. Log files collet customer care data, billing data, and performance data. The information tables are viewed using an Application Server facility.

Data Server
The Data Server is the main storage for mPlatform data.
It is an Oracle RDBMS database holding information on subscribers, services and applications.
It performs data-centric operations like creation, retrieval, updates and deletion of subscriber and system data.

Nokia Advertising Services
Nokia mPlatform Advertising is a part of the Nokia mPlatform Solution. It provides operators and service providers with all the tools to become a key player in the advertising business. It provides added value not only to the subscriber and the advertiser but also to the ad broker, media agency and content provider. Nokia mPlatform Advertising includes all the needed elements for processes involved in advertising like campaign management, inventory management, targeting, ad delivery, tracking, reporting and billing.

Nokia mPlatform Solution Integration Interfaces
The concept for the integration interfaces and interface hierarchy for mPlatform is identified in the next figure. The three main interfaces include connections to operators, subscribers and applications.

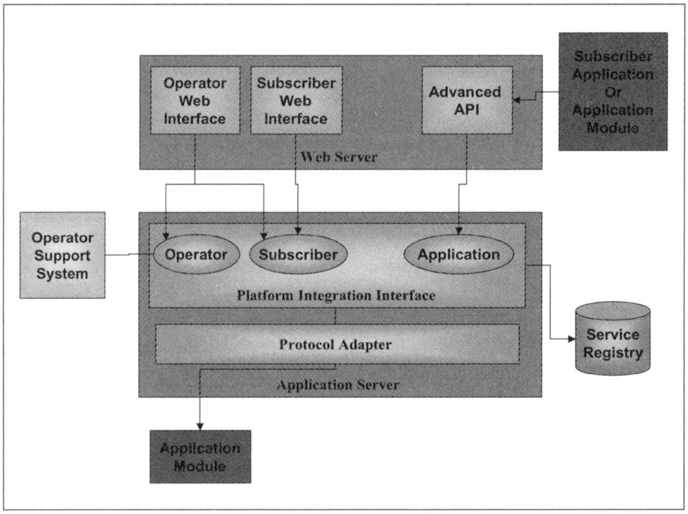

mPlatform Integration Interfaces

Advanced API
The Advanced API offers the possibility for developers to integrate their applications into the mPlatform Solution. Utilization of the Advanced API reduces the time to integrate new applications or to remove old ones. The Advanced API also reduces the dependencies between the mPlatform Solution and the applications.

Most changes in the mPlatform architecture will be transparent to the application.
The Advanced API protocol is based on SOAP/XML over HTTP. The Advanced API also features extensible data objects that enable applications to share data during a subscriber session.

Subscriber Interface

The Subscriber interface provides services that enable subscribers to perform navigation and personalization.

Operator Interface

The Operator Interface provides provisioning and maintenance services.

Nokia Commerce Services

Nokia will offer an open platform for all parties, including operators, banks and credit card companies, service providers etc. We think that various business models will exist through alliances between different players. Nokia´s commitment to open platforms ensures entrance and maximum benefits for all of the parties.

Nokia will offer different solutions for different needs in mobile commerce: the wallet application, as a terminal based solution, for easy online transactions and the Nokia Payment Solution, as a server based solution, for payment service provision.

Nokia Commerce Offering includes, e.g., Wallet Application and Nokia Payment Solution, as described thoroughly in the Mobile Commerce chapter.

Nokia Operational Services

Operational Services comprise the integration with operator or service provider customer care and billing systems, which is required to be able to charge the subscribers for service usage and to enable service and subscriber provisioning.

The charging and rating of Mobile Services is managed by the Nokia Charging Center (NCC). The NCC enables rating and charging of GPRS access and WAP and SMS content for both postpaid and prepaid subscribers. It provides an easily adaptable and flexible framework that allows operators and service providers to offer mobile data services without customising existing Billing and Intelligent Network Systems.

Using NCC's graphical pricing tool, new services can be offered and existing services modified in a timely manner. Rating is done against user definable event attributes such as data volume and sessions.

For prepaid subscribers, NCC provides operators and service providers the ability to bar and unbar access and content as a subscriber's account balance crosses credit thresholds.
NCC has integrated interfaces for most relevant Nokia network elements. This allows operators and service providers using Nokia equipment to decrease the time it takes to integrate NCC into their network.

In a similar manner, the Operational Services covers the integration to the operator's existing service provisioning tools.

Nokia Artuse™ Profile Directory

The Profile Directory will initially enable operators to manage subscriber profiles for WAP and Multimedia Messaging services. In the future, other mobile services that can benefit from profile information will also be supported. In addition, the Profile Directory has an open profile recovery interface, so it can be used with non-Nokia products.

With its interfaces to the customer care and billing system, the Profile Directory enables operators to easily download existing subscriber data to the Profile Directory database, add new subscribers, as well as handle mass downloads of subscriber data.

As a network independent product, the Profile Directory supports all major networks such as GSM, TDMA, CDMA, GPRS, and up-and-coming 3G networks

Nokia Artuse™ Mobile Terminal Management Server

As mobile devices become more powerful they also become much harder to manage. The provisioning of mobile handsets is one of the most important features of a mobile infrastructure. The Nokia Artuse™ Mobile Terminal Management Server (MTMS) is targeted at Mobile Operators and Service Providers that want to ensure that mobile phones are configured correctly. MTMS also writes phone capability information to the Nokia Artuse Profile Directory enabling other products like the Nokia Artuse MMS Center to use this information to recognize the data handling capabilities of a terminal and act accordingly.

The key driver in a successful mobile Internet business lies in launching a stream of content and revenue rich applications for customers today. With the Nokia middleware, an operator or service provider can be the fastest to market. The operator or xSP can continuously launch a host of seamless and user friendly services over diverse channels, terminals and networks. All these can be accessed by the customers with a single log in.

To ensure quick implementation and delivery, Nokia middleware is designed to be technology independent. Open standards support smooth customization and the development of third party applications and services. Nokia middleware also provides smooth evolution from 2G to 3G.

To maximize revenue from the very beginning, modular elements can be combined with the existing systems and infrastructure. Service delivery, location based services, billing and tariffing, connectivity functions and others are all available today to satisfy the requirements of the mobile world.

The Nokia mPlatform Solution is a platform for offering mobile services by bringing together software solutions from application vendors and operators.

- End-users can access the services with a single log in, and many of the services can by customized according to personal preferences.
- Developers have an open programming interface allowing them to rapidly create applications that meet market needs.

The Nokia mPlatform Solution is independent of the operator's network configuration. Therefore, it can be installed and updated easily. Nokia provides consulting, integration, and care services via its knowledgeable personnel, alliances and key partners.

The Nokia mPlatform Solution is a comprehensive solution for operators and service providers, who want to offer seamless, user friendly mobile services over diverse channels, terminals and networks. The Nokia mPlatform Solution gives the operator or service provider the ability to offer the latest value added mobile services for subscribers with a single log in to access the services. These can be customised to bring the user personalised information wherever and whenever they want.

Nokia mPlatform provides the operator or service provider with access and enabling functions. Furthermore, Nokia provides applications and open interfaces, all with subscriber and service profiles. Operators and xSPs can trust Nokia to provide a solution that brings applications now as well as enabling easy ways to customise and quickly develop more advanced applications. In addition, Nokia provides full operational support, allowing the system to be managed effectively with the ability to charge as necessary.

Nokia mPlatform comprises modular network elements, taking care of everything from IP security to the customer care and billing. Modules can be combined with the operator's or xSP's existing infrastructure, giving a complete solution for diversifying or establishing a Mobile Internet business.

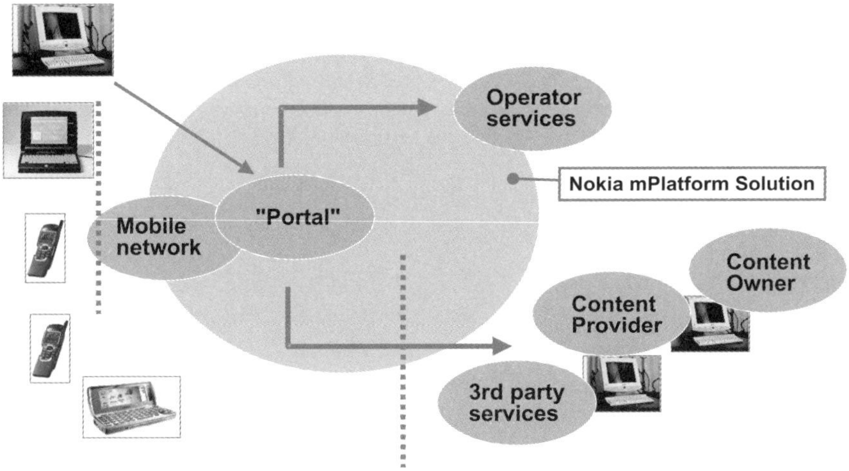

Building a trust position with the Nokia mPlatform Solution.

The Nokia mPlatform Solution enables the operators to achieve a sustainable trust position with mobile users and 3rd party content providers.

The Nokia mPlatform Solution offers means for creating and launching new applications frequently and providing open application interfaces and common functionalities. These enable the applications to work seamlessly together. They allow application developers, without knowing every detail of the underlying networks, to implement a wide range of applications. Applications developed by 3rd parties not only utilise network functionality, but also add value to the operator's networks. Eventually, operators are capable of differentiating themselves on the market and increasing the usage of their networks.

The Nokia mPlatform Solution enables the effective exchange of information between consumers and various applications and services. For a consumer using a mobile phone or a personal computer, mPlatform provides a single point of access to various applications and content located on the Internet. Also, mPlatform provides information for the consumer for self-care and self-provisioning. For applications, mPlatform provides an Advanced API supporting open standards such as HTTP, XML, and SOAP.

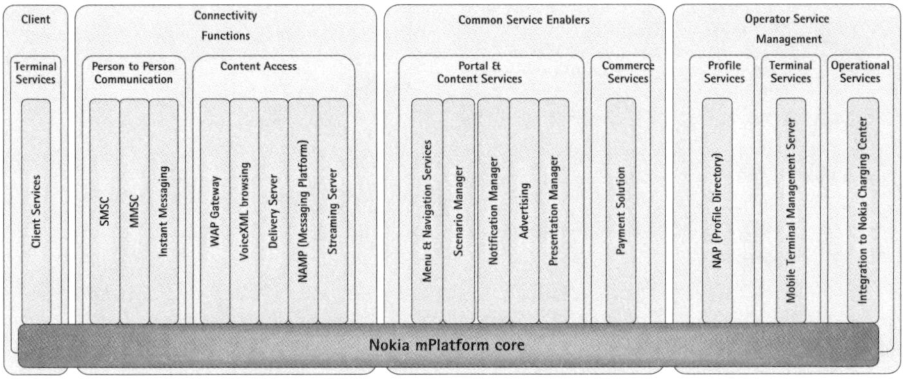

Nokia mPlatform Solution

The Nokia mPlatform Solution offers service providers an integrated service provisioning platform with open access to value-added functionalities. The mPlatform architecture is modular so that:

- mPlatform can be customised for the Service Provider's environment
- new and enhanced mPlatform features can be introduced frequently
- scalability and availability support the fast growth of the Service Provider's business
- new technologies, for example, new access technologies or handset types, can be introduced.

IP Convergence Based on SIP

In the future mobile world, we will see the merging of the two most successful phenomena of the last decade - mobility and the Internet. The Internet has given us the ability to access all kinds of services, information and entertainment through one common interface. These services in turn are created and offered by almost endless numbers of service developers, and most importantly, by the end users themselves. User-generated content will also drive development in the mobile Internet era. This is made possible by the open Internet protocols, standard extendable service creation languages and the modularity of the network. The result is personalised communication technology that enables people to shape their own mobile world.

Mobile phones have made us familiar with the concept of being reachable at any time and with the ability to reach other people or services quickly. Combining the Internet's open service creation and its vast amount of content and services with the reachability and immediacy of mobile networks creates the mobile Internet - something much more than the Internet we know today. The Mobile Internet means personal trusted services wherever, whenever, both time- and location-dependent and yet at the same time independent.

As network systems converge towards All-IP, we shall see an even greater multitude of services, as communication services, in addition to services such as e-mail and browsing, will be also be realised with IP. Person-to-person communications will still be perceived as the most important service by end-users, but the modes of communication will change.

Service creation for the mobile Internet and communications services is based on an open content format definition, such as XML (Extended Markup Language) and IETF defined protocols such as SIP (Session Initiation Protocol).

SIP was originally targeted for setting up communication sessions, but it has been extended for Instant Messaging and Presence. The same methods and tools used to create web services and content will also be used for creating rich-call and messaging services.

The creation and integration of these SIP based communication services will be much faster and easier than those based on intelligent network protocols. Operators will be able to create their own service packages and utilise the best of the third-party services.

Applications in All-IP can no longer be viewed as standalone. A rich experience for the end-user is created using several components interacting with each other.

Service mobility means accessibility of a person's personalised services through any access network or terminal. Another aspect of service mobility is reachability; services follow you wherever you go, so that e.g. messages reach you at all times. Personal reachability is achieved with protocols such as Session Initiation Protocol (SIP) and mobile IPv6.

The business model for the future mobile market will be a convergence of the Internet model, the content publishing business and the mobile information service model. Mobile operators are in a key position to exploit the new business opportunities brought about by the mobile Internet. A mobile portal will complement service provision as an end-customer interface and also as the marketplace for digital content. As the new roles and competition in the mobile Internet world emerge there will be a fundamental shift in the ownership of the customer's business.

An operator's key assets in this new business model are: the provisioning and billing infrastructure; control of the end-to-end Quality of Service for operator-hosted services; a large end-customer base with the subscription and profile data of the users; a recognised and established mobile brand; mobile network-specific enablers such as location and presence information; established dealer channels and, naturally, the mobile network infrastructure itself. Also, operators have existing working relationships with mobile terminal vendors, giving them deep insight into the opportunities provided by terminals.

In the mobile future as Nokia sees it, consumer behaviour is driving the development of applications and services. Nokia will limit the complexity of the technical environment supporting these applications and services from the end-user. Users simply do not want to be concerned with the underlying technologies. In addition, Nokia supports open technologies, standards and relevant initiatives that support and facilitate the deployment of global technologies and applications and stimulate market growth.

A Day in the Life of...

Mark Smith, Marketing Manager with a major European company, is about to leave home for work in the morning. As he goes briefly through the morning paper while having breakfast, he turns on his mobile phone and sets it as his preferred communication device. Now all his incoming calls are forwarded to it. At the same time, his presence information is updated, partly automatically and partly manually, into "At home, Not available for 'Work Group', Available for 'Friends' and 'Family'".

In the middle of his breakfast he remembers that there is no meeting room reserved for the afternoon's meeting. He tries to contact his secretary, but her Dynamic Contact Card shows that she is not available. Perhaps she is having her breakfast in the same way as Mark. However, the DCC gives Mark an option to send an instant message, leave a voice mail or see his secretary's homepage on the Net. Mark leaves a voice mail, takes a final sip of coffee and leaves for work.

Introduction

The promise of new revenue from third-generation networks comes essentially from SIP based services, where the mobile Internet is much more than simply making Internet-based content mobile. When All-IP becomes a reality, we will have these mobile Internet services up and running, being widely used as part of our everyday life and most importantly, generating revenue. What All-IP networks will provide on top of that is a revolution in communications services, service mobility and the speed of creating these services.

Mobile networks contain a lot of useful information that can be used to make applications more intelligent; location information, presence and micro-payment capabilities, to name a few. Towards IP convergence, all this will be even more important, as communications services will also be IP based. This, in turn, enables the same success factors that have caused the Internet explosion to be used in creating communications services. Using these capabilities and making the applications interact and share information with each other with the help of an intelligent middleware platform will enable all the new and exciting mobile Internet applications.

Even in the mobile Internet world, users will still perceive person-to-person communications as the most important service, and from the user's perspective, the primary benefit of IP in communications is integration and interaction of services along with network-independent addressing. Voice, video, instant messaging and presence are all components of personal interactive communications, which can interact with each other to create services that are both beneficial and easy to use. Therefore, the ability to use one common protocol for all the communication services will offer significant benefits.

In the following sections, we will highlight the four key issues providing the operators with the possibility of added revenue when implementing All-IP networks, person-to-person communications, service mobility, service interaction and service creation. We shall also discuss the concepts of presence, rich call and messaging as different components of personal communications and the technology behind these services.

IP Convergence

Service Creation is Fast and Easy

The phenomenal growth of the Internet and the browsing paradigm (WWW) has put the spotlight on IP-based networks and the inherent advantages of IP technology.

In All-IP networks, the service creation platforms are also based on IP protocols. All services, including communications services, are created with Internet-based protocols and languages. The benefit is that when all services are created with open and standard IP protocols, the creation and integration of these services is much faster and easier than those based on telecom protocols.

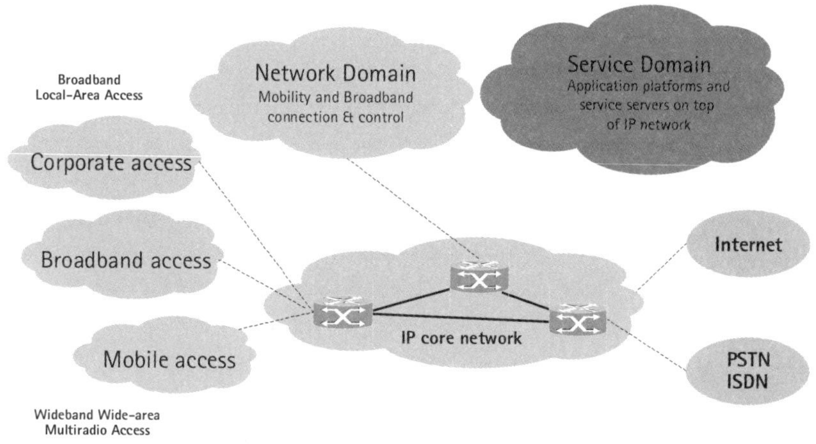

Network architecture for mobile All-IP.

Operators profit from the endless amount of competing innovative Internet service-creation companies and developers. The operators can create their own service packages, but also provide the best services available from small software houses or world-class portals.

Yet, the operators are in the best position to offer added value to their subscribers. First, they have the subscriptions and profile data of the users. Second, they own the provisioning and billing machinery. Third, they control the end-to-end QoS for services. And fourth, they have access to the mobile network-specific enablers such as location and presence information with the user's permission.

Service creation will occur in three identified domains:

- Network operators are able to offer services that fully exploit the capabilities of their IP network and service capability servers integrated to the solution.
- Third-party service providers or corporations can use SIP capabilities by interfacing with the operator network server using an external application server. It is also possible for a third-party service provider to operate SIP application server.
- End-users are able to create services by using the capabilities offered by an operator or a third-party service providers. For example, a web interface with capabilities of building services using a simple Call Processing Language could be offered to end-users for the purpose.

Since the underlying elements of SIP are similar to HTTP, creating network-based services such as time-dependent call forwarding is quick and straightforward. Millions of software developers can design and implement new dynamic SIP-based voice services as quickly and easily as they develop web pages.

By not requiring major hardware upgrades to application servers, but rather enabling new software-based services using SIP, service providers can reduce the time associated with deploying new features from months to days. This means ever-improving communications services for the subscribers.

Service Mobility Increases User Benefits

Regardless of the efforts to create global standards for network access, there will always remain alternative methods for accessing services. Differences such as wide and local area mobility vs. fixed access methods will remain. Nokia supports a high degree of access independence. The different network domains of wide-area cellular networks, Intranets, and the Internet will remain. Although from a broad perspective all networks converge into a single IP network, different networks inherently have some characteristics that differ from each other. Securing the best quality interaction between networks is needed for seamless end-to-end services.

Service mobility creates a unified service experience.

Service mobility means easy accessibility to a person's personalised services and their personal reachability through any access network or terminal. User Interfaces and services are independent of time and place, but are at the same time customised according to the terminal used and are aware of place and time. Our not so distant future vision is that sessions (e.g. a videoconference) can be started on a PC, carried on with a mobile terminal and continued with a DVB-T screen at home. When all services (both browsing-type mobile Internet services and person-to-person communication services) are IP based, they can be offered through any access network - the operator can create services on one platform and offer them via multiple access networks, thus greatly increasing the addressable customer base and serve the existing base better. Services that are accessible wherever, whenever and through whichever device will increase customer loyalty and reduce dissatisfaction. Service mobility and personal reachability is achieved with SIP and Ipv6 Mobility. IPv6 gives a device a universal address, enabling it to be reached from any network.

Service Interaction Opens New Business Opportunities

The convergence of the publishing model of the Internet and the service delivery model of the mobile Internet will put operators and third-party service providers in a position where service provision brings new business potential.

Applications in All-IP can no longer be viewed as standalone. A rich experience for the end-user is created using several components interacting with each other. From the user's perspective, the primary benefit of IP in communication is the integration and interaction of services. Voice, video, instant messaging and presence are all different parts of personal interactive communications, all of which can also interact with each other to create services that are both beneficial and easy to use.

Historically, service integration has been rather difficult in mobile networks, since they have all required their own protocols, standards and service domains. In the All-IP era, service integration is easy, since services are created within one domain. It is easy to integrate location-based applications with m-commerce applications, m-commerce applications with calendar applications, calendar applications with location-based applications and so on. The possibilities are endless.

Voice and data applications have been rather hard to integrate in the same network due to the different standards and bearers. In All-IP networks, the operator needs to take care of only one network, while SIP remains the enabling protocol.

New IP Convergence Paradigm will be Based on Previous Generation's Legacy

In Nokia's view, IP convergence builds on well-established service paradigms, e.g. SMS and MMS, by adding new functionality and new content and service types in user-understandable steps. Because consumers can relate to the new services as enhanced service experiences from previous networks, the barriers for adopting them will be significantly reduced, leading to rapid take-up and high penetration.

The market will develop towards All-IP services as an evolution, having some technological discontinuity in between technology shifts. Whereas messaging was introduced as the main application in the 2G era and browsing and rich call in the 2.5/3G, real-time services will be the focus of attention in the All-IP era.

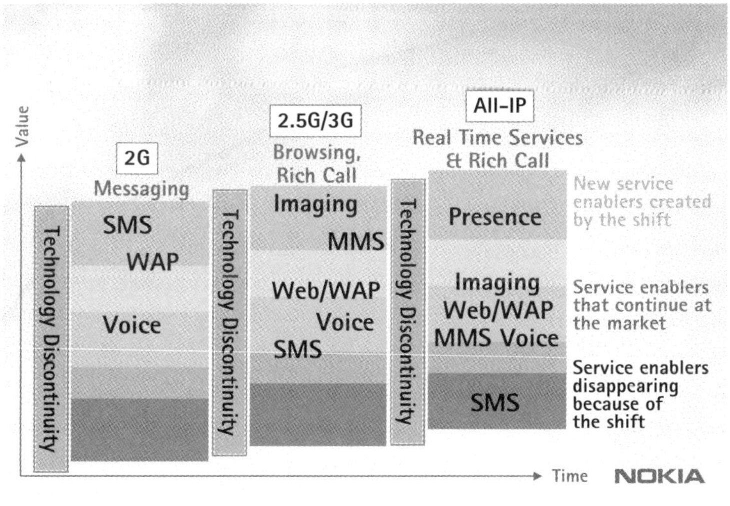

Market development is evolutionary with some technological redundancies along the way

In the development of third-generation mobile networks and beyond, the focus has shifted towards the IP paradigm. In particular, the work so far has focused on the adoption of IP transport, IP mobility and VoIP protocols (e.g. SIP) in mobile networks.

In future mobile networks, IP will allow the provision of multimedia services and of other services, such as instant messaging, web browsing, etc. in an integrated fashion. In particular, users will no longer be forced to see the cellular network as capable of providing only voice services or to be bit pipes to access the Internet. With respect to this vision, current attempts to evolve the service infrastructure have simply tried to apply the current concepts and services to basic services, e.g. voice services, without considering the importance that service integration will assume.

Before we can set the objectives for the new service platform of the All-IP era, we must first consider the basic assumptions behind the changing service paradigm. An IP-based service architecture, capable of supporting multimedia and service integration, will be a key element of future cellular networks. Also, current service architecture models, as they are evolving at present, do not meet the requirements for service integration, from a technology point of view. Therefore, a more radical and distributed IP-based technology architecture approach is needed. In practice this means the reuse of existing IETF protocols, when possible, which will allow for an easier deployment and acceptance of the service architecture, both in the cellular and IETF environments.

Technology Behind IP Convergence: SIP

The Session Initiation Protocol (SIP) will be the standard signalling protocol/mechanism to support multimedia sessions in third-generation networks. This choice ensures that future IP-based communication services like voice services will be truly Internet based. SIP is defined by the IETF Session Initiation Protocol working group in RFC2543 and is very different from traditional signalling methods used in telecommunication networks. It shares much with existing Internet protocols such as HTTP and SMTP (protocol header formats).

Like HTTP, SIP is a text-based client-to-server protocol. SIP was designed to establish, modify and terminate multimedia sessions or calls. Clients send requests to servers and servers send responses back. A request invokes a method on the server. However, SIP is different from HTTP in the way that an SIP terminal (User Agent, UA) may act both as a client and as a server. Therefore "client-to-client" communication is possible. The fact that it is text based makes SIP relatively easy to implement and debug. SIP is also an extensible and modular protocol. The modularity helps to implement SIP clients in different device categories, from small devices with limited capabilities to desktop computers with full features.

SIP service architecture follows the Internet's successful design. Services are developed in the application layer and the network is used exclusively for data transport. The SIP service architecture also takes into consideration the growing power of the terminals by moving more responsibility to communication end-points. The ultimate goal vision is to push service logic to the end points leaving service providers and end-users to develop and control the services. As a consequence, standard interfaces to develop services are needed and are provided by service development tools familiar from the Internet world such as SIP/CGI (Common Gateway Interface; an API protocol) and SIP/Servlet (Java Servlet) servers on the application server side. The XML-based SIP Call Processing Language (CPL) can be used by developers or even by end users themselves (e.g. by clicking preferences in an XML sheet) to create SIP services. Actual web developers will be able to use their experience to develop and integrate new communication services

SIP also enables IN- and ISDN-type services as well as user mobility. SIP supports instant conferences, adding parties to full mesh and enabling multicasting and uses similar global URI- (Uniform Resource Identifiers) based addressing as Internet e-mail; SIP:user@domain, where "user" is a user name or a telephone number and "domain" is a domain name or a numeric address.

Consumers benefit from SIP by having the same address while choosing the most suitable interface, e.g. mobile terminal, PC, PDA, etc. While SIP is the generic protocol over all access networks, consumers may use the most capable access network available, e.g. in terms of price, bandwidth, usability, etc. Services are also personalised, which makes the user experience appealing.

The primary advantage that can be seen with the SIP proposal is the way in which it can integrate voice, video, and other interactive communications services with instant messaging and presence in an All-IP service concept manner.

SIP will greatly enhance person-to-person communications by combining multiple media types and communication methods with presence and community information. With SIP, adding "richer media" (video streams, live video or application sharing), presence (availability, preferred media type) or community information is straightforward as all the services use the same communication protocol.

SIP in Messaging

There are several possibilities to implement messaging services with SIP. SIP provides methods for message distribution for different types of content depending on the size of the message attachment. SIP can be used for creating separate TCP/UDP sessions for bigger messages. SIP messages can also contain user information such as a caller image or ringing tone. This allows the operator to implement different services around messaging applications, as explained below.

Messages can be sent as an in-line attachment in SIP signalling, which is a useful scheme for small messages. For example, a user may send a picture in the message showing the subject of the telephone call. For large content, this can be enhanced so that small messages are carried in-line, and bigger messages are only referenced in SIP signalling for retrieval. The user may have set in his profile whether to fetch the content instantly on screen or to send it to a mailbox for later retrieval.

Another way is to use SIP only to establish a new uni-directional data session. This way, when the messaging server receives a message targeted for a mobile client, it sends a SIP signalling message to the client specifying that a uni-directional data session, using codec X (e.g. jpeg) is needed.

Protocols for instant messaging and presence are being standardised in IETF. The IETF SIP working group has enhanced SIP to include the necessary capabilities (i.e. extensions to the protocol) for instant messaging and presence.

SIP and Ipv6

Ipv6 is relatively easy to implement in SIP addressing. Combining the two, highlights personal mobility using a unique personal identity, which enables us to originate, receive and access services from any terminal and any location. From the end-user's point of view, Ipv6 and SIP enable ubiquitous communication in person-to-person, as well as person-to-machine environments. For services this means increased availability and usage, while more devices are capable of being online.

SIP and Java

SIP and Java are closely linked together. Java applications can be transported as an SIP payload, and Java applications running in a mobile terminal can use SIP-based services in communications. A simple use case would be distributing a mobile game, checking availability for gaming, setting up the game and playing against another person over SIP. In a game such as chess, the moves can be transported using SIP. In games requiring more "bandwidth", SIP may be used for signalling a TCP/UDP connection.

SIP and HTTP

SIP and HTTP stacks have a lot in common. Advanced SIP implementation usually provides a simple HTTP stack. One example of HTTP and SIP co-operation is that after some SIP signalling, a large file is fetched over HTTP using the HTTP GET method. Moreover, an SIP client (e.g. mobile device) may host a simple HTTP server and offer files etc. so that other parties may download information from the client (embedded web server).

SIP works well in conjunction with the Web (MIME, URI, DNS). In fact, a lot of SIP services could be originated by web pages (click to call, web phone book...). SIP services can also use the resident client web/WAP browser to present information (call presentation service, messages, pictures...). SIP also enables "multicast Web", e.g. sending web pages on the top of SIP.

SIP and H.323

SIP will be implemented in the terminals, whereas H.323 conversion will be executed in the network gateways. SIP and H.323 can be used together. SIP can locate the called party and determine its capabilities, including H.323. H.323 is then used to connect the two parties. Unfortunately, there is currently no specification on translatiion between the two. Conversion is made more difficult by the multiple versions of H.323 (v1, v2, v3).

There might be SIP/H.323 gateways in All-IP mobile networks. However, it is unlikely that there will be cellular H.323 terminals on the market, and it is likely that eventually SIP will overcome H.323 in fixed and WLAN implementations.

Why SIP is the Protocol for IP Convergence?

The management and operational costs of running a communications service are vastly reduced, since new functionalities, such as presence and instant messaging can be added easily to an existing SIP network, avoiding the need to run and build a completely separate, parallel network to support these features. Equipment costs are reduced, since only one set of proxy and location servers are needed, rather than two. System capacity is improved, as resources can be shared across many services. Infrastructure investments in firewalls, mobility services, etc. that have been put in place for supporting SIP can also be reused. Service providers have invested intellectual capital in SIP; reusing SIP means these providers have a much shorter learning curve.

In short, by implementing SIP, the operator gets the whole array of IP-based services in the same network that are easy to implement, manage and control. New service development and introduction are made rather simple when compared to other protocols.

Applications in IP Convergence

IP will greatly enhance person-to-person communications by combining multiple media types and communication methods with presence and community information, not only person-to-person, but also instant multipoint-to-multipoint, i.e. conferencing. With IP, adding "richer media" (video streams, live video or application sharing), presence (availability, preferred media type) or community information is straightforward, as all the services use the same communication protocol (IP) and signalling (SIP).

Rich Call

Rich calls are voice or video conversations supported with concurrent access to an image, data or other value-added information. The key in the rich-call concept is the end-user experience: the value and potential of combining telephony with information to provide enriched services.

By joining the fixed and mobile worlds, SIP will enable true service mobility and access independence. It will also support rich calls, or concurrent access to images, data or other value-added information during a single session. This will add "see what I mean" capability for subscribers through a combination of voice, mail, web browsing, instant messaging voice over IP and other services. If SIP is the engine for rich calls, presence is the ignition, enabling subscribers to initiate the services and select from a range of different communication modes.

Most of the services do not require voice-over IP or IP telephony service machinery; they can be realised by combining data with circuit-switched telephony or video-telephony. Still, IP telephony or IP multimedia is expected to emerge as the rich-call technology.

SIP provides essential elements for rich call before the call, during the call and after the call. With SIP it is easy to implement, for example, dynamic phonebook and click-to-talk features, which enable easy connection establishment. During the call establishment itself, SIP enables sending a rich clip, which can be a picture with a call subject or a special ringing tone. During the call it is easy to share documents and generally implement information push features. Furthermore, during the session tear-down, the same kind of call termination messages are possible as in call establishment. And finally, after the call it is still possible to send an instant message containing, for example, an advertisement.

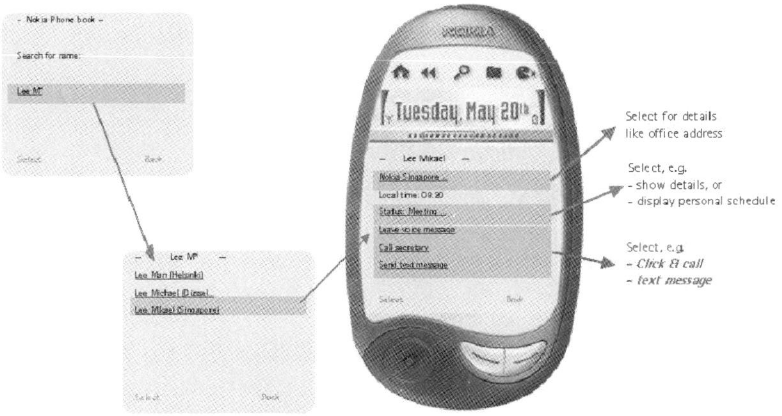

Presence

As mobile communications become more personal, presence will add a whole new dimension to staying in touch. Tomorrow's subscribers will know when and how they can reach their friends, even before they make a call.

With presence, mobile terminals will provide dynamic status information about subscribers, including their call/connection status, identities, terminal capabilities, location and availability. The result will be a visible network personality, one that signals valuable information about the subscriber's reachability. In the fixed world, this information supports buddy lists and other similar services, which let Internet users know when selected friends are also on-line.

With presence, mobile subscribers will be able to show their status (busy, away from my desk, on holiday...) to callers and suggest the most appropriate mode of communication (chat, instant message, e-mail, etc.).

As rich call is largely about combining different communication modes (telephony, messaging, browsing), presence information should also be seen as a user interface to rich-call services rather than just Instant Messaging. Presence is an application in itself, but is also an enabler and enrichment for other applications such as rich call, messaging and browsing. Also, presence enables call completion so that the caller may choose the best communication method with the callee, thus generating traffic in new ways that was hitherto not possible.

In its simplest form, a basic presence service might allow a user to publish a personal greeting that is visible to a caller prior to initiating the call.

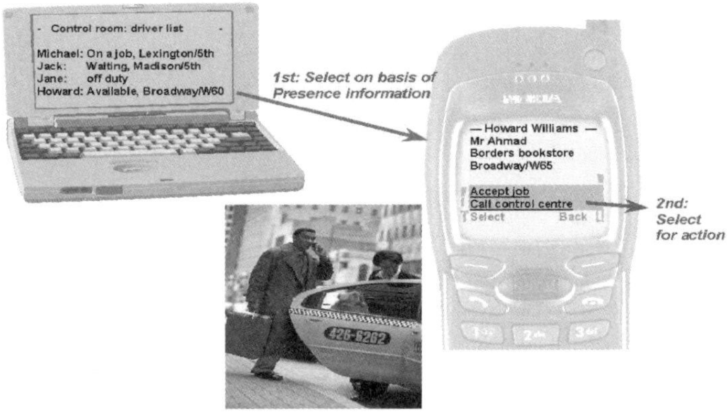

Presence is a central constituent in numerous rich-call services. Presence information can be shown to rich call users and thus influence their behaviour (service and application selection) and also be used in the network to control the rich call in a similar way to the supplementary services of today.

Messaging

Mobile messaging today predominantly means SMS, which has become a tremendous success over the past years, and voice messaging. As an evolution from SMS, Multimedia Messaging (MMS) will enrich the messages with multimedia content, boosting mobile-to-mobile messaging to even higher levels. With Multimedia Messaging it is possible to combine text with richer content types, photographs, images, audio and voice clips, and eventually also video clips, and send these combinations in the same simple manner as SMS today.

Instant messaging is an extremely popular application in the fixed Internet. Enhancing today's SMS, IM enables messaging that is enhanced by presence information. Users are aware of each other's status, which provides better service and supports new types of use scenarios. Messages can be sent real-time to the receiving party whose on-line status can be checked with the presence feature. With SIP, the proven SMS business model may be implemented with new technology.

With SIP, instant messaging can be seamlessly integrated with conversational communications thus creating the concept of ubiquitous communications, seamless usage of various real-time communications media and messaging services depending on the presence information of the parties.

Mobile instant messaging combines the success stories of SMS messaging as being always reachable, and Internet-based instant messaging, which uses the presence information of the parties. SIP-based instant messaging paves the way for Integrated Messaging solutions, allowing users to combine the benefits of e-mail, MMS and instant messaging and choose the appropriate communication method according to the presence information of the receiving party.

Common Critical Elements for IP Convergence

Quality of Service is Part of the Service Concept

In short, Quality of Service (QoS) means providing a consistent, predictable data delivery service. In other words, satisfying customer requirements. QoS refers to the ability of a communication system to provide an appropriate transport service for delivering the relevant traffic stream types of different users. It can sometimes be difficult to define the exact technical parameters required to ensure such delivery, as end-user's perceptions of "service quality" may differ from one user to another. QoS only manages bandwidth according to application demands and network management settings, and in that regard it cannot provide certainty if sharing is involved. Hence, QoS with a guaranteed service level requires resource allocation to individual data streams. Clearly, Quality of Service is an end-to-end issue, as it affects the terminal as well the network solutions. Any QoS assurances are only as good as the weakest link in the "chain" between sender and receiver.

In most GSM networks, coverage, price and service rollout were the only way to differentiate between service providers as the services themselves were standardised. Services in the All-IP era are becoming more demanding in terms of data rates, IP packet delay, consistent data throughput capacity, etc. Applications have different QoS needs (e-mail, browsing, streaming, two-way voice/video, etc.). More often applications are based on quality, reliability and timeliness assurances. In particular, services that use voice and streaming media are quite demanding in this sense. Operators have the possibility to differentiate themselves from their competitors by offering different kinds of service packages that have different kinds of QoS levels. The requirements for transport services vary according to application stream types and end users.

Examples of these service-level packages are:

- Gold, Silver and Bronze web browsing users
- Different QoS (max. bandwidth, packet priority) for different ISPs
- QoS profiles for wireless corporate intranet users

In the future, it is possible to broaden the scope of these QoS packages into a more application-driven one. QoS may be a parameter of the application and the actual service is created around that application. Then, if a consumer buys a certain service, he is able to do so whether he belongs to a certain service-level or not (provided his terminal and access mode support the required QoS).

3GPP (Third Generation Partnership Project) has defined four different traffic classes. Those are conversational, streaming, background, and interactive classes. The following table outlines the classes with service examples.

Conversational	**Streaming**	**Background**	**Interactive**
Telephony	Streaming video	Calendar update	Signalling
Video Telephony	Streaming audio	Mail download	Interactive games
VoIP	Video surveillance		Web browsing

The operator has the possibility to offer different Quality of Service packages for different market segments. For example, in the "Business package" all four traffic classes are allowed. The operator guarantees high bit rate, for example, for broadband live video and, calendar synchronisation and e-mail, downloads have the highest traffic handling priority. Further, the "Free time package" could allow only voice calls (VoIP) and the background traffic class. The maximum bit rate could be 64 kbit/s, but the guaranteed bit rate would be slightly lower. Also, only the lowest traffic handling priority would be allowed.

In summary, Quality of Service does not create additional bandwidth. Instead, it manages the existing bandwidth in such a way as to allow for efficient transportation of traffic through the network allowing for predictable delays, delay variation (jitter) and packet loss rates.

IP Convergence Opens New Revenue Possibilities

It is important to point out that IP convergence is not a threat to operators' revenue streams, not now or in the future. The All-IP infrastructure enables new services, as well as new flexible billing models, that were not possible before. The same charging model, i.e. price per minute and per message, can be maintained for voice telephony and messaging, i.e. for non-real-time applications. Initially, we may see more package-based billing or service combinations. Package-based billing is the most suitable scheme for real-time applications, such as SIP-based instant messaging.

The operator may provide different levels of service packages and combine them with different Quality of Service levels depending on their customer needs. This in turn enables more personalised services, increased customer loyalty and decreased customer dissatisfaction. The service packages may be comprised of the following individual services:

- store & forward extension and storage space,
- Quality of Service levels,
- transcoding service in the network (e.g. JPEG-to-GIF),
- voice mail,
- maintaining a presence server and the possibility to create groups
- instant messaging and
- offering Mobile Location Services.

There is no single billing and charging model available for all operators. This is one of the strenths of IP convergence. All-IP services enable better and endless possibilities for operators to differentiate from competitors with different service and billing concepts.

The operator may also offer and charge for SIP Call Processing Language (CPL) services. For example:

- "If the size of the in-line content is bigger than 10 KB and it is from an unknown sender, then forward it to mailto:john@yahoo ",
- "Attach Star Wars theme as a ring tone for my outgoing call attempts, and me.jpg picture"
- "Pass only urgent priority calls through", e.g. my boss, secretary and wife
- "If caller is Gregor Schmidt, send soccerresults.html file to him" (in SIP payload or by mail)

These services can be uploaded dynamically by the users or the default configuration may be used.

The operator may also offer and charge for a transcoding service. For example, if a sender sends a JPEG image to a mobile user and the service knows that the receiver only supports GIF, it may transcode it into another format.

It is also possible to charge for maintaining a presence server and virtual groups. Users may create their own friend groups or virtual communities. The operator's location data information may also be utilised here (e.g. authentication, etc.).

A future vision may well be that the basic messaging based on All-IP protocols, e.g. SIP, could be free and only advanced services that require additional network support or additional information which only the operator can provide, are charged for.

Future Outlook

Consumer behaviour is driving the development of applications and services. The key to commercial success lies in understanding end users, their lifestyles and attitudes, and in creating the product-service combinations that match their expectations and needs. Those who deliver winning product categories and platforms with the right technologies will be the victors of the 3G world.

In the quest to further develop and facilitate the next generation of products and services, Nokia will develop a technical architecture that enables high-quality, seamless interoperability between key applications, network environments and user identity/addressing. However, we will limit the complexity of the technical environment supporting 3G applications and services; users simply do not want to be concerned with the underlying technologies.

Nokia pioneers the open technologies, standards and initiatives behind the terminals and applications for the new generation. We are identifying the relevant communication modes, defining the key technologies required to support them and playing a highly active role in the drive to develop a common wireless platform.

Nokia has a vision of a future mobile world, where not only voice but also a significant part of data communications are wireless. In that environment, mobility and the Internet are combined, offering seamless services for end-users.

In the mobile future, services will be scaleable and accessible, wherever and whenever, based on end-user needs. Uniform services will be available through all types of access options, and will always be optimised for the terminal. Nokia's ultimate target is seamless interoperability between technologies, thus enabling seamless services.

IP Convergence Will Happen

In the Nokia vision, the future is about seamless connectivity to Internet-based services. This creates an increasing demand for bandwidth, connectivity features and economy that cannot be served by the circuit-switched mobile networks in the present form. The majority of traffic will be packet based in 4-5 years, leading the mobile telecommunications world to reinvent itself via Internet Protocol (IP). Implementing this vision requires knowledge of the mobility business - both infrastructure and terminals - which Nokia truly has. 3G is all about services and Nokia All-IP is about delivering those and more in even better ways.

Applications in IP convergence cannot be viewed as standalone applications. A rich end user experience is created by using several components that interact with each other. The primary advantage that can be seen with IP is the way in which it can integrate voice, video and other interactive communication services with instant messaging and presence. In conclusion:

- IP multimedia will greatly enhance person-to-person communications by combining multiple media types and communication methods with presence and community information.
- The key in the rich-call concept is the end-user experience; the value and potential of combining telephony with information to provide enriched services.
- Presence is a service in itself, if seen externally by others, but also an enabler of and enrichment to other services.
- Instant and multimedia messaging can be seamlessly integrated with conversational communications thus creating the concept of ubiquitous communications.

Nokia has defined an IP-based service architecture based on the SIP paradigm and promoting integration of IP-based services (e.g. VoIP, multimedia, e-mail, instant messaging, web browsing, WAP, etc.) over an access-independent IP-based mobile network. The focus will be on SIP-based mobile networks.

Working with other Internet players, Nokia is committed to supporting objective and open standards for new services. Interoperability will be crucial, to ensure a truly open mobile Internet environment. With its strong understanding of end users and services, Nokia is well placed to take advantage of this concept, which promises to change the nature of connectivity into 3G and beyond.

M2M - A New Opportunity in the Wireless Data Business

We have recently witnessed a series of fundamental changes in communication culture and technologies. Almost twenty years ago, the introduction of mobile phones freed personal voice communication from location. Subsequently, customers learned to satisfy some of their other communication needs by using data services on their phones. A wireless connection to the Internet was introduced for mobile phone users, expanding their communication options even further.

The next step in the evolution of communications is extending this connectivity beyond the person. To enable the vision of a mobile information society and true independence of location, we have to enable communication between individuals, devices and systems. This paper introduces Machine-to-Machine, or M2M solutions and outlines the business opportunities they offer. In the context of wireless data, M2M solutions are an ideal bridge from 2.5G business to the 3G world. The service-focused business models of tomorrow may already be here in the form of M2M solutions.

Nokia supports open technologies in M2M business and will actively drive the market towards global M2M standards. Our partners in the M2M business will benefit from our long experience and competitive position in the cellular marketplace. Furthermore, the highly advanced Nokia M2M Platform, including GSM terminals and the M2M Gateway, will benefit corporates, cellular operators, application developers and service providers in creating and maintaining industry-leading M2M solutions.

Wireless Data Market Evolution

As we move rapidly towards the third generation (3G) mobile world, more attention than ever is being paid to the wireless data market. While the core of the business remains in the saturated and very competitive voice market, there is a growing demand to look for alternative revenue streams. To achieve future growth in the diverse data business will require completely new strategies and business concepts. SMS (short message service) has already evolved from a means of personal communication to a popular service delivery medium and a real revenue source for innovative cellular operators. The growing popularity of the Wireless Application Protocol (WAP) is another good example of how technological innovations influence the process of service evolution.

But we have only taken the first steps towards the future mobile information society. The wireless data market is continuously growing and it would be shortsighted to believe that web browsing is all that the wireless data business will be about. Content is the real driving force behind the business. Strong service and application concepts are required as the foundation of successful business cases before the data market will really take off. In addition to the horizontal applications relying on web content, more and more specialised services with tailored content will be introduced.

As the number of different wireless data uses and applications increases, it can be confusing to group together the full variety of applications from web browsing to remote device management. A more extensive picture of the business may be attained by analysing wireless data services and applications based on end-user identity, network environments, access methods and interaction modes. When concentrating on communications and the means of data access, three types of wireless data services can be identified that all have unique requirements for service creation and strategy development. Next figure describes these three segments: Messaging, Browsing and M2M.

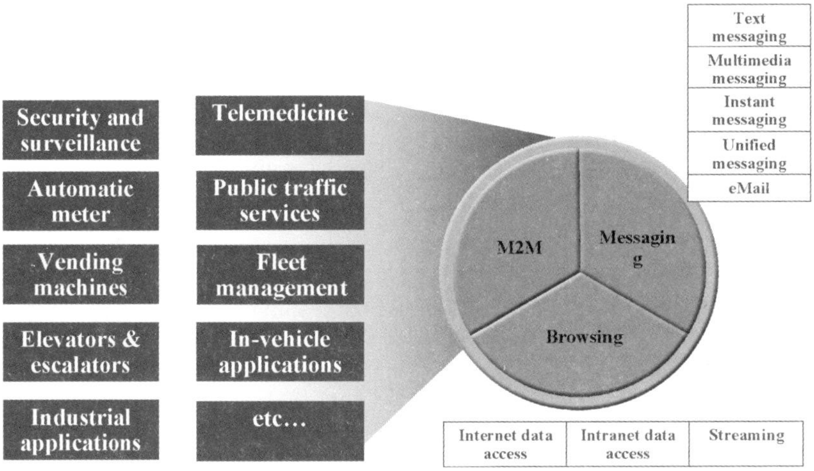

Wireless data segments.

This paper outlines the fast evolving business area of M2M communication. In this context it is to be understood that M2M stands for a much broader business scope than that of telemetry or telematics, even though both terms are often used to refer to M2M business in general.

Although M2M applications are discussed as a separate group here, it should be kept in mind that they may be combined with other applications to create solution packages that best satisfy overall customer needs.

M2M Applications

M2M refers to the terms machine-to-machine, man-to-machine and machine-to-man. This segment covers applications in which wireless data is used to create a link between systems, remote devices or locations and individuals. This link is typically used for collecting information, setting parameters, sending indications of unusual situations or taking care of an on-line transaction.

New M2M applications are continuously emerging, and they may serve almost any business area and physical environment. It is fair to say that only imagination limits M2M applications. However, in order to give a concrete picture of the total business potential, the best known M2M applications are presented here and grouped based on their uses and application environments.

Telemetry

- AMR (utility meters)
- parking meters
- industrial metering
- elevator control
- vending machine control

Telematics / in-vehicle applications

- driver navigation
- driver safety
- vehicle diagnostics
- mobility services
- traffic information

Public traffic services

- traffic information
- electronic tolling
- schedule synchronisation
- optimising road usage
- speed cameras
- changing traffic signs

Industrial applications

- process automation

Security and surveillance

- access and mobility control
- surveillance cameras
- asset monitoring

Sales & payments

- vending machines
- POS terminals
- gaming

Fleet management

- cargo tracking
- route planning
- order management

Telemedicine

- remote patient monitoring
- remote diagnostics
- equipment status tracking
- staff scheduling

Public services

- information kiosks
- web access points

Characteristics of M2M Solutions

The M2M business is about creating solutions aimed at improving existing businesses or creating new opportunities. M2M combines telecommunication and information technology to automate processes, to integrate all company assets with its IT system and to create value-added services to be used by the employees or customers of a company.

M2M solutions are often vertical applications tailored to serve the specific needs of one company or corporate body. The ultimate goal is to increase the performance and competitiveness of the company, which may be achieved through increased efficiency or better service. Although some of the M2M solutions are internal to a company and their focus is on improving processes, an ever-growing number of M2M solutions also aim to serve the customer base of the company.

M2M solutions should be designed to grow along with the company and its needs. Technology development or changes in the business environment may create a need to introduce new applications or services. The following example illustrates an M2M solution that develops over time and finally combines both company internal applications and customer services. It even generates new business.

An M2M Solution for a Vending Business

The first target of the solution is to minimise the costs of vending machine operation by using remote monitoring, data collection and configuration of the machines. Maintaining optimum stock in each machine, proactive maintenance, immediate fixing of malfunctions and changing prices remotely are just few of the alternative ways to improve efficiency that this telemetry application provides.

The M2M solution in the vending business may be expanded by providing the customers with the possibility of mobile payment. In addition to getting rid of the trouble and cost of handling coins, mobile payment increases sales. The mobile payment solution can be implemented today in co-operation with a mobile operator, and the evolution of mobile handsets will enable new concepts in the future. Creating a direct link to the end user through his or her personal handset will for the first time make it possible to build customer profiles and provide personalised services and offerings to customers based on their individual needs.

The next step of the M2M solution in the vending business could be expanding the business scope of a vending operator. The M2M-enabled machines could be used to provide information or content - e.g. news or games - that are financed by a third party content provider. The same content may also attract people to the machine and make them spend an extra dime for a snack or a bottle while there.

Similar types of opportunities can be found in almost any segment or environment. The solutions just have to be tailored to serve the needs of each specific business. M2M solutions should not be seen as narrow vertical applications, isolated from the external world, but rather as tools to integrate devices, applications, people and companies to become interactive parts of the mobile information society.

M2M Business Opportunities

Corporate Bodies

The obvious beneficiaries of M2M applications are corporate bodies, the customers of M2M solutions. Companies are constantly looking for new ways to increase efficiency, cut costs and improve customer service to gain a competitive edge in their core business. The possibilities raised by M2M solutions are already widely recognised in several industries. When competition in the traditional business area is getting more intense and eating into profits, expanding the business - for example, in terms of service provision - with the help of an M2M solution may be the way to maintain profitability.

An M2M solution integrates all the assets of a company in one complete system, including those in the field and even around the world. Labour costs can be minimised when needless site visits for service, maintenance, device updates and so on, can be avoided thanks to on-line connections with all devices. Manpower can thus be used more productively. The amount of downtime can be minimised or even eliminated, and the risk of fraud can be diminished by emptying coin boxes in time or getting rid of cash as a payment method. Sometimes legislation sets requirements that can be best fulfilled by introducing an M2M solution.

Many businesses will find that M2M provides the way to make their operations more profitable than ever before, because M2M solutions both improve competitiveness and enable them to focus on core business development. Thus, there are plenty of untapped business opportunities for providers that understand these needs and are able to offer turnkey M2M solutions for the corporate sector.

Cellullar Operators

With the voice market saturating and competition in the cellular business tightening up, cellular operators are looking for new revenue streams. For an operator, M2M solutions provide a new, appealing opportunity to strengthen its position in the profitable corporate sector and to expand its business scope towards providing total solutions.

From the operator's point of view, the M2M business case is ideal in many ways. The traffic can often be concentrated in the off-peak hours to balance capacity utilisation, and much of the traffic flow is predictable. With M2M solutions it is possible to expand the customer base to new segments, and win all cellular traffic of new clients by bundling traditional telecoms services with company-specific solutions. Furthermore, M2M solutions make it possible to generate significant additional revenues with a minimum of initial investment.

Those operators that choose to take the role of a solution provider in M2M business, rather than that of a data carrier provider, have a huge opportunity to drive the fast growing M2M market. The natural position and relationships of a cellular service provider are an extremely good basis from which to enter the M2M business. M2M solutions also complement the existing service portfolio very well. The first movers have already shown that a clear strategic focus and competence development in the M2M business will be generously rewarded.

Service Providers

Perhaps the least noticed and understood interest group in the M2M business is the service providers. Nevertheless, service providers play a critical role in developing and expanding the M2M business and their significance will only increase with the growing demand for M2M solutions.

The business scope of an M2M service provider may vary considerably. Probably the best known business model is that of an ASP (application service provider), which manages and hosts software on a basis of renting or leasing it to companies wishing to minimise the cost of their IT resources and equipment. In addition, M2M service provision may include application development, system integration, MVNO (mobile virtual network operator) activity or content provision or a combination of any of these.

M2M service providers may play the role of a cellular operator, but as the demand for M2M services increases there is a need for a broader spectrum of service providers. M2M service provision is an outstanding business opportunity for existing ASPs that may use M2M services in expanding their portfolio for existing customers or to penetrate new businesses. Also, companies currently acting in completely different fields, such as manufacturing, may find M2M service provision gives them a way to expand their business scope and fully capitalise their business understanding. It could raise their profitability to a whole new level.

Value Systems in M2M Solution Creation

The overall business trend of partnering and focusing on core competencies also applies to the M2M business. Specific knowledge of the business environment, communication technologies, SW development, HW integration, customer care and billing are required to create a successful M2M solution. Thus it is always produced by a joint effort by experts in different fields. The variety of contributors involved in creating an M2M solution is presented in next figure.

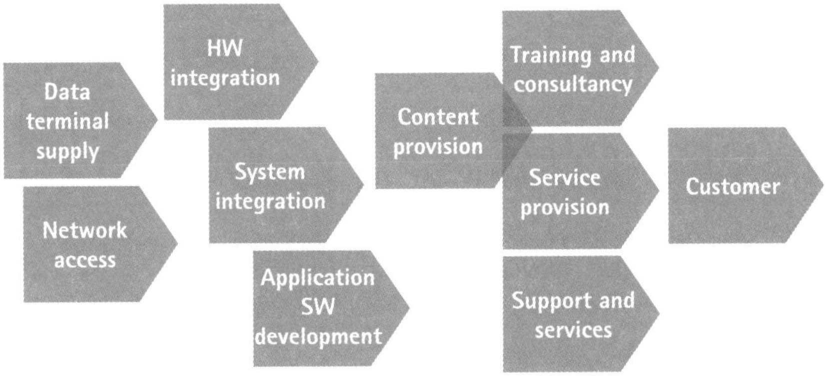

M2M Value System.

As described earlier, the variety of M2M application segments is almost limitless. The solution requirements and optimal divison of roles between parties within the value systems differ accordingly. Some roles, such as those of service provider or end user, are simply irrelevant in some cases. Furthermore, one company may adopt several roles within the value system if the required competencies are in place. The optimum division of value system shares must be defined individually for each M2M solution.

One of the greatest challenges of the M2M business seems to be mastering the complexity of the value network in providing total solutions. Customers do not want to purchase the solution piece by piece from different companies and instead of reliable components they want end-to-end solution reliability.

Therefore, the winners of the M2M race will be those companies who succeed in establishing and managing partner networks with a top competence portfolio for creating end-to-end solutions.

Requirements of Succesful Solutions

The requirements of a successful M2M solution are reliability, upgradability and affordability. Today it is not always easy to meet all these requirements, and even though M2M business is in a phase of rapid growth, even more focus will be needed to fulfill these three basic requirements. Application development and system integration are often defined as the most problematic areas of M2M solution creation. So, developing efficient tools in this area is one of the key issues in meeting M2M solution requirements.

M2M solutions will typically link a company's remote assets (such as vending machines, vehicles and other equipment) to its existing IT infrastructure. Therefore, expertise in both cellular and IP technologies will be required for effective application development. The reliability of a solution can be guaranteed only by taking an end-to-end approach to development work. In addition, some applications set higher requirements for security than others, such as in the case of mobile payment or transferring price information. Efficient information security measures such as data encryption and authentication are needed to provide adequate security.

Upgradability means that the customer must be able to take advantage of the latest technology, whether in terms of lower equipment costs, higher data rates or more affordable operating costs. However, solution upgrades cannot lead to expensive re-development cycles, because this is likely to lead to the loss of a big part of the benefits achieved with the solution. Therefore, easy upgradability of both SW and HW components is one of the key elements of an M2M solution.

The factors affecting the affordability of a solution are the costs of application creation, system components, operation and upgrades. The optimum cost can be reached only by considering all these factors; optimising only one of them may be achieved at the expense of the others. Application creation and upgrade costs may be minimised by selecting reliable products with efficient development tools and standard, open interfaces. Operational costs may be minimised, for example, by using different data bearers flexibly and by negotiating bulk tariffs with cellular providers.

Market Estimate

The M2M business is in a phase of fast growth. The total number of M2M connections is estimated to exceed 100 million during year 2004. The benefits of wireless data in creating connectivity and all-IP solutions are being used more and also more widely by companies as the awareness of the possibilities increases. On the other hand, new technologies and more appropriate M2M products and platforms enable the development of applications and services that have not been possible or have been too difficult to implement before. The market share of GSM connections in M2M market is estimated to grow from 30% to 60% during 2001-2005.

Fulfilling the requirements described in previous chapters is the key to M2M market growth. This will happen while more intelligent products are being introduced and more competent players with extensive M2M strategies are entering the business and forming alliances. M2M solutions are becoming a part of any company's everyday life.

During the next few years, much of the market growth will come from after market solutions, where M2M connectivity is introduced to existing systems and products. The nature of the market is however shifting towards line-fit manufacturing. An increased number of products and systems will be equipped with data connectivity from the outset. The demand for retrofit solutions will nevertheless remain and generate part of the total volumes in the future.

GSM - The Optimum Technology For M2M Solutions

It is obvious that solution functionality and user experience are the things that really matter to M2M customers; they simply do not want to be concerned with the underlying technologies. But when designing an application or a solution, what constitutes the best technology must to be considered carefully. This decision could have a far-reaching impact on operational costs or the possibilities for expanding the solution. There are several alternatives from landlines to proprietary data technologies. It has become evident that, where available, GSM is most often the optimum technology for M2M solutions.

GSM is a global, proven system with considerable benefits compared to other technologies. The history of openness and thorough standardization of GSM enable the creation of truly global M2M solutions. GSM technology roadmaps are created jointly with all the major players in the GSM industry to guarantee the same level of stability and global compatibility in the future.

The GSM market has faced a huge growth during the past decade and the trend is continuing. Consequently, the prices of GSM technology and communication are decreasing, benefiting M2M solutions from the beginning. In addition, the high number of GSM products and communications providers makes customers independent of a single supplier or network operator and increases their negotiating power.

A great benefit of the GSM solution is cheap and easy installation. No cable digging is needed and equipment with GSM connectivity can be easily moved from one location to another. For mobile applications, such as those in a vehicle, wireless technology is the only alternative. Further, high-level data services with full security features are already available in GSM networks and products, and there is a strong focus on developing them further to suit the needs of the consumer market.

The most important factor of GSM technology is that it enables M2M business entry now. GSM is also rapidly becoming the world's most popular 3G evolutionary route. Advanced solutions may be created immediately based on existing technology, and these solutions and services will create an easy evolution path to 3G business.

Nokia Operator Wireless LAN

Networks are increasingly being used to store all types of information, which presents its own challenges to the concept of mobility. Users are traveling and teleworking more yet still demand access to their corporate information systems no matter what their location.

Nokia Operator Wireless LAN effectively makes a user's desktop portable, allowing access to multiple media. Nokia sees this freedom of mobile laptop connectivity as one of the main building blocks of the emerging Mobile World. In this concept, mobile business subscribers will be the first target segment to have high-speed, low-cost Internet and intranet access wherever they are. Wireless LAN will be a main access technology providing solutions for high-speed wireless IP laptop connectivity.

With wireless LAN, mobile operators will be able to expand their business to mobile laptop users. Many GSM business subscribers will already have a laptop and will form a ready market for the new service. Offering the fastest wireless indoor data access, up to 11 Mbit/s, Wireless LAN will be a strong attraction. High-speed data access will also be extremely convenient for all subscribers to use.

Nokia Wireless LAN offers SIM-based authentication, which is a very simple way to authenticate and provides the means to combine with GSM/GPRS/3G billing systems. The Nokia solution also provides end-users with global roaming of Wireless LAN terminals, similar to today's GSM handset roaming.

With Nokia Operator Wireless LAN, indoor access can be made available in a wide variety of locations, the main ones being hotels, airports, railway stations, business parks, campuses, and corporate buildings - locations where business subscribers most need high speed laptop connectivity.

High Speed, Low Cost Wireless Internet Access

One of the definitions of the Mobile World is that everyone will have high speed, low cost access to the Internet wherever they are. However, this access will not be based on a single 3rd Generation technology, but rather a range of technologies that give people the greatest choice of how and where they get this access. Wireless LAN will be one of these technologies but it should not be seen as merely a wireless extension of existing Local Area Networks - instead it would be a vital component in the world wide communications network, providing solutions for high-speed wireless IP laptop connectivity.

Base stations capable of providing Wireless LAN, known as Access Points, will typically be located at places where business people, the core potential users of the technology, will congregate. These are indoor areas such as airports, hotels, conference centres, restaurants and cafes and will provide typical cell ranges of 20 to 100 meters. They will also be installed in offices. Wireless LAN will allow business users to check data from their company intranet, send and receive e-mails, save documents and take part in videoconferences, among many other applications.

Mobile business GSM users are expected to form the first target market for Nokia Operator Wireless LAN. They are already familiar with the convenience of voice and improved reachability, and increasingly, data mobility and are expected to welcome the opportunity to receive high-speed data directly into their laptops - even when they are abroad.

Reaching the Mobile Laptop User

Defining the Standard

Although the first commercial wireless LAN systems using the 2.4 GHz frequency band were introduced some seven years ago, the standard defining Wireless LAN, IEEE802.11, has had a difficult gestation period. Many years passed before agreement was reached on the first version of the standard, providing transfer rates of one or two Mbit/s. The standard also approved two incompatible approaches, "frequency hopping' and 'direct sequence', causing many in the industry to lose faith in IEEE802.11 and leading to the proliferation of yet more incompatible systems.

However, IEEE802.11 is now enjoying a major resurgence, thanks to the ratification of IEEE802.11b. This defined a single solution offering 11 Mbit/s wireless data communication. The performance advantage of IEEE802.11b over all the other wireless LAN technologies in the 2.4 GHz band gave it a sufficient advantage to draw most of the industry back into a common approach.

This renewed confidence in the standard was underlined by the introduction of Wireless LAN products by industry leading companies and the establishment of WECA (Wireless Ethernet Compatibility Alliance). This industry group, currently comprised of 120 companies and which includes Nokia as a founding member, promotes IEEE802.11 High Rate systems under the brand name WiFi (www.wi-fi.org). WECA conducts tests to ensure that systems produced by different vendors are compatible with each other. Compatibility challenges have now been overcome.

Typically, a WiFi certified wireless LAN product provides the following functionality:

- The Wireless LAN Card is an extended type II PC Card with up to 11 Mbit/s data rate
- Radio technology: 2.4 GHz, IEEE 802.11b compatible direct sequence spread spectrum, 13 channels
- Coverage area: Outdoor max. 400 m radius, office 20 m-100 m radius (typically)
- Wired Equivalent Privacy (WEP) with up to 128-bit secret key encryption over the air
- SW drivers for at least Windows operating system versions

Great interest has also been shown in wireless LANs operating on the 5 GHz band. This high frequency band is able to support even higher data speeds than IEEE802.11b. Two standards are vying for acceptance in this band: HiperLAN/2 in Europe and IEEE802.11a in the US and other countries. Interest in this band stems partly from the higher data rates available and partly from concerns about cross interference on the 2.4 GHz band.

However, 5 GHz systems are currently expensive and have high power consumption and it will probably be some time before 5 GHz products will be comparable in costs to current 802.11b systems.

Making Secure Connections

Security is a major consideration when developing and implementing any wireless system. Potential users must know that their data is secure. The IEEE 802.11b standard defines WEP (Wired Equivalent Privacy) security features, enabling RC4 data encryption with a key length of up to 128 bits. This encryption works over the 2.4 GHz air interface and protects the user data from eavesdropping. This WEP security is, however, considered insufficient in the areas of encryption and security key management. The IEEE 802.11 Enhanced Security Task Group (Tgi) is working on a new WLAN security scheme with the following main objectives:

1) To define a new strong encryption method
2) To integrate the new encryption method into access management systems, which can allow practical deployments of systems with thousands of users

Because today's business users and companies are very cautious about remote access and data security, it is recommended that the Nokia Operator Wireless LAN user applies separate Virtual Private Network (VPN) software to create an encrypted end-to-end tunnel between the user's terminal and the corporate network. Nokia IP Security Solutions offer a comprehensive line of products for security applications including VPN and firewalls. However, the VPN system is independent of Nokia Operator Wireless LAN and any company specific VPN solution can be used.

SIM + Wireless LAN = New Business for Mobile Operators

Charging for Wireless LAN services is another important issue. The Nokia Operator Wireless LAN solution provides operators with the means to turn a technology into a new business. With Nokia Operator Wireless LAN, charging is performed in a similar manner to GSM, using a GSM SIM card. The new Nokia Wireless LAN card supporting up to 11 Mbit/s high speed data access, offers an integrated SIM card reader. Nokia is the first vendor to introduce this unique feature in the Wireless LAN terminal market. The benefit of using a SIM card is that the identification and authentication of the user to the network is automatic and can be integrated into the billing system of a mobile phone operator.

Secure SIM card based authentication also provides global operator-to-operator roaming. This works in a similar manner to and as simply as in GSM, providing end-users with high-speed laptop connectivity even abroad.

The Nokia Operator Wireless LAN terminal also uses the SIM card to authenticate the public Wireless LAN access network. The authentication information is sent to the mobile operator's core network via a dedicated gateway and verified using the existing cellular functions. Combined with cellular SIM based user authentication and billing characteristics, the Nokia Operator Wireless LAN solution allows the mobile operator to implement fast wireless Internet access for roaming laptop users with a cellular subscription.

Nokia Operator Wireless LAN Scope

The Nokia Operator Wireless LAN system consists of two separate parts: the public wireless LAN access zone (hotspot), which provides the actual radio access network, and the cellular operator network side, which offers SIM authentication, roaming and billing services.

A single operator will typically have many wireless access zones in different public locations, such as airports, hotels and corporate sites. These are connected to the IP core network, which provides traffic transmission and relays the authentication and billing information to the cellular network.

The Nokia solution consists of wireless LAN cards for the terminals, wireless access points, an access controller and an authentication server.

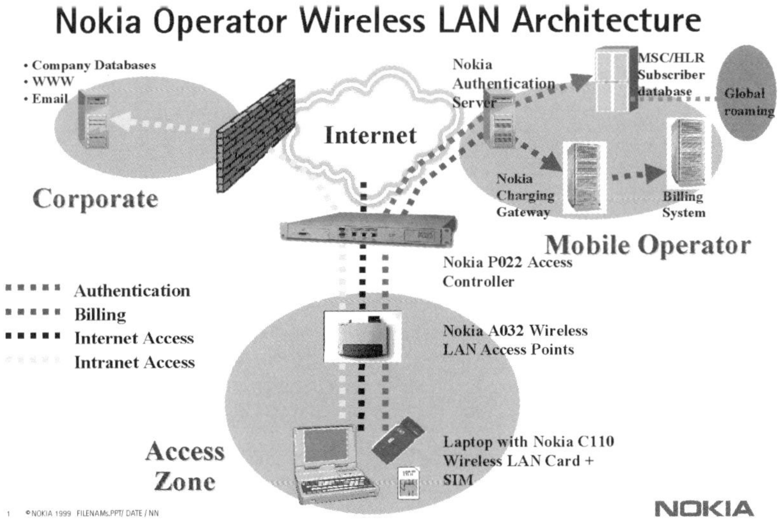

Access Controller

The access controller controls access from the radio access network to the Internet services. It allocates IP addresses to the terminal and also authenticates the terminal before the connection to the Internet is established. It relays the authentication messages between the mobile terminal and the authentication server. It also collects the billing records and sends them to the authentication server.

The access controller monitors each incoming and outgoing IP packet and only permits traffic between the authenticated terminals and the Internet. It supports several types of alternative charging mechanisms, such as time based, packet based or flat rate billing.

Authentication Server

This product provides GSM-based authentication services to the access controller, as well as collecting charging information. It provides authentication and billing services for a number of access controllers simultaneously. It uses industry standard RADIUS authentication protocol to carry the SIM authentication.

Wireless LAN Access Point

The radio access network is provided using IEEE 802.11b compatible wireless LAN Access Points, which are connected to a separate public access zone LAN segment. The Access Points provide a data rate of up to 11 Mbit/s and incorporate advanced management features.

Careful radio network planning ensures good performance and large coverage. For this purpose, Nokia provides dedicated Wireless LAN radio network planning software, offering an easy way of making site surveys and finding the best locations for access points.

Mobile end user Wireless LAN terminal

The Wireless LAN mobile terminal inside the laptop consists of a Nokia Wireless LAN card and access software. All the data traffic from the terminal to the network goes over the Wireless LAN link to the public access controller that performs the authentication.

The Nokia access point supports all IEEE 802.11b standard compatible Wireless LAN terminals. Currently, the Nokia 11 Mbit/s Wireless LAN card is the first terminal solution with an integrated SIM card reader. The access software provides the SIM authentication capability that is part of the product.

Seamless indoor mobility

The IEEE 802.11 Wireless LAN standard supports seamless mobility between different access points that are connected to the same LAN. This feature is also deployed in Nokia Operator Wireless LAN. In a typical wireless access zone installation, all the Access Points within the same building are connected to the same LAN, allowing the user to wander freely with his laptop inside the building. As the user moves to the coverage area of a new Access Point, the Wireless LAN network automatically switches the connection from the old Access Point to the new Access Point and the user notices no difference.

Behind the Scenes

The Nokia Operator Wireless LAN solution works in the following way: When the public Wireless LAN service is available, the cellular subscriber simply inserts his SIM card into the Wireless LAN terminal and starts the access software. The user can also configure the wireless LAN card to automatically log into the service. For convenience the mobile operator can provide the user a separate SIM card for WLAN services.

The operator Wireless LAN network components carry GSM SIM authentication messages over the Internet to the cellular core network, which verifies the user and grants access. This approach supports normal GSM roaming. After the proper authentication, the Wireless LAN network authorizes IP connection for the user terminal and starts collecting charging information. As the user disconnects from the network, the charging information is relayed via the cellular infrastructure to the home operator, who adds the charges to the user's normal (GSM/GPRS/3G) bill.

Benefits

Benefits for the operator

- Allows the operator to enter a new market segment and extend their service offering to corporate solutions.

The Nokia Operator Wireless LAN solution extends the operator's corporate product portfolio from handsets and GSM/GPRS/3G access to wireless broadband laptop solutions and global Wireless LAN roaming.

- Allows cellular operators to become ISP type businesses by offering high-speed wireless (IP) data access for mobile users and corporations.
- Subscriber management and billing based on a SIM card (same as GSM SIM).

Allows operators to sell Wireless LAN access as a complementary value-added service for existing cellular subscribers.

- Seamless operator-to-operator roaming utilizing existing GSM core network roaming capabilities.

The Nokia Operator Wireless LAN solution relays SIM authentication messages to the cellular network, which verifies user validity. In this scheme, both home subscribers as well as roaming partner operators' customers may access the same public Wireless LAN radio access network. Authentication requests and charging are always relayed to the home operator via the mobile operator's core network. The SIM enables reliable Wireless LAN subscriber authentication, authorization and accounting, as well as roaming between Wireless LAN networks from various operators.

- The Nokia Operator Wireless LAN solution offers the mobile operator an easy and cost-effective way to enter the new mobile laptop market.
- A License free wireless access solution with minimum investment costs.

As Wireless LAN uses unlicensed frequency bands, it offers a cost-effective alternative for high speed IP indoor data services in most countries. The solution can be used to offer either new wireless IP services in a limited coverage area or to complement the operator's network in public indoor hot spots.

- Brings significant savings in system investments, as Nokia Operator Wireless LAN utilizes existing cellular core infrastructure functions such as user management and billing.

Benefits for the end user

- Fast and secure wireless indoor laptop connectivity for mobile business users.

The Nokia Operator Wireless LAN solution provides mobile professionals with reliable, fast and secure public wireless IP access. This allows them to synchronize their e-mails and establish remote access to their corporate data at access speeds comparable with fixed Ethernet while travelling. The fast connection brings significant benefits and timesavings to the mobile laptop user.

- High speed wireless access to Internet services and corporate data in public places such as airports and hotels.
- Secure wireless IP access to corporate networks.

The Nokia Operator Wireless LAN solution supports all legacy VPN solutions, allowing the user to establish a highly secure end-to-end encrypted IP tunnel to his corporate network via public access zone.

- Can provide a single bill for both GSM and Wireless LAN usage for the end-user.
- Significantly improves business performance, while allowing employees to access corporate Intranet and e-mail systems, from all over the world.

Conclusion

The world of wireless communications is changing. The increased penetration of mobile phones and laptop PCs in the corporate environment, together with the ever-increasing pace of the business world and the increased demand for perfect reachability, have created a new business opportunity for the mobile operator to cover new market segments with Wireless LANs.

Nokia public Operator Wireless LAN technology will allow the operator to enhance its service portfolio for business data users, to generate more revenue and to enter completely new market segments, such as corporate indoor IP wireless access solutions. Early and quick adaptation of the Operator Wireless LAN solution is extremely important for the operator due to the unlicensed nature of the technology and the need to limit the number of radio access networks in the same building, for example, at airports.

Nokia Operator Wireless LAN is an innovative and unique solution. Providing significant benefits for the operator, it offers a competitive alternative to meet new mobile laptop user requirements and complement public wireless IP service offerings.

The deployment of SIM authentication and existing mobile operator subscriber management and the billing infrastructure dramatically decreases required system investment costs. Even more importantly, it allows the mobile operator to sell public Wireless LAN access as an add-on service for their existing mobile business customers and thus increase the revenue from the business segment.

Wireless LAN represents a great opportunity for operators to grow their business significantly, while using existing networks. The Nokia Operator Wireless LAN solution is available today and will turn that opportunity into a reality.

Mobile Commerce

By 2002, one billion consumers globally will have a mobile phone and will not leave home without it. Mobile phones evolve towards Personal Trusted Devices - the ultimate digital wallets - and will be used for a wide range of mobile transactions in the remote, local and personal environments, i.e. over the digital mobile network, in the physical proximity of and in conjunction with other devices belonging to the same user, such as a PC or set-top box.

Transactions conducted with Personal Trusted Devices include mobile banking and trading, as well as payment and ticketing. Frequent use will create user habits from which mobile services will benefit. It is worth noting that mobile phones already outnumber personal computers.

The mobile transactions market has begun with remote transactions on the Mobile Internet. Soon, concepts like Smart Covers will facilitate the development of local transactions in a fast, easy and secure way. This opens up huge opportunities for the ultimate digital wallets.

A major part of mobile commerce consists of the purchase of different types of digital content that in most cases, ends up in being consumed in the mobile phone. Consumers want to personalize their mobile phones with ringing tones, graphics and picture messages from Club Nokia. Games, new game levels, downloadable phone applications, as well as music and video feeds will soon follow.

Mobile Services

The first step for a mobile transactions service is often the Mobile Extension - mobile access to existing (Internet) services. A typical example is mobile banking, where customers can conduct money transfers and trade stocks using their mobile terminal. Essentially, the available services are the same as the bank's Internet services. Services become far more accessible and are no longer restricted to a specific location. The mobile device also provides ease of use, as it can be used in seconds, unlike a PC.

Successful PC e-commerce sites offer personalized user interfaces and so should mobile sites. The more consumers have to scroll through the information layers to find what they need, the more likely they are to give up before getting there. Ultimately, service personalization means offering products or services that users require, at the right time, in the right place. The PC can deliver this promise only partially, as it is not always connected, e.g. when the consumer is traveling. The fact that a mobile phone can deliver services and a connection at all times, gives it an undisputed advantage as a commercial platform for conducting e-business.

A unique personalization feature of the mobile phone, enabling services at the right time and in the right place, will be location-sensitivity. The offered information or transaction opportunities are based on the user's location, tracked, e.g. by using e-OTD or other mechanisms. A typical example is displaying the lunch menu of a nearby restaurant to a hungry tourist.

It is essential for wide acceptance among the end users that the service offering is suitable for the mobile context. It is not enough to make mobile extensions of current Internet services, since not all of them offer added value in the mobile context. The mobile phone does have its limitations, like a small display. Nonetheless, factors like mobility, availability, personalization and ease of use with appropriate value-added services easily overshadow the limitations of a pocket sized device. In mobile commerce, it is time for service providers to focus on suitable service and content provisioning.

Open Standards and Interoperability

In order to avoid market fragmentation, it is crucial that the architecture is based on open non-proprietary standards that are actively supported by the industry. A security system based on the Wireless Application Protocol (WAP), featuring a wireless extension of the Public Key Infrastructure (PKI), is an outstanding solution to address the challenge. WAP/WPKI is being specified in the WAP Forum, a coalition of hundreds of companies.

Nokia has been active in pioneering industry initiatives to drive the development of a common framework for mobile commerce solutions. Examples of the key industry inititives include the MeT Initiative, the Mobey Forum and the WAP Forum.

Security

Mobile commerce has to be based on secure applications and solutions to gain trustworthiness. Strong security is in the interests of the service provider and the consumer. The security consumers experience depends on the security services provided by the entire value chain. The security chain is as weak as its weakest point. The challenge is to implement a security scheme so that it remains convenient and simple to use.

An unfortunate consequence of inconvenient security methods is low mobile service usage. The static PC-centric user environment helps consumers cope with multiple usernames and passwords by storing them in secret files on the PC or on electronic post-it-notes below the monitor. This simply does not transfer well to mobile use due to the limited keypad, screen and challenging environment. In short, complex security solutions may create barriers that discourage consumers from going 'mobile'.

Becoming a Trusted Brand

Mobile commerce carries a powerful new dimension for customer service and is a huge source of substantial revenue. However, it is "a big cake with many eaters". Many players want to have a control point somewhere in the transaction process, although the trusted brand position seems to be the most wanted position.

When making mobile transactions, the consumer must decide who to do business with. In the mobile commerce value chain we can distinguish four different brand selections to be fulfilled either by one or by several companies.

- Merchant brand = Who do I want to buy from (e.g. e-tailer, retailer)
- Trust brand = Whom do I trust to handle the payment (e.g. bank, credit card company, service provider etc.)
- Portal brand = Who links me to the merchant (e.g. mobile portal, Web site)
- Connectivity brand = Who provides me with reliable connections (e.g. network operator, Service Provider)

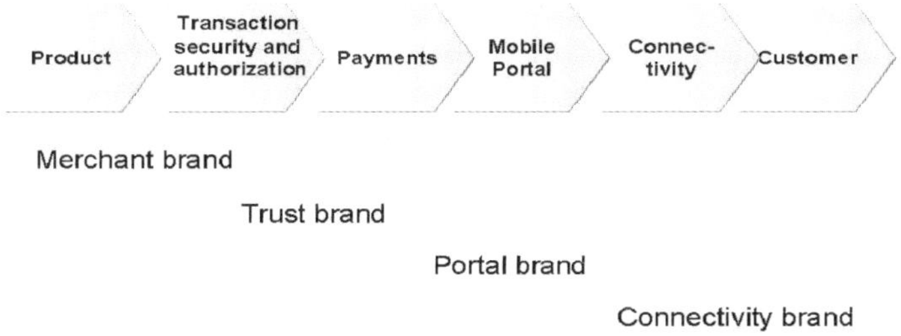

The merchant and connectivity brands fall naturally into the territories of merchant and mobile network operator, respectively. The portal space is more open, as practically all companies in the value chain have the chance to operate there.

The development of the Trust Brand position is still in its infancy and creates significant opportunities for new business and all value chain players alike. From a market perspective, three camps are well positioned to fulfil the trust brand role:

- Merchants
- Network operators
- Financial institutions
- other independent players

We believe that various models, and their combinations, can coexist in the global marketplace. However, cooperation between the players is essential for the success of mobile commerce. The suitability of a particular model to a given market depends on how various players shift in the value chain, government regulations and attitudes and the adoption rates of the required technology.

The merchant can become a Trust Brand. Large merchants such as department chains, e-tailers or e-marketplaces already operate as "retail convenience deployers" by offering extra services to strengthen customer relationships. They have strong brands of their own and technology-based offerings can build brand equity as an innovator. Payment schemes already exist in conjunction with loyalty programs and extending them towards mobile payments is quite natural. The downside for the merchant is the required large investment in technology and the need to take financial risks.

The mobile network operator already has established consumer relationships, as well as key infrastructure, such as billing systems. Their large customer bases and strong marketing skills provide an easy entrance into mobile business, especially for those who do not have security or strategic concerns about moving the technology outside the merchant's own control. A drawback is that it in order to the entire customer base, agreements must be negotiated with several mobile network operators. Another challenge is whether the operator brand carries enough weight to be a trust brand in consumers' minds.

Banks and other financial institutions enjoy the benefit of having the most trusted brand position. Consumers are willing to let them handle not only their mobile commerce payments, but also all or most of their personal finances. The mobile extension creates a new cost-efficient channel towards their customer base. Indeed, banks already possess extensive customer bases, large corporate/merchant customer bases as well as volume-proof technology systems. These assets lead to the opportunity of becoming a financial transaction intermediary, i.e. Trust Brand. This means that ultimately, existing bank account and credit/debit card agreements can be extended to the mobile commerce space. The challenge is often in the rigid convenience requirements from the mobile user mass market. While much of the existing infrastructure can be utilized, some processes may need to be streamlined in order to be less bureaucratic.

The Role of Wireless Application Protocol (WAP)

The Wireless Application Protocol (WAP) is the Internet based global standard that allows mobile devices to interactively access content and applications from the Internet and corporate intranets. WAP creates an open global environment for wireless applications. WAP is network, bearer and manufacturer independent. WAP specifications are published by the WAP Forum, which is a nonprofit industry association open to all. The WAP Forum is widely represented by various members of the mobile commerce value chain, positioning it as the most efficient body to deliver the required specifications.

WAP arose from the need to find a way to effectively take into account the critical constraints of the current wireless world: limited bandwidth, challenging conditions of use, specific graphical user interface and the processing characteristics of a mobile phone. The very same constraints will become a reality in the impending future of mobile commerce. Therefore, we believe that WAP is the best available choice of platform for the sustainable evolution of mobile commerce.

The evolution of mobile networks towards the third generation will bring increased bandwidth and the always on capability of packet-switched networks. Simultaneously, the processing capacity of handsets will steadily develop and the screen size, and consequently the graphical user interface will become more advanced. Due to the strong industry momentum behind WAP, we believe that its specifications will evolve alongside the evolution of the market.

WAP Security and WPKI

The current specifications of WAP address the PAIN-criterion of a secure system.

Criterion	What it means	in WAP
Privacy	Only the communicating parties are able to understand the contents of the transferred information.	Wireless Transport Layer Security (WTLS) Class 1 Encryption of mobile networks such as GSM
Authentication/ Authorization:	The communicating parties can be sure that the other party is who it claims to be	WTLS Classes 2 and 3 WAP Identity Module (WIM) in the mobile terminal Ability build application-level security such as passwords/user names PIN codes of mobile phones
Integrity:	The receiving party can be sure the received information is exactly what the other party originally sent.	WTLS Class 1 Digital signature using WMLScript Crypto Library and WIM
Non-repudiation:	Once someone has received a message the sender of the message cannot deny having sent it.	Digital signature using WMLScript Crypto Library and WIM

The security aspects of the WAP 1.2 and WAP 1.3 specifications are fundamentally built on the Public Key Infrastructure (PKI). It is a certificate-based cryptographic system that utilizes asymmetric algorithms based on key pairs (public, private) associated with each party. A certificate is a data structure that binds the identity of the certificate holder to a public key. Certification authorities (CAs) act as a clearinghouse issuing security certificates and ensuring their authenticity.

Originally, the PKI emerged on the wired Internet, but its adoption by consumers has been slow. While the technological merits of PKI are undisputable, the PC-centric user environment allows for substitute application-level security methods. Consumers have accustomed themselves to managing multiple usernames and passwords in a number of ways, i.e. writing them in secret files on a PC or onto a Post-It-Note below the monitor, and they see little added value in changing their habits. The perceived low added value has a direct correlation to how much they are willing to pay, e.g. for PC smart card readers that are relatively expensive to manufacture.

On the other hand, the mobile environment is completely different. The limited keypad and challenging usage situations while on the move mean PKI will deliver much greater added value, since substitute methods are clearly inferior. Furthermore, the cost dynamics are attractive in the markets, in which consumers pay for the terminal purchase themselves. Many future consumers will probably not even realize that they are purchasing the functionalities of a smart card reader. If you ask them, they are out to buy a phone. Therefore, the mobile phone will outpace any other card reader solution in market penetration speed. Once consumers become accustomed to PKI usage through their mobile phones, it can be forecast that, e.g. the authentication to Internet commerce will be done with the mobile terminal. Therefore, we also see mobility as a key driver for Internet PKI adoption.

The extension to the mobile environment, called Wireless PKI, Mobile PKI or WAP PKI, covers the infrastructure and the procedures required to enable the trust provisioning needed for authentication and digital signatures for servers and clients. The WPKI specification will address certificate enrollment and lifecycle management - more specifically - the creation, distribution, verification and revocation of the certificates. The key elements of the WAP/WPKI system are:

- Mobile phones with WIM (WAP Identity Module in which secure data is stored) support
- WAP Gateway enhanced with certificate-based identity validation capability
- Registration Authority (RA) for certificate enrollment
- Back-end PKI infrastructure with access to Certification Authority (CA) infrastructure

Certificate Enrollment in WPKI

The back-end PKI infrastructure is pretty much the same for both wired PKI and wireless PKI. The main difference is in certificate enrollment. The process of enrolling a Certification Authority (CA) certificate needs to be re-worked to accommodate an efficient mass-market roll out.

In smart card PKI-solutions, the issued certificate is fixed to a specific user identity through customized enrollment. For example, in order to get a government-issued ID chip card, one goes to a police station personally with a passport in hand and fills in an application. After that, a certificate is issued.

Customized enrollment is not suitable for the mobile commerce mass market. To maintain high security standards, it would mean that the WIM card issuer would have to mobilize and train its whole service network for such a security operation or conversely ask its customers to travel long distances to get the task done. Neither approach is very appealing.

In the wireless PKI solution, anonymous key pairs are pre-installed in WIM cards with a corresponding PIN (Personal Identity Number) code. A manufacturer certificate ensures the keys' authenticity. In the first instance of secure service usage, the Registration Authority (RA) of the WPKI validates the credentials and requests the CA to create and send a certificate thus binding the anonymous key pair to a specific user identity. An example of the process is given below.

- The registration authority prompts the user with a request form
- The user fills in a form that might request for example, a user ID and a shared secret - like a one-time password (e.g. from the existing WAP service agreement)
- The user 'signs' the response with the private key on the WIM (activated with a PIN code)
- RA validates proof of possession of the private signature key (POP) and proof of the subscriber's identity (POI)

- RA sends a certificate request to the CA
- CA issues the certificate

The WAP specification on certificate storage allows for different options. The cleared certificate can be stored on the WIM card or in a directory.

Wireless Transport Layer Security (WTLS)

Within the WAP protocol, transport level security is specified using a protocol known as the Wireless Transport Layer Security (WTLS) protocol. WTLS is conceptually and functionally equivalent to Secure Socket Layer (SSL), also known as: Transport Layer Security (TLS), the popular security protocol used on the traditional Internet. WTLS invisibly encrypts and decrypts information sent between a WAP client and a WAP gateway so that a third party cannot decipher the communication. The protocol also ensures the integrity of a communication so that the recipient of secure information can verify that the content has not been altered in transit.

The WAP 1.1 specification has already defined three classifications for WTLS :- Class 1, Class 2 and Class 3. The default level, WTLS Class 1 allows for an anonymous, secured channel between the WAP client and the WAP server. While the WTLS Class 2 includes Class 1 features with Server authentication. WTLS Class 3 includes Class 2 features and client authentication.

The WAP Identity Module (WIM)

The WIM, i.e. Wireless Identity Module, stores and processes information and security keys needed for user identification and authentication. Additionally, it has the ability to perform cryptographic operations. Its main function is to enable digital signatures that are required for authenticated mobile transactions.

A digital signature is the electronic equivalent of signing a receipt. It is the element needed that can replace the need for visual checks of ID cards, hand-written signatures and paper receipts. Legislation in many countries such as the U.S., Germany and the U.K. has already been or is being developed towards accepting a digital signature as a legally binding transaction. Digital signature technology can fulfill the requirements of authentication and non-repudiation, which are key conditions in establishing the merits for legally binding commercial transactions. The digital signature is executed in the WAP application security layer. It is specified in the existing WAP 1.2 specification, approved in December 1999. It is based on the WML Script Crypto Library. This specification describes security functions, which can be devised using WML Script.

WIM has been defined as a separate tamper-resistant hardware device, such as - but not limited to - a smart card or a SIM card. Technically, a WIM can be implemented in several ways. The main differences in implementation came from business and usage considerations.

WIM functionality can be stored on the operator issued Subscriber Identity Module (SIM) card. However, the mobile network subscription functionality is separated from the other applications that require authentication and signature capabilities. Standardized features and interfaces are used to ensure interoperability between technologies from a different manufacturers and SIM cards from different operators.

WIM functionality can also be stored on a second smart card in the terminal. The Dual Chip approach means that a second separate SIM-sized smart card is placed semi-permanently in the phone. It is removable and issued independently of the SIM card, e.g. by a mobile commerce service provider or financial institution. Its strength is that it separates network subscription from the other applications thus creating more latitude in terms of business models. Naturally, this requires phone designs that have a possibility of storing two SIM cards.

Eventually, customer requirements and demand will have an impact on which of the proposed WIM solutions will be taken into use. Since the WIM-o-a-SIM card solution is the most evolutionary one and uses the existing phone designs, it is likely to be the first to achieve larger consumer usage. However, it is good to bear in mind that from a service provider's technical perspective, the implementation of WIM is irrelevant. As long as it adheres to WAP specifications, the server-end execution is exactly the same.

Mobile Commerce Solutions

Nokia will offer open platform for the use of all parties, including operators, banks, and credit card companies, service providers etc. We think that various business models will exist through alliances of the different players. In each case Nokia´s commitment to open platforms ensures entrance & maximum benefits for all of the parties.

Nokia will offer solutions for different kinds of mobile commerce needs. Examples of the solution offering include the wallet application, as a terminal based solution, for easy-to-use online transactions and the Nokia Payment Solution, as a server based solution, for payment service provision.

Wallet Application

The wallet inside a mobile phone one of the essential building blocks of a Personal Trusted Device - offering consumers a visualization of the digital wallet and a trusted user interface.

The first release of the wallet application enables users to make convenient mobile transactions via WAP browser. The application is capable of storing protected personal information, like credit card details, inside the terminal, also making it possible to provide and use virtual cards flexibly. It simplifies storage and retrieval of personal information in mobile transactions. By enabling Internet type payments, where card information is transferred to the Internet merchant, wallet makes it possible for wireless providers to adapt their e-business to the mobile world very easily.

The wallet application menu contains three different modules: Cards, Personal notes and Settings. The Cards module is used for storing personal card information, such as payment card (credit, debit, etc.), loyalty card and dispatch note information, for different usage purposes. The card details consist of card info (name, number, etc.), account info (billing address, etc.) and shipping info (shipping address, email, phone number, etc.). From the wallet the user can fetch the required information via WAP browser and easily fill in the required fields. The Personal notes is a notebook where the user can store sensitive private information, like passwords, usernames, etc. From the Settings user can switch the wallet code request on and off and change the wallet code when wanted. The data stored inside the terminal is protected with a special wallet Code (PIN) that is used to authenticate the user to the application.

The wallet application is based on existing open technologies, like WAP and ECML (Electronic Commerce Modeling Language) standard. ECML is an open Internet standard for digital wallets and online merchants that is already commonly used in the Internet world. Its main function is to facilitate automatic exchange of transaction information from the client wallet to the server application. ECML does not only transfer credit card information; it can also be used whenever the applications need a common method for transferring e.g. address information

ECML is security mechanism-independent, and can/will be integrated to other transaction protocols and security elements when available.

http://www.ecml.org/

How to use wallet for mobile transactions:

1. The customer browses the merchant's WAP-service and selects the item to buy
2. The customer selects wallet payment and the merchant's server sends the payment request (billing data form in ECML format) to the customer
3. The user selects the payment card from the wallet menu (behind the wallet code)

4. The user fetches the billing and shipping details from the wallet to complete the form and the form is transferred back to the merchant
5. In the case of more costly purchases, the merchant may require the consumers to digitally sign the purchases whereupon the signing request is sent to the user
6. If the phone supports digital signatures, the user signs the purchase with the signing PIN
7. The merchant's server sends an acknowledgement or receipt to the customer

Nokia Payment Solution

Nokia Payment Solution is a platform that enables payment service providers - operators and service providers who wish to boost up the mobile marketplace - to mediate payments between three main parties of mobile commerce: financial institutions, merchants and consumers. The Nokia solution is a platform that collects, manages and clears payments initiated from mobile phones and other web-enabled terminals through various payment methods, such as credit and debit cards, operators' pre-paid or post-paid systems (phone bills), and a virtual purse. The virtual purse is an integrated account in the Nokia Payment Solution for every customer.

The concept of the Nokia Payment Solution is derived from the need of payment service providers to position themselves in the emerging mobile commerce value chain. Establishing a strong position in customer authentication, clearing of macro and micro payment consolidation, payment service provider gain a sustainable position in the mobile commerce value chain.

The solution is a network technology independent solution utilizing SMS, WAP and web technologies.

Nokia Payment Solution consists of three major components: Nokia Payment Server, Nokia Payment Channel Manager, and Nokia Virtual Purse application.

Nokia Payment Server manages the transaction flow between consumers, merchants and the clearing channels. It records all payment events, validates consumer and merchant identities, and processes the transactions towards financial institutions and billing systems.

Nokia Payment Channel Manager manages the SMS based communication between the customers and the Payment Server. The merchants can initiate the push of SMS payment contracts to the Payment Channel Manager. The Payment Channel Manager sends a SMS payment contract to the customer. The Payment Channel Manager initiates the payment processing over the Payment Server and returns the payment confirmation to the content provider.

In addition Nokia Payment Channel Manager analyses the HTTP headers in the message flow between the merchant or content providers and the consumer terminals. Whenever HTTP requests for priced content are intercepted, Nokia Payment Channel Manager initiates the payment transaction and redirects the consumer to the Nokia Payment Server. At the end of the payment transaction, Nokia Payment Channel Manager releases the requested HTML/WML page and redirects the consumer back to the merchant application. Nokia Payment Channel Manager controls the delivery of the requested content and requests the removal of the payment transactions in case the delivery of the digital content was not successful.

Nokia Virtual Purse is one of the clearing channels of Nokia Payment Server and holds a virtual account for each registered consumer. Consumers can fill their Virtual Purse from any other available payment methods.

Mobile Commerce Fora

MeT (Mobile Electronic Transactions) Initiative (http://www.mobiletransaction.org)

In April 2000, Nokia, Ericsson and Motorola, the three leading mobile manufacturers, announced the Mobile electronic Transaction (MeT) initiative that aims to establish a framework for secure mobile transactions. Since it was founded, Panasonic, Siemens, Sony and NEC have joined the initiative as new sponsors and over 50 other mobile e-commerce stakeholders, such as operators, financial institutions, service and content providers, and security companies have joined as Associate Members.

The MeT initiative is global and open. Open briefings, workshops and Associate Member Summits have been held to inform and share ideas in developing the framework in the right direction.

The MeT initiative is leveraging on and supporting existing open standards such as WAP when appropriate, but also specifying new ones when needed. A specific concept or functionality might be missing or may not be viable, for example, for implementation in the mobile environment. MeT supports open standards and a fair and non-discriminatory implementation environment. MeT also takes a heuristic view in specifying viable concepts and technologies and how they will work together as a seamless solution.

An important target of the initiative is to specify the core functionality of the Personal Trusted Device (PTD). A PTD is a mobile phone incorporating a comprehensive set of security functionality supported by security elements and thus capable of performing secure transactions. Many services, such as ticketing for example, also require a transaction database. To facilitate secure mobile transactions of all kinds, the functionality needs to be independent of the underlying technology whether it is GSM, TDMA or W-CDMA. Another important aspect to consider is the obvious, but often forgotten need for the functionality to be implementable in mobile phones. The mobile phone environment still differs distinctly from the PC environment having its own limitations and advantages.

The MeT objective of ensuring a consistent user experience independent of device, service and network has been very warmly welcomed by the market. A good example of a consistent user experience is the Automatic Teller Machine (ATM) market. Regardless of the ATM manufacturer, the user experience is similar, however, the user interfaces may differ from each other. Another good example is the control of cars. It does not really matter, which car you step inside of and start driving. Even if the brand is unknown to you, it is relatively easy to use the control devices. Once you have built a mental model of how a car should work, you are able to drive one even if the car is unfamiliar. The mental model for PTD's should encompass all tasks required to operate a PTD, i.e. PTD initialization, User authentication and Digital signing. A consistent user experience increases the trust for services, especially the ones in which money is handled. A consistent user experience needs to ensure that the indication of a secure environment should be visible for the user all the time, e.g. by an icon, in such a way that an external service cannot mimic it. Icon usage should, in other words, be consistent.

Release 1 of the MeT specifications was published in February, 2001 and release 2 is expected in the first quarter of 2002.

Mobey Forum (http://www.mobeyforum.org)

The Mobey Forum (pronounced Mo-Bay) is a financial industry-driven forum, whose mission is to encourage the use of mobile technology in financial services. The formation of the Mobey Forum was publicly announced by the world-leading financial institutions and mobile manufacturers on the 10th of May 2000. Currently, there are three categories of Membership:

1. Founder Members, those who founded the Mobey Forum
2. Membership (open to Financial Institutions and Mobile Phone Manufacturers)
3. Associate Membership is open to all others.

The Mission

The mission of the Mobey Forum is to encourage the use of mobile technology in financial services - such as payment, remote banking and brokerage. It aims to do this by:

- raising the awareness of mobile financial service implementations
- facilitating the open provisioning of mobile financial services
- identifying business considerations and working to obtain the interoperability of the technical and security requirements for the mobile finance industry, in order to promote competition
- acting as an active liaison between various standardisation forums in the mobile industry and the financial industry, so as to promote competition.

The Mobey Forum Workgroups

Business Workgroup

The objective of this workgroup is to investigate business models and to elaborate on business requirements for m-commerce applications.

- By giving input to the Requirements and Technology workgroup
- By promoting pilot projects
- By coordinating activities with other m&e-commerce bodies
- By elaborating on recommendations and best practices

By creating business requirements and models for the financial industry.
The list of items that will be addressed include: Payment systems and m-billing; Payer's and payee's m-commerce infrastructure; Trust infrastructure for m-commerce; Financial applications on the chip card in the mobile phone

Requirements and Technology Workgroup

The scope of this workgroup is to obtain interoperability of technical and security requirements of the financial industry for mobile services and technology.

- By exchanging information on future mobile technologies and their impact on mobile financial transactions
- By exchanging information on future requirements on financial services
- By creating technical and security requirements of the financial industry for mobile services and technology

Rules and Regulations Workgroup

The scope of this workgroup includes:

- Acting as active liaison between various standardisation forums in the Mobile Industry and the Financial Industry. This includes, but is not restricted, to the WAP forum and the Global Platform.
- Encouraging members to participate in the relevant standardisation bodies.
- Collecting and studying legalisation aspects and consequences for the Financial Industry. This includes studying the competition aspects, legislation concerning electronic identification, electronic signature and its archiving aspects etc.
- Patents and intellectual property rights, also concerning the work of the Mobey Forum and other related issues. The Workgroup can give statements regarding internal disputes of the Mobey Forum if requested by the board.

Marketing Workgroup

The Marketing Workgroup is responsible for acquiring new members and building awareness of the forum and its goals within the financial services and m-commerce industries. This will be done:

- By planning and coordinating publicity operations,
- By planning, coordinating and taking care of public relations materials and events,
- By producing and distributing information according to the decisions of the workgroups and the Board of Directors,
- By acquiring research on the issues involved in the workgroups and/or on the Board of Directors,
- By developing and carrying out marketing operations according to a Marketing Plan approved by the Board of Directors.

Forum Nokia – Community for Third Party Developers

Open technologies and industry initiatives are together supporting the transformation of mobile phones from voice-only devices into multipurpose data terminals that will bring a myriad of new usage contexts, balanced efficiency, and increased personal choice for users across the globe. At Nokia, we have a long tradition of emphasizing openness and interoperability. Our support for open standards and platforms enables independent software developers to participate in the development work with Nokia.

To support that work, Forum Nokia (www.forum.nokia.com) was established in 1995 to provide third party developers with online services. The Nokia 9000 Communicator was the first mobile phone allowing third parties to develop applications for it. Hundreds of software development kits were distributed via Forum Nokia and tens of applications were available for the first Communicator in 1996.

Today, Forum Nokia boasts more than half a million registered members. Free registration enables software developers across the world to get acquainted with the opportunities of application and service development for Nokia's broad base of mobile devices. Mobile technologies, such as the Symbian Operating System (OS), WAP, Bluetooth, SyncML and Smart Messaging, are supported in Forum Nokia. Since all of Nokia's products are based on widely adopted standards, the fruits of the development work can be easily modified to benefit the entire industry. A good example of this is the Nokia 9210 Communicator, a multi-task mobile device targeted to meet the varying needs of mobile professionals and corporate users. Using Symbian's OS, the latest communicator with a color-display is an excellent platform for software developers to design add-on functions and applications easily and effectively. Once the sales of the communicator begun, the application development awareness grew, which in turn made application development accelerate. The availability of add-on applications has a positive effect on sales and the market for applications grows.

The history of PCs illustrates this point: computer systems were closed in nature until the 1980s. Application developers had to pay royalties to the patent owner before they could work on their own programs. It wasn't until IBM and its DOS operating system that an open platform came along. The PC revolution picked up speed and many software developers seized the opportunity.

Who Are The Developers

The companies making software for Nokia wireless platforms are not necessarily the same companies who have participated in the PC software industry. It is difficult - almost impossible - to describe a typical developer.

The registered members in Forum Nokia are individual developers or they come from small, mid-sized and large companies, they might be programmers as well as marketing managers or students, their interest in application development might be just amateur enthusiasm or they might have professional ambitions and profit responsibility. Their background is either in traditional software development or they have realized the boost in mobile phone usage and that has influenced them to start mobile application creation. The divergence is a richness, different brains hatch different application ideas and one thing is certain - the interest is global.

Nokia's Role in Wireless Standard Initiatives

Nokia keeps challenging the brains by creating more and more intelligent technologies for the mobile world. Together with other industry players, Nokia participates in the forums by standardizing mobile applications, enabling protocols and drives certain wireless platforms to become de facto platforms for application development. Sub-sets of established technologies, such as Java, have been shaped to become suitable for wireless devices. Unnecessary functionality and Application Programming Interfaces (API) have been stripped and the mobile edition profiled for mobile has been completed. Since the development work has been a result of co-operation between most of the big mobile phone manufacturers, the deployment of these standards expands quickly.

The Challenges of Wireless Application Development

A difference from traditional application development is that the services are not dependent on location or time, which allows the developers to create true value adding services. On the other hand, the challenge is to make the services usable taking the limitations of the mobile devices into account. The user interfaces consist of rather small screens with limited color palettes and varied navigation utilities. The interaction methods vary, there are a diverse amount of softkeys, a touch screen with pen input, voice steering, a joystick or a 8-direction navigation button. The processors are not as powerful as in desktop computers and the performance and capacity in mobile networks is not always sufficient.

However, network evolution and the development of technologies and protocols optimized for the wireless world, has elevated the popularity of mobile applications. The advantages of wireless platforms include the fact that in pocket-sized mobile devices an application can be launched in a few seconds and the possibility to create applications that can communicate wirelessly.

Application Types

Applications can be categorized by the way they use network functionality.

Standalone applications are installed into the mobile device's memory provide additional functionality to the mobile device. Standalone applications can be simple games or office applications such as calculators, stopwatches or spreadsheets.

Client-server applications use a network connection only occasionally, for example when retrieving information from a server based database. Packet data is the optimal connection method, since information is both fed and retrieved - but only occasionally. By adding the possibility of acquiring the necessary information automatically from a network, applications can be really useful. Applications in this category could be news tickers that retrieve selected information a few times a day and push it to the user's screen.

Communicator type products have also triggered a market for specially tailored applications. A customer company has certain specific requirements for an application and often when a client-server solution is in question, the Communicator platform is an optimal solution - a pocket sized device with diverse communication capabilities and an open operating system and attractive set of APIs.

The application category that most often needs a continuous network connection is mobile Internet applications. The usage type emphasizes browsing and receiving information over feeding it. Applications that rely on a continuous network connection can provide quite an attractive user interface to information online.

Forum Nokia Developer Services

In Forum Nokia, development in all these application categories is supported in many ways. We do not simply provide the technical instruments, but also share our knowledge in the design phase and offer various marketing services for the completed product. We also help the developers to connect with customers and vice versa - before and after the application is completed.

Development is stimulated in many ways. Occasionally, Forum Nokia arranges application contests. The contests are often defined to have a certain theme, for example "Mobile Internet" or applications optimized for the Communicator. The winners from around the globe have explored new business opportunities with the contest award as the trigger.

Technology sections

The Forum Nokia web site is divided into different technology sections. From each section the developers can find documents and downloadable files. Today, the supported technologies include Java, WAP, XHTML, MMS, Smart messaging, Bluetooth and SyncML. Additionally platforms, such as Symbian and GEOS are supported. Some Nokia products, Nokia Card Phone and Nokia M2M also allow third party application development, which is supported in dedicated sections. Information of Windows™ based application development for Nokia handsets can be found via the Nokia PC Connectivity SDK developers link. Development possibilities for server side APIs can be accessed through Nokia developer alliances.

Tools

In order to ensure quick and easy application development, Nokia provides developers with programming tools - Software Development Kits (SDK) and other toolkits. A professional tool includes both the application programming environment and an emulator for the wireless device. The emulator is an essential part of the development environment - the developer does not have to possess every single phone model for functionality testing, but they can acquire a new model emulator by downloading it from the Internet. Another advantage is that the emulators can be published and distributed before the actual product appears on the market, enabling consumers to start using the tested applications and services the same day they purchase their new mobile phone. The tools are provided on a CD-ROM or they can be downloaded from the Forum Nokia web site. Nokia is co-operating with several leading programming tool providers to include a developer environment for Nokia mobile phones into the professional programmers tools. The tool houses bring in their expertise in building the debugging and other basic programming necessities, while Nokia's part is to arrange all extra functionality needed for wireless application development and Nokia product-specific features. For example, the installation of finalized applications into Nokia phones is made easy this way.

Documentation

Development tools are often accompanied by thorough documentation. Programming tutorials are often required as well as guides for using the tools. While the latest technologies, such as XHTML, have in-built user interface sensibility, adjusting the content automatically to the screen in question, to achieve best appearance and usability, some optimizing for different user interfaces, interaction methods and screen sizes is needed. For that purpose, Nokia provides detailed information about different user interfaces, API documentation and a collection of best practices for wireless application development. The same rules that are valid in the computer world do not necessarily apply in the wireless environment, where the efficiency of information access is an asset, but the user should still experience the service design as being attractive and visual.

Technical Support

Even good documentation, however, is not always enough. In order to solve technical problems that might endanger smooth implementation, the developer must be provided with technical assistance. Forum Nokia has and a large, capable and enthusiastic community, with thousands of daily visitors. While some developers wish to develop their application in secrecy, others are willing to share tips in the discussion boards, and at the same time learn from others.

While a discussion board for technical support is sufficient for most developers, some companies wish to solve their technical problems together with Forum Nokia's technical assistance team. To guarantee the best possible case handling, Nokia has limited the access for such technical assistance services to the developer alliance level. In the alliance, developer companies also have access to advanced APIs and other services.

Application Testing

Once the application is ready, it may get an approval certificate from Nokia. The Nokia "OK" process includes application screening and technical testing. In the certification process, the application is checked for its suitability to Nokia values. Depending on the technology and application type, testing might also include investigation of some of the most critical usability aspects of the applications user interface, digital application signing and "stress testing". In this latter part, the application will be run in tight circumstances on the device, such as with low memory, with incoming calls, faxes, etc. to determine whether it will interfere with the operation of the device. Digital signing is possible for certain applications, such as installable Symbian applications.

The target product must support the software certificate. After this applications installation package, its sis-file, will be digitally signed. Thus, the user will know where the application originated and the device will not warn the user of an unreliable source, when the application is installed. The application is - depending on the technology - tested either by an external liaison on company or in house. The developer can also book testing time in laboratories around the globe. Nokia is also studying additional methods: testing utilities integrated into development tools and internet based testing. After passing the Nokia "OK" Process, the developer may use the Nokia "OK" logo on the applications marketing material.

Application Distribution

Passed Nokia "OK" tests are a requirement for entering some of the distribution channels that Nokia provides. Application distribution directly to end users is enabled via Club Nokia, Nokia's customer retention program. For network operators and service providers, different channels apply. Some attractive consumer applications have also ended up on sales package CD-ROM and they might also be integrated into Nokia's products.

Developer Alliances

Developers who wish to create applications in closer relationship with Nokia can apply for the Nokia developer alliances. For the alliance member companies, Forum Nokia offers a wide set of exclusive services, including advanced application development tools and tutorials, technical support and access to advanced APIs both for the terminal and the server side. The developer newsletters ensures that the information of new development possibilities is delivered rapidly and efficiently.

Additionally, alliance members are entitled to marketing and business services, boosting visibility for both the company and their software applications. For example, mobile internet application developers in the alliance are automatically linked to the application demonstration server, which is one of the tools for network operators to select applications into their service offerings. Applications developed by alliance level companies have priority position when Nokia selects partners to showcase their products in different events. The alliance is also an important match making tool: members are recognized as being active and talented as well as having long term plans to profit from wireless application development.

Benefits for the Consumers

Third party applications and services give consumers a chance to personalize their mobile device's content and appearance by selecting individual ringing tones and operator logos, bookmarking their favorite services and installing extra applications. If we compare this with computers, there probably aren no two units with the exactly same applications and bookmarks in the whole world! This freedom of choice has now been given to mobile phone users.

Benefits for the Network Operators

For network operators, the large Forum Nokia developer community provides a flexible research and development environment to complement their own departments for service development. Since there are members from over 200 countries, development in many languages is guaranteed. Nokia works closely with network customers and match making with the developers has led to many successful partnerships. The opening of advanced server side APIs for Nokia's trusted developer partners gives the operators a possibility to offer their customers a series of smart services that utilize, for example, the location information received from the network. Another way for network operators to differentiate themselves from competitors is to provide invoicing information fetched from network based billing systems. Such services have proven to be so useful for users that they decide to stick with the operator providing smart services, even when a competing network operator changes their pricing to be more attractive for consumers.

Towards the Third Generation

The third generation networks bring more convenience way using wireless services. The increased data transfer speed can be utilized in many ways. Firstly, it allows new types of applications and services, carrying multimedia information: pictures, sound and even streaming video. Secondly, the applications can be run seamlessly from a server. A service that is not used daily is not required to be installed into memory, but can be downloaded in a couple of seconds when needed. Actually, the user does not even notice that the network connection is established. The storage limitations are no longer a problem even for memory consuming multimedia applications. Processor evolution is also good news for application developers as well are appearing digital convergence products, where different digital devices are integrated into one piece solutions.

Independent developers, each with their own niche market in mind, are helping Nokia to shape the impact that mobile devices have on everyone's lives.

Club Nokia

Club Nokia is an online community and loyalty program designed to help new and existing Nokia phone owners to make the most of their Nokia phone.

The mobile phone is no longer simply a voice communications device, but a very central part of many people's lives, containing their personal information such as calendar and contacts, and a channel for receiving information and entertainment. Club Nokia is a way for us to enhance the user's Nokia phone experience.

Being a member is free and easy. Prospective members can join online via Club Nokia web (http://www.club.nokia.com) or wap (http://mobile.club.nokia.com) or by posting a form that comes with their new phone. Registered Club Nokia members enjoy a wide range of exclusive benefits like instant service and support, sneak previews of new Nokia products, special priority offers, access to the Club Nokia web site, Mobile Club (the Club Nokia WAP extension) and much more.

All Club Nokia services are localized. When a member logs in for the first time via a generic address (www.club.nokia.com or http://mobile.club.nokia.com), the member is guided to a localized site. From that point on, the local Club Nokia handles all correspondence. Thus, Club Nokia services vary from country to country and depend on which Nokia product the enduser has.

Development Principles for Club Nokia Services

Every time a new service is developed for Club Nokia, a strong terminal linkage is built into the service. This is done in order to enhance the value for the end user, so that the terminal can be smoothly customised to match the end-users lifestyle and preferences. Services are designed simultaneously with new terminals, and they leverage the ability to infulence the terminal design.

The aim of Club Nokia service design is to create a seamless end user experience and offer convenient payment methods. An example of such an experience is embedding service access menus in terminal user interfaces.

As the services are based on open technologies, this opens up the possibilities for developers to create different types of content and applications to cater for different end user needs and market segments.

An example of a service embedded in a terminal menu

Club Nokia Digital Services Overview

Club Nokia's digital services cover imaging, music, games and personal information management services. These services manifest themselves to the end user in different ways, today we see ringing tones, operator logos, animated screensavers, picture messaging, game level downloads and high score sending. The end user also has a possibility to possibility to compose own ringtones and create graphics.

In the future, the services will take full advantage new open technologies such as MIDP, XHTML, midi,

Multimedia Messaging Services (MMS) and SyncML. Location based services will also open new possibilities in service creation.

In the content delivery arena, digital rights management will have a key role in enabling secure content delivery to mobile terminals. This is a key enabler that will be built into the service offering.

Club Nokia web provides all Club Nokia members with online storage where they can store their personal data. The storage space can be used for, e.g., personal ringtones and user graphics. These can further be downloaded to the member´s own Nokia phone or sent to friends.

Club Nokia Community

In addition to the Club Nokia digital services descibed above, Club Nokia has community and customer relationship functionality built in. Club Nokia offers information about Nokia products, local Nokia events and campaigns, and information about mobile communication in general. Club Nokia careline services and Club Nokia instant services are designed to make phone maintenance more convenient.

Club Nokia Careline provides technical assistance and information about Club Nokia activities.

When members' phones need to be repaired, they can take it to a Club Nokia Service Point. Members can find their nearest Service Point by looking in the Welcome Pack or checking from the Club Nokia Website. Club Nokia Instant Service is exclusive for Club Nokia members. A membership card and proof of purchase should be shown when visiting a Service Point.

The Club Nokia services are continuously expanded and developed further by Nokia teams. The latest services can always be found at http://www.club.nokia.com

5.
Epilog

Epilog

The Mobile Internet is not conventional Internet accessed from a mobile device. We will not spend our time browsing Internet pages for content as we do today, although this will still be possible. Instead, we will use services and applications to access content, make transactions, do business, link up with friends and family, play games, watch videos and listen to and download music. More important than anything else, we will use the Mobile Internet to help control our lives and give us more time to do the things we choose to do, regardless of our location.

The new environment is about enabling people to shape their own mobile world through personalized communication technology, but it is also an environment where companies make business by matching the consumer needs with their service portfolio or product offering. To ensure the success of the created services in the Mobile World they must be highly user-frienly. Actually, the user should be able to ignore the underlying technologies and purely to enjoy the richness of supplied services no matter what the access method is.

Focusing on the services, the key to commercial success depends on three main factors see the next figure. First, understanding consumer needs, their lifestyles and attitudes, and consequently being able to provide a matching combination of service portfolios and products. Second, readiness of the business system to enable service consumption. And third, understanding that as the consumer needs and business systems evolve, the technical architectures need to match them accordingly.

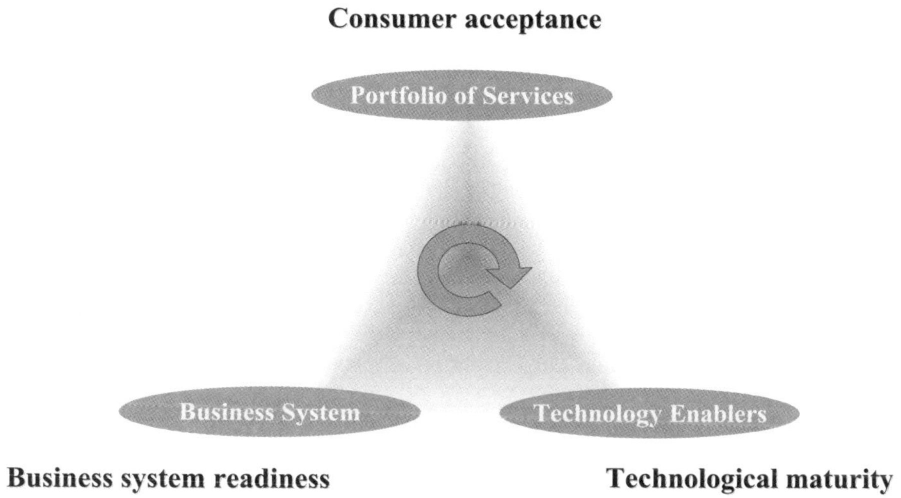

Enabling mobile services.

The principal factor for enabling service consumption is consumer acceptance of the portfolio of services. Nokia has a good understanding of the mobile device consumer segments, the general attitudes and needs of these segments and the characteristics of technology adoption in each of them. This knowledge is applied to new services and through product and user interface categorization it enables an optimal user experience and ease of use in each of the segments.

The second main factor for enabling service consumption is to ensure that a healthy business system exists. Understanding the profit-making logic on the mobile and web domains, and applying this knowledge to the innovation possibilities in the mobile world will combine the best of these domains. To ensure the flow of new services that match the evolving consumer needs, it is important, for example, to independent software vendors, to unlock the innovation potential through solid business models. This consumption of new services benefits the operators through volume and through higher average revenue per user.

The third main factor for enabling service consumption is to understand the technology life cycles and maturity levels. All technology choices should emphasize service enabling - technical architectures can be perceived as end-to-end service enablers.

The key emphasis of this book has been to present the technology viewpoint to enabling the mobile services. Having a common framework and understanding this technical foundation will be important when companies interact with each other and consider introduction of new services for the consumers. To facilitate this, Nokia has taken a global initiative to develop a comprehensive technical architecture for the Mobile Internet.

The initiative, Nokia Mobile Internet Technical Architecture, aims to provide seamless interoperability between all interaction modes, any network environment and with any type of access. The ultimate objective of the initiative is to create a user-friendly Mobile Internet experience for everyone. This will be done by identifying the relevant communication modes, by defining the key technologies required to support them and by driving industry participation to develop a common Mobile Internet platform.

In this book the most essential MITA elements and methodology were presented, and a selected set of the technologies for the Mobile Internet and Nokia's end-to-end solutions for the Mobile World were described.

In the Web Domain section, we introduced the basics of Internet protocols and version 6 with mobility extensions in relation to ad-hoc networks sections. To address an increasing demand of end-to-end controls in the Web domain, the latest QoS technologies of these controls were presented, as an example. Furthermore, to extend the openness in the Web domain, the basics of the Linux operating system were portrayed. Moreover, the implications of voice over IP with SIP for service creation and the implications of JAVA for developers were introduced. The semantic web technology related to browsing were looked into in relation to future technologies and challenges.

In the Mobile domain section, access technologies were presented as the building blocks of the Mobile domain. Security and the technologies related to it are of vital importance when creating a trusted environment for the end-user and these matters were also studied in more detail. General services for a mobile user, such as the multimedia messaging service and location based services were covered. Additionally, the implications of SIP for the mobile domain were looked at from the Presence, Instant Messaging and person-to-person communication perspectives. Technologies linked directly to mobile terminals were also presented.

These two sections presented two domains that should work seamlessly together, i.e. meet the Mobile World challenge. The way Nokia has solved this question was presented in the section that describes Nokia solutions for the Mobile World. Therefore, the network evolution, end-to-end quality of service, the virtual private network proposal and the future of mobility were given an indepth approach. More tangible solutions such as mobile terminals, machine-to-machine and end-to-end product applications and services were also discussed in this section. As presented in this book, Nokia's response to the ever growing demand of new applications and services is Forum Nokia. With this community of third party developers the aim is to ensure the continuous evolution and development of services.

By looking at the different elements that have been described in this book regarding the challenges of Mobile World, one realizes the breadth of multiform issues we are dealing with. By presenting current Nokia solutions we attested how widely Nokia has taken all these issues into close consideration and is going towards the Mobile World with MITA being the guide. Since openness and the standardisation of technologies benefits all players in the field, this book introduces many of the prevailing standardisation committees in which also Nokia actively participates.

All in all, we are in middle of IP Convergence driven technology evolution from three network environments: mobile networks, Internet and Intranets to Mobile Internet. Role of the end-to-end solutions, new technologies like, SIP and Web services, together with increasing programmability in mobile devices leads us to the end-users' requirement driven personalized Mobile World. To achieve this goal of Mobile World mobile devices, networks, and service solutions must work together as an end-to-end unity and developers must have an access to develop new services into it.

An open solution benefits all; profitable business scenarios call for interoperability, short development cycles, large volumes and most of all global reach. Unless there is a commonly accepted architectural solution, the markets will be fragmented as well as requiring separate parameters, and where total volume is much smaller than in a single global market.

Nokia Invites all developers, service providers, operators and other industry players to participate in further clarifying and specifying the market requirements and technical solutions required to provide the Mobile Internet Technical Architecture needed for a superior end-to-end services.

Appendix

3GPP - A Global Initiative

The technical and political debate over 3rd Generation Mobile Systems was reaching a boiling point at the end of the 1990's. The ITU dream of a single globally common 3G standard, IMT 2000, had fallen through due to the political situation, and the new aim was to produce a family of standards sharing the same architectural principles. The commercial advantages of global markets would not compensate the losses in technical investments for the three largest camps, the American, the European, and the Japanese, if they had to give up their technology. Each of them had developed their technology mostly independently of each other.

However, some regional standardisation organisations did have similar technical views, and could possibly agree on a common standard. For example, WCDMA radio technology was being developed both in Europe and Japan. Discussions on creating a common standardisation partnership was started early among the supporters of WCDMA. Finally, 3GPP was formed in December 1998, to produce full set of UMTS (Universal Mobile Telecommunication System) specifications.

3GPP was to a large extent formed around a compromise between the Japanese and the European telecom companies. The key enabling agreement was that the Japanese accepted the European born GSM based Core Network and the Europeans accepted WCDMA radio, which was then viewed as mostly Japanese technology.

The European and Japanese companies managed to also lobby part of the American and Korean standardisation community to join the original partnership. Later China has also joined, and brought its flavour to the technology.

3GPP is a well defined, agreed and committed joint effort among its partners. The scope is to produce specifications for 3G, based on the evolved GSM Core Network and Universal Terrestrial Radio Access (WCDMA). These specifications, with minor or no regional modifications, are in general terms respected as standards in the regional partner organisations. Moreover, the regional standardisation organisations have submitted the 3GPP specification to ITU as an IMT 2000 family member.

Organisation

3GPP constitutes of partners, members and observers working jointly in a functional organisation where technical specifications are developed. Following figure illustrates the structure and relationships in the organisation.

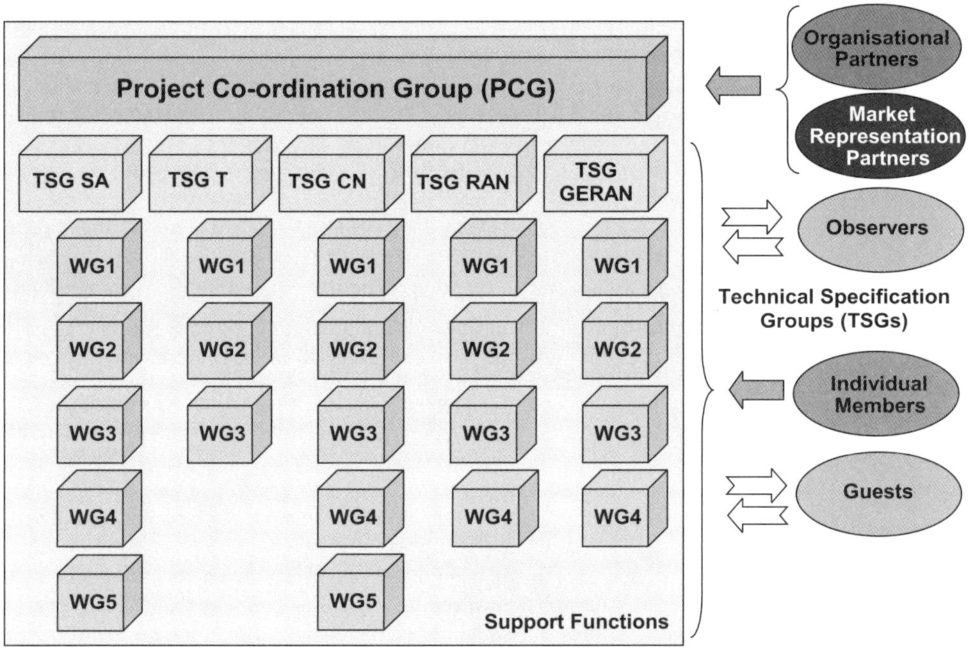

3GPP Organisation

The functional organisation of 3GPP consists of a Project Co-ordination Group (PCG) that administers the work of Technical Specification Groups (TSGs), who subsequently control the Working Groups (WGs). The partners and members participate in the work according to their defined roles, partners being responsible for the overall co-ordination, and individual members contributing to the technical work. A special support function is provided to help the management and production of standard documents. The work may be followed by some observers or guests that might in the future become partners or members. The following chapters provide more details on the function of each organisational block.

Organisational Partners

The Organisational Partners (OPs) are the partnering regional standardisation organisations that have authority to develop standards in their region. They participate in the work mainly at the PCG level, and also meet independently based on need. Their responsibility is to define the future of 3GPP, in the sense of general policy and strategy. This is carried out by maintaining the scope of 3GPP documented in the Third Generation Partnership Project Agreement. Also, the human and financial resource questions, as well as creation or termination of TSGs are the responsibility of the OPs. The OPs can also handle procedural appeals handed to them.

Together with the Market Representation Partners (MRPs), the OPs are also responsible for handling new partnership applications, and ultimately the dissolution of the whole 3GPP organisation.

The following regional standardisation organisations are 3GPP partners (use the provided internet links to find more information):

- ARIB (Association of Radio Industries and Businesses) (http://www.arib.or.jp/arib/english/). ARIB is responsible for radio system research development and standardisation in Japan. It is accredited by the Minister of Public Management, Home Affairs, Posts and Telecommunications.
- CWTS (China Wireless Telecommunication Standard group) (http://www.cwts.org/cwts/index_eng.html). CWTS is the national standardisation body responsible for defining, producing and maintaining wireless telecommunication standards in China.
- ETSI (European Telecommunications Standards Institute) (http://www.cwts.org/cwts/index_eng.html). ETSI's mission is to produce telecommunication standards in Europe and elsewhere. ETSI is the main European telecommunication standardisation organisation, and its standards are widely used within and outside Europe.
- T1 (http://www.t1.org/). Committee T1 is an ANSI (American National Standards Institute) accredited organisation defining standards for interconnection and interoperability of telecommunication networks and end users at various interfaces. T1 is one of several ANSI accredited organisations working in this area in North America.
- TTA (Telecommunications Technology Association) (http://www.tta.or.kr/english/e_index.htm). TTA is responsible for a broad range of telecommunication standardisation activities including multimedia, mobile communication, and conventional telephony. TTA is accredited by law in Korea.
- TTC (Telecommunication Technology Committee) (http://www.ttc.or.jp/e/index.html). TTC is responsible for research and standardisation of protocols for interconnection of telecommunication systems and end users in Japan.

Market Representatives

The Market Representation Partners are industry forums and consortiums that have been invited to join 3GPP, and have agreed to commit to the 3GPP scope. Their role is to provide guidance to the process so that the standards meet the market requirements, and they mainly work together with the OPs at the PSG level to manage the overall direction of 3GPP work. Currently 8 market representatives have signed the 3GPP agreement (Internet links are provided for finding more information on these organisations):

- 3G.IP (http://www.3gip.org)
- GSA (Global Mobile Suppliers Association) (http://www.gsacom.com)
- GSM Association (http://www.gsmworld.com)
- IPv6 Forum (http://www.ipv6forum.com)
- MWIF (Mobile Wireless Internet Forum) (http://www.mwif.org)
- UMTS Forum (http://www.umts-forum.org)
- UWCC (Universal Wireless Communications Consortium) (http://www.uwcc.org)
- WMF (Wireless Multimedia Forum) (http://www.uwcc.org)

Observes and Guests

3GPP is aiming to produce globally applicable telecommunication standards, and one way to ensure this is the ability to expand to new geographical areas, hence mechanisms facilitating the introduction of new partners and members have been defined.

The prospective new partners that have the qualifications to become an OP may become Observers to familiarise themselves with the work. The participation rights of observers will be decided on case by case basis by the existing OPs.

While following the work, observers are encouraged to contribute to the work, in particular by bringing up regulatory issues that might lead to a situation where the 3GPP specification is not applicable to the area governed by that observer.

The following three standardisation bodies have become observers (Please note the web links for further information):

- ACIF (Australian Communications Industry Forum) (http://www.acif.org.au/splash.html)
- TIA (Telecommunications Industry Association) (http://www.tiaonline.org)
- TSACC Telecommunications Standards Advisory Council of Canada (http://www.tsacc.ic.gc.ca/)

Similarly, a mechanism is available for prospective Individual Members. The OPs may assign a right for a company to follow and contribute to some part of the technical work as a guest for a limited period of time.

Individual Members

3GPP membership is open to those members of Organisational Partners that are willing to support 3GPP, contribute to its work, and to utilise the results. The Individual Members are typically operators, vendors and consultancy companies that are planning to become players in the 3G market. Many member companies are large global enterprises, and are therefore members via many OPs.

The Individual Members are the carrying force of the technical work, which is driven by contributions from the member companies. The members also make their experts available to fill the chairman positions in the TSGs.

Support Functions

The support functions are provided by the Support Team that is based on the ETSI premises in Sophia Antipolis, France. They are responsible for the management of the technical meetings and documents. The support team prepares meeting minutes from regular 3GPP technical meetings, and does the final editing of the output documents.

Project Co-ordination Group

The Project Co-ordination Group (PCG) oversees that the technical work is done according to the rules and principles of the partnership. PCG follows that the timeline of specification production is fulfilled. PCG appoints the TSG chairmen based on elections in TSGs and also allocates other resources for TSGs. PCG may also act as a body of appeals from Individual Members on technical or procedural matters.

Technical Specification Groups

Structure of the Technical Work

The technical work in 3GPP is carried out in Technical Specification Groups (TSGs) that are responsible for creating, approving and maintaining the technical specifications and reports in their responsibility area. The work is organised into two hierarchy levels, where each TSG controls several Working Groups (WGs). The detailed technical work takes place in the WGs, whereas TSGs carry the main responsibility for the approval of documents, discussion on new Work Items, and work co-ordination within the TSG.

The following sections introduce the TSGs and their WGs.

TSG Services and System Aspects (TSG SA)

TSG SA is responsible for all system wide aspects of the 3GPP system. This includes the definition of the system architecture, assignment of functionality into sub-systems, and defining the main information flows. TSG SA is also responsible for all aspects of services and bearer capabilities. Furthermore, security, network management, speech coding and charging topics are studied and specified by TSG SA.

In addition to its very central technical work, TSG SA has the responsibility for system wide co-ordination of all technical work in 3GPP. All other TSGs report to TSG SA, and it is often referred to as the de facto plenary of 3GPP. This arrangement makes TSG SA the most important technical forum of 3GPP, since it already makes the technical matters well prepared for the PCG.

The detailed work is carried out in the following five working groups:

- TSG SA WG1, Services: Stage one service work is carried out in this WG. This includes defining the framework and requirements for services, service capabilities and features. The aim is to be as independent of the protocols as possible. The results of the work are used by the other TSGs and WGs as basis for their protocol work.
- TSG SA WG2, Architecture: This is one of the most influential WGs in the whole 3GPP, since it is responsible for defining the system architecture, including the high level capabilities of the whole system. The decisions often affect many other WGs, and other WGs often turn to TSG SA WG2 asking for guidance on technical questions that seem to have system wide aspects.
- TSG SA WG3, Security: This WG is responsible for all security aspects in the 3GPP system including: analysis of threats, development of security mechanisms and algorithms, and security requirement setting for protocols used in various interfaces. The work is often done together with WGs defining protocols and architectures.
- TSG SA WG4, Codec: The codec WG is responsible for defining the codecs for speech, audio, video and multimedia, and setting the corresponding Quality of Service (QoS) and channel coding requirements for other WGs. The work is done taking the end to end efficiency and interoperability aspects into account.
- TSG SA WG5, Telecom Management: The Telecommunication Management Network (TMN) for the 3GPP system and all other technical and co-ordination aspects of management are the responsibility of this WG.

TSG Terminals (TSG T)

TSG T is responsible for all aspects of Terminal Equipment (TE) interfaces. This includes TE performance specifications, and specifying USIM and its interfaces. TSG T is also responsible for end to end service interworking, and service capability protocols, including messaging.

TSG T has the following three WGs:

- TSG T WG1, Mobile Terminal Conformance testing: This WG is responsible for creating specifications on how to test that a Mobile Terminal conforms to 3GPP specifications. The work is further divided into studying radio frequency measurements, Electromagnetic compatibility (EMC) and signalling.
- TSG T WG2, Mobile Terminal Services & Capabilities: This WG specifies the Services and Service capabilities the Mobile Terminal will deliver in accordance with the 3GPP specifications. This includes applications, features and interfaces in the terminal. Messaging is one important application being looked at in this WG.
- TSG T WG3, Universal Subscriber Identity Module (USIM): This WG is responsible for the development of USIM internal and USIM to Mobile Terminal interface specifications. The work also includes the aspect of SIM card roaming between different systems.

TSG Core Network (TSG CN)

TSG CN is responsible for the GSM/UMTS CN, which is an evolution of the GSM and GPRS CNs. This organisation was inherited and moved from ETSI to 3GPP. It is the European component of the 3GPP compromise.

TSG CN is responsible for the layer 3 protocol between the User Equipment (UE) and the CN. It includes Call Control, Session Management and Mobility Management. Also, interfaces with the CN and the interconnection to external networks are in the scope of TSG CN.

The work in TSG CN is carried out in five WGs:

- TSG CN WG1, MM/CC/SM (Iu): This WG specifies the Call Control, Session Management and Mobility Management protocols between the UE and the CN over the Iu and radio interfaces. The SMS protocol is also included. The work currently includes SIP, that is used for session initiation in the IMS (IP Multimedia Sub-system).
- TSG CN WG2, CAMEL: WG2 is responsible for specifying the stage two and stage three protocols for CAMEL (Customised Application for Mobile Network Enhanced logic).
- TSG CN WG3: Interworking with external networks. This group specifies the interworking function needed towards both the terminal in the 3GPP system, and the terminal in the far end in external networks. The work also covers the specification of bearer capabilities both for circuit and packet switched services, especially data services.
- TSG CN WG4: MAP/GTP / BCH/SS. This WG is responsible for stage two and three specification of protocols within the CN. These protocols relate to Supplementary Services, Basic Call Processing, Mobility Management within the Core Network, and the Bearer Independent Architecture. Part of this work includes applying existing standard protocols for use within the CN.
- TSG CN WG5, OSA: The UMTS system uses Open Service Access (OSA) to facilitate service implementations. CN WG5 is responsible for specifying the Application Programming Interfaces (APIs) for OSA.

TSG Radio Access Network (TSG RAN)

TSG RAN specifies the newly defined UMTS Terrestrial Radio Access Network (UTRAN). When the work of TSG RAN was started, a lot of information from ETSI and ARIB were merged to form the first set of draft specifications.

UTRAN interfaces with the WCDMA radio interface to the terminals and connects to GSM/UMTS CN via the Iu interface.

TSG RAN includes four WGs:

- TSG RAN WG1, Radio Layer 1 specification: WG1 is responsible for all physical layer aspects, such as physical channel structures, mapping of transport channels to physical channels, physical layer multiplexing, channel coding, error detection, spreading, modulation, physical layer procedures, measurement provision and UE physical channel capabilities.
- TSG RAN WG2, Radio Layer 2 specification and Radio Layer 3 RR specification: This WG is responsible of all radio interface protocol aspects in radio interface L2 and the radio part of L3. Part of WG2 work is also applicable to GERAN.
- TSG RAN WG3, Iub specification, Iur specification, Iu specification and UTRAN O&M requirements: RAN WG3 is responsible for all protocol work in the UTRAN terrestrial interfaces Iu, Iur and Iub. In addition the WG is responsible for overall UTRAN architecture, and transport of O&M information between Node B (denotes the UTRAN Base Station) and OMC (Operations and Maintenance Center).
- TSG RAN WG4, Radio performance and protocol aspects from a system point of view - RF parameters and BS conformance: This WG produces requirement specifications for Radio Link, System performance, and Radio Resource Management. In addition, BS conformance tests are specified in this WG.

TSG GSM/EDGE Radio Access Network (TSG GERAN)

The 3GPP system is based on GSM CN technology, and the corresponding work has been transferred completely from ETSI to 3GPP. Since the division of GSM work to two different places created a lot of confusion, it was later agreed by the partners that the development work of all GSM specifications can be moved to 3GPP. The remaining part includes radio interface aspects, that are now specified in TSG GERAN.

This organisation also emphasises the fact that GERAN is a prospective radio access technology for 3G services, and bringing it into 3GPP strengthens the evolution of current GSM operators in that direction.

TSG GERAN is responsible for specifying all aspects of the GSM/EDGE radio access network and its internal and external interfaces, including management and conformance testing. TSG GERAN has the following 5 WGs:

- TSG GERAN WG1, Radio Aspects: This WG is responsible for EDGE (Enhanced Data for GSM Evolution) radio interface L1 and RF, and the Ater interface (between CCU (Channel Codec Unit) and the TRAU (Transcoder and Rate Adapter Unit)).
- TSG GERAN WG2, Protocol Aspects: WG2 is responsible for defining the various protocols in GERAN including radio interface L2 and L3 RR (Radio Resource), A, Gb and Abis interfaces.
- TSG GERAN WG3, Base Station Testing and O&M: GERAN WG3 has responsibility over the Base Station conformance testing and GERAN specific O&M matters.

- TSG GERAN WG4, Terminal Testing - Radio Aspects: This WG is responsible for creating specifications for GERAN terminal conformance testing in the area of radio interface L1 and RLC/MAC (Radio Link Control/Medium Access Control).
- TSG GERAN WG5, Terminal Testing - Protocol Aspects: The responsibility of this WG is to create specifications for GERAN terminal conformance testing for all protocols above RLC/MAC, and for higher layer services.

Working Procedures and Methods

3GPP was established to produce documentation very quickly, by maximising the usage of modern working methods like electronic document handling, and by allowing fast decision making at an appropriately low level of the flat organisation. In fact, 3GPP has been successful in this respect. During its first year, 3GPP produced the first release of the specifications, Release 99, containing over 300 TSs and TRs. Together they contain well over 10,000 pages of high quality technical specification. This is quite an achievement even if much of the documentation was based on existing documentation, like the adopted GSM CN specification series.

3GPP work, like any tightly scheduled standardisation work, is committee work where often more than 100 people of over 20 nationalities meet to discuss and debate over technical, and sometimes not so technical matters. The input contributions are presented in Temporary Documents (Tdocs), a few hudred of which are often submitted for each meeting. Most meeting days are well over 10 hours long, making the weeks rather intense.

Needless to say, a well defined process is needed to allow anything sensible to be produced in a bearable timeframe. The 3GPP Working Procedures and 3GPP Working Methods are fundamental documents for the operation of 3GPP; They define how 3GPP works.

The immediate responsibility to drive the work in each group lays on the chairman. The chairmen are selected via elections for two year terms. The candidates are volunteers from the Individual Member companies. The chairman tries to have all committee decisions made by consensus, and voting is used extremely rarely in dispute situations. This is possible because of the generally very good working atmosphere. Every group seems to have their own way of handling difficult situations. Some groups have made it their pride not to have to vote, which shows excellent joint work spirit.

The WGs meet 4 to 8 times a year, normally five working days each time. The WGs produce new versions of Technical Specifications (TSs) and Technical Reports (TRs) for TSG approval. Liaisoning between the groups is essential during the process for coherent specification. To facilitate information exchange, the TSGs meet four times a year. The PCG meetings are held twice a year.

In addition to the scheduled meeting cycle, ad hoc meetings are often held to progress the discussions, especially at the WG level. Each group also has its own e-mail list, where discussions are held and Tdocs are distributed before the meetings. The 3GPP internet pages (http://www.3gpp.org) are used efficiently to share information, e.g. meeting invitations and all meeting documents. In fact, the 3GPP UMTS specifications are openly available via these pages.

In practical terms, the specifications are developed based on written contributions from meeting participants. Initially, when a new specification is produced, an editor is selected to progress the work based on meeting agreements. At this phase the contributions are in free written form, e.g. proposing a principle to be adopted, or text to be added or removed. When the specification reaches a certain state of maturity, the text is frozen. Corrections and additions to frozen specifications are applied through a formal Change Request (CR) process that on the one hand, forces any change to be well premeditated, and on the other, allows the easy monitoring of the extent and nature of the necessary changes.

New work areas are controlled by the TSGs and the PCG via the specified Work Item (WI) handling structure. WI is a generic name for new work areas, and WIs are categorised hierarchically into features, building blocks, and work tasks. A feature can be viewed to represent a system level expression of the work, whereas building block represents the corresponding functionality in a physical or logical entity, and the work task identifies the needed technical work in one of the 3GPP WGs.

Since the first release, Release 99 that was finalised in March 2000, 3GPP has already produced Release 4 in June 2001, and is working towards the completion of Release 5 in early 2002. Note that the release numbering was changed from production year to the document version number. The main scope of the Release 99 was to introduce the new radio access network, UTRAN with WCDMA radio interface. Release 4 includes major enhancements to the circuit switched side of the CN, and Release 5 adds a totally new subsystem, the IP Multimedia Subsystem (IMS) that handles connections to the Internet, making UMTS an integral part of MITA.

References

- Third Generation Partnership Project (3GPP) Partnership Project Description, 04/12/1998, slideset of 47 slides.
- Third Generation Partnership Project Agreement, 04/12/1998, 12 pages.
- Third Generation Partnership Project 3GPP Working Procedures, 17/07/2000, 25 pages
- 3G TR 21.900 V3.3.0 (2000-06), Technical Report, 3rd Generation Partnership Project; Technical Specification Group; Working Methods, (3G TR 21.900 Release 1999), 28 pages.
- http://www.3gpp.org

IETF

The Internet Engineering Task Force (IETF) is the home of the TCP/IP suite of protocols. IETF is not defining any protocols for link layers, and has only limited coverage of the presentation formats used over the Internet. However, more or less everything falling in between is in the scope of IETF standardization, including protocols such as the Internet Protocol itself (IPv4 and IPv6), IPSEC, DNS, TCP, SCTP, SIP, RTP, etc. The programming interfaces for the protocols are usually not defined by IETF.

IETF work is conducted in Working Groups operating within one of the defined Areas. Each Working Group has a Charter and a set of Work Items. When a Working Group completes its Work Items & fulfills its charter, it can either attempt to recharter or to go dormant. Going dormant is a sign of success - the working group has solved the problem it was formed to solve. The Working Groups are overseen by the Area Directors (who collectively form the Internet Engineering Steering Group - the IESG) and the Internet Architecture Board (IAB). Currently, there are nine different areas in the IETF: Applications Area, General Area, Internet Area, Operations and Management Area, Routing Area, Security Area, Sub-IP Area, Transport Area, and User Services Area.

When members are interested in starting a new working group, they can form a 'Birds of a Feather' (BOF) session (taken from the saying, 'Birds of a feather flock together.'). If the BOF gathers sufficient interest and is thought to be working on a useful and solvable problem or problem area, a working group may be formed from the BOF. It is important to note that working groups are chartered to solve a particular problem in a reasonable amount of time.

The Internet Assigned Numbers Authority (IANA, http://www.iana.org) administers protocol parameter numbers and values for all IETF protocols. IANA, IAB, and IESG are chartered by the Internet Society (ISOC, http://www.isoc.org). ISOC also funds the RFC editor function of the IETF.

IETF and its Working Groups are open for anybody to participate. Participation is done on an individual basis, not on a per company basis. Contributions to the IETF are documented in Internet Drafts and the work takes place on WG email lists. Face-to-face meetings scheduled three times a year are reserved for reporting on the status of the Working Groups and solving any outstanding issues that could not be solved on the mailing lists. All IETF contributions are governed by the Intellectual Property guidelines specified in RFC 2026 section 10.

Internet Drafts (IDs) are submitted to the IETF for temporary online distribution. The lifetime of an ID is six months, after which the draft is removed from the online repository if it has not been revised during that period. The numeric suffix before the file type extension identifies the ID revision, starting from '00'. When all technical issues in an ID have been solved, the Working Group chairs can call for a Working Group Last Call, for comments on the draft from the working group. After the comments have been satisfactorily addressed, the ID is sent to the IESG for IESG last call. The IESG last call is the final call for comments on the ID from the general Internet participation. After all open issues are solved, the draft is sent to the RFC Editor for publication as an RFC.

RFCs (Request for Comment) are permanent IETF publications. Usually the RFC text has been available as an Internet Draft before publication as an RFC. RFCs are never modified once they have been published. New revisions of the RFCs are published with a new RFC number, which obsoletes the older one. The old one will be available indefinitely, however.

Both IDs and RFCs are available online through many sources, for example at the IETF web site (http://www.ietf.org). All IETF documents are published in ASCII format with the .txt suffix. The same text can also be made available in the PostScript format (.ps).

There are four classes of RFCs: Standards Track, Informational, Experimental, and Historical. Standards Track RFCs specify IETF protocol standards. Informational RFCs provide useful information about issues of interest to the IETF community, while Experimental RFCs specify protocols and practices for experimental use only. Protocols not recommended for further use can be retired from the Standards Track to Historical status.

The Standards Track includes three maturity levels. Each protocol starts on the Proposed Standard status when approved by the IESG. When independent, interoperable implementations exist and operational experience is available, the standard can be revised and elevated to Draft Standard status. Elevation to Draft Standard is a strong recommendation of the usefulness and value of the protocol. It should be noted that if the protocol implementation requires the use of any licensed technologies, the requirement for independent implementations also includes independent exercises of the licensing processes. Internet Standard (or just Standard) is the final status on the Standards Track, requiring significant implementation and successful operational experience. Standards will get a number on the STD series in addition to a RFC number. For example, the Domain Name System specifications have the STD label STD0013.

The list of current Internet Standards Track documents can be found from STD1, currently RFC 2900 (http://www.ietf.org/rfc/rfc2900.txt). It should be noted that while there are thousands of RFCs, there are less than a hundred standards on the STD series.

ITU

The International Telegraph Union (ITU) was founded in 1865, when 20 European states signed the first International Telegraph Convention. The first telegraph regulations dealt with common rules to standardize equipment, operating instructions and common international tariff and accounting rules. During the first 20 years the ITU concentrated only on telegraph issues, but in 1885 also matters concerning telephony were taken under discussion. A long leap was taken forward when wireless telegraphy studies were started in 1903 in the ITU and the first International Radiotelegraph Convention was signed in 1906.

The International Telephone Consultative Committee (CCIF) and the International Telegraph Consultative Committee (CCIT) were created in 1924 and 1925, respectively. The International Radio Consultative Committee (CCIR) was established in 1927. In 1932 the International Telegraph Union decided to combine the above-mentioned Telegraph and Radiotelegraph Conventions to form the International Telecommunication Convention. It was also decided that the name of the Union would be changed to the International Telecommunication Union.

After the United Nations was established the ITU became a United Nations specialized agency for telecommunications in 1947. The following year the headquarters of the ITU were transferred from Bern to Geneva. In 1956 the CCIF and the CCIT were merged into the International Telegraph and Telephone Consultative Committee (CCITT).

The Plenipotentiary Conference held in 1989 set up the Telecommunications Development Bureau (BDT) in order to give technical assistance to the developing countries. At the same Conference it was decided that the High Level Committee would be created to carry out an in-depth review of the structure and functioning of the Union. The recommendations of the High Level Committee were presented at next Plenipotentiary Conference in 1992. The creation of three sectors was decided: the Radiocommunication Sector (ITU-R), the Telecommunication Standardization Sector (ITU-T) and the Telecommunication Development Sector (ITU-D) into which functions previously carried out by the International Frequency Registration Board (IFRB), CCIR, CCITT and BDT were integrated.

Most of the work in the ITU is currently done in the three Sectors established in 1992. The next big changes in the management, functioning and structure of the ITU are expected in the Plenipotentiary Conference of 2002. These changes will be based on proposals made by the Working Group on ITU Reform (WGR). This group was established by the Council in 1999. The mandate for establishing the WGR had given by Plenipotentiary Conference in 1998. The function of the WGR is to review the management, functioning and structure of the Union and the rights and obligations of Member States and Sector Members.

The ITU has established the World Telecommunication Policy Forum (WTPF) in order to provide a forum where ITU Member States and Sector Members can discuss and exchange views and information on telecommunication policy and regulatory matters, especially on global and cross-sectoral issues. Although the WTPF shall neither process regulatory outcomes or outputs with binding force, it shall prepare reports and opinions for consideration by the members and by ITU working groups. The three WTPFs held so far are the WTPF '96 Global Mobile Personal Communications by Satellite (GMPCS), the WTPF '98 Trade in Telecommunication Services and the WTPF2001 Internet Protocol (IP) Telephony.

Structure of the ITU

ITU is an intergovernmental organization, where governments are represented through their telecommunication Administrations (Member States). Other entities are called Sector Members. These include Regional Operating Agencies (ROA), Scientific Industrial Organizations (SIO), regional and international organizations, etc. The ITU membership includes 189 Member States, 661 Sector Members and 17 Associates (May 2001).

The structure of the ITU is given in the Constitution of the International Telecommunication Union. The Constitution together with the Convention form the basic instruments of the ITU. The modification and approval of these documents are the responsibility of the Plenipotentiary Conference. It convenes every four years and is composed of delegations representing the Member States. It is the highest decision making body in the ITU and it determines the policy, structure and activities of the ITU. The next Plenipotentiary Conference will be held in Marrakesh, Marocco from September 23rd to October 18th 2002.

During the four year period between Plenipotentiary Conferences, the main policy-making body in the ITU is the Council. It is composed of Member States elected by the Plenipotentiary Conference. The duties of the Council are stated in the Constitution and the Convention of the ITU. It is stated among other things that "The Council shall consider broad telecommunication policy issues in keeping with the guidelines given by the Plenipotentiary Conference in order to ensure that the Union's policies and strategy fully respond to the constantly changing telecommunication environment and shall prepare a report on the policy and strategic planning recommended for the Union, together with their financial implications". The Council shall review and approve the biennial budget of the Union and consider the budget forecast for the two-year period following that budget. Each year the Council shall consider the report prepared by the Secretary-General on implementation of the strategic plan adopted by the Plenipotentiary Conference and shall take appropriate action. In addition, the Council shall perform any duties assigned to it by the Plenipotentiary Conference. One of these duties is the invitation for the Council to establish the Working Group on ITU Reform (WGR).

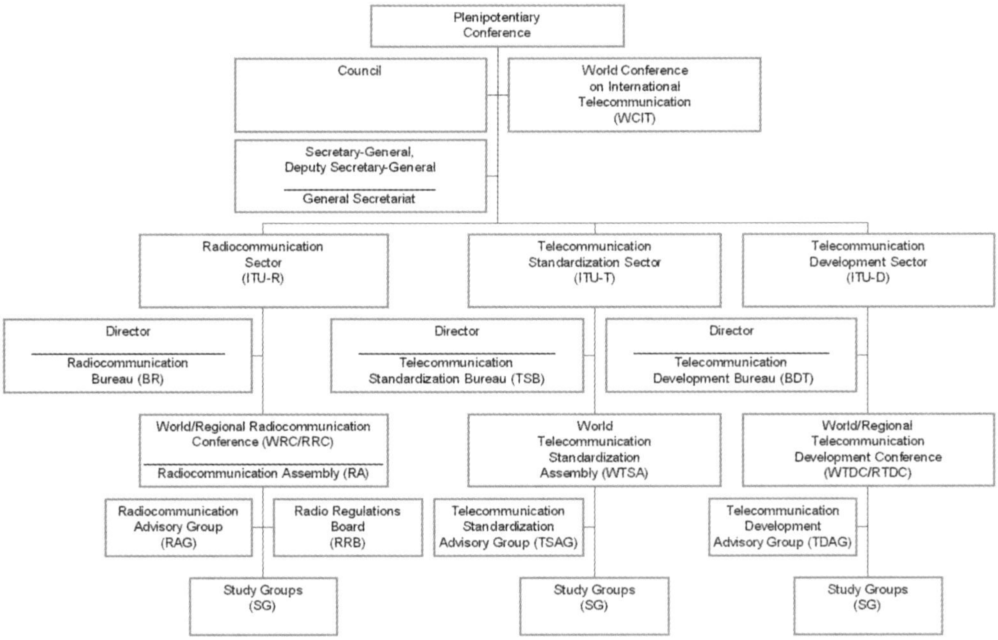

The structure of the ITU

The Sectors of the ITU

There are three Sectors in the ITU, namely, the Radiocommunication Sector (ITU-R), the Telecommunication Standardization Sector (ITU-T) and the Telecommunication Development Sector (ITU-D). Although the structures of the Sectors are quite similar as can be seen in figure 1., the scope and activities of the three Sectors are different.

Radiocommunication Sector (ITU-R)

The mission of the Radiocommunication Sector is, inter alia, to ensure rational, equitable, efficient and economical use of the radiofrequency spectrum by all radiocommunication services, including those using satellite orbit and to carry out studies and create recommendations on radiocommunication matters.

The World Radiocommunication Conference (WRC) and the Radiocommunication Assembly (RA) normally convene every two to three years and the meetings may be associated with each other in order to improve the efficiency and effectiveness of the Radiocommunication Sector. WRC may revise the Radio Regulations, address any radiocommunication matter of worldwide character, and instruct the Radio Regulations Board (RBB) and the Radiocommunication Bureau (BR). In addition it may determine questions for study by the Radiocommunication Assembly and its Study Groups. The Radio Regulation Board considers reports of unresolved interference investigations carried out by the Bureau and formulates Recommendations. It also provides advice to Radiocommunication Conferences and Radiocommunication Assemblies.

The Radiocommunication Assembly is responsible for the structure, programme and approval of radiocommunication studies. The RA approves and issues ITU-R Recommendations and Questions developed by the Study Groups. It is also responsible for approving the programme of technical work, which is carried out in Study Groups.

The Radiocommunication Advisory Group (RAG) reviews priorities, programmes, operations, financial matters and strategies related to the Radiocommunication Assemblies, Study Groups and the preparation of Radiocommunication Conferences. It also provides guidelines for the work of Study Groups.

The technical work in order to draft ITU-R Recommendations is carried out by Study Groups. At present there are seven Study Groups.

ITU-R Study Groups

SG 1 - Spectrum Management
SG 3 - Radiowave Propagation
SG 4 - Fixed-Satellite Service
SG 6 - Broadcasting Services (terrestrial and satellite)
SG 7 - Science Services
SG 8 - Mobile, Radiodetermination, Amateur and related Satellite Services
SG 9 - Fixed Service

Studies in Study Groups are driven by contributions from its membership. Contributions are discussed at meetings, with assessment made by Rapporteurs and the management team of the groups. There are several levels of meetings including those by Study Groups, Working Parties, Rapporteurs etc.

The Working Party 8F (WP8F) in ITU-R is responsible for the overall system aspects of IMT-2000. WP8F has developed the vision of IMT beyond IMT-2000. WP8F is in close collaboration with the ITU-T Special Study Group on "IMT-2000 and Beyond" in order to develop a common ITU vision of systems beyond IMT-2000. It has also been agreed that increased collaboration with external organizations is needed in the development of systems beyond IMT-2000.

Telecommunication Standardization Sector (ITU-T)

The main functions of the ITU-T is to study technical, operational and tariff questions and to issue recommendations on them with a view to standardizing telecommunications on a worldwide basis. Technical or operating questions specifically related to radiocommunication are the responsibility of the Radiocommunication Sector. Currently, in ITU-T, the topics of high priority are IP-related issues, IMT-2000 and tariff and accounting issues. Other key topics include optical fibre transmission technologies, access networks, multimedia, TMN, signalling, numbering and addressing and global interconnectivity and interoperability.

The Membership of ITU-T (August 2001) includes 189 Member States, 454 Sector Members (180 ROAs - Recognized Operating Agencies, 232 SIOs - Scientific and Industrial Organizations including manufacturers, 42 Others (Regional and International organizations, etc.)) and 19 Associates.

The World Telecommunication Standardization Assembly (WTSA) is organized every four years to consider specific matters related to telecommunication standardization. It defines the general policy for the Sector, establishes the Study Groups and approves their work programmes for each study period of four years and appoints the Study Group Chairman and Vice-Chairmen. In addition the WTSA considers the reports of the Study Groups, approves, modifies or rejects draft Recommendations and considers the reports of the Telecommunication Standardization Advisory Group (TSAG). The WTSA may assign specific matters within its competence to the TSAG.

The Telecommunication Standardization Advisory Group acts between WTSAs reviewing priorities, programmes, operations, financial matters and strategies for the Sector. It also follows up on the accomplishment of work and provides guidelines for the Study Groups. The last WTSA, held in October 2000 in Montreal, Canada, authorized TSAG to take action in some new areas. For instance, TSAG may create groups with a short lifetime to address items that require rapid reaction. It may also restructure and establish ITU-T Study Groups and assigns chairmen and vice-chairmen in response to changes on the telecommunication market.

Standardization work in ITU-T is carried out by 13 Study Groups (SGs), where Recommendations are developed on the basis of the study of Questions (i.e. areas for study). Studies in the Study Groups are driven by contributions from its membership. Contributions are discussed at meetings (Study Group, Working Party, Rapporteur, Joint meeting), with assessments made by Rapporteurs and the management team of the Groups.

ITU-T Study Groups

SG 2 - Operational Aspects of Service Provision, Networks, and Performance
SG 3 - Tariff and Accounting Principles Including Related Telecommunications Economic and Policy Issues
SG 4 - Telecommunication Management, Including TMN
SG 5 - Protection Against Electromagnetic Environment Effects
SG 6 - Outside Plant
SG 9 - Integrated Broadband Cable Networks and Television and Sound Transmission
SG 11 - Signalling Requirements and Protocols
SG 12 - End-to-end Transmission Performance of Networks and Terminals
SG 13 - Multi-Protocol and IP-based Networks and their Internetworking
SG 15 - Optical and Other Transport Networks
SG 16 - Multimedia Services, Systems and Terminals
SG 17 - Data Networks and Telecommunication Software
SSG - Special Study Group on "IMT-2000 and Beyond"

During the operation of the CCITT (1956-1988) all the Recommendations that had been readied during the four year period between Plenary Assembly meetings were approved by the Plenary Assembly. The approved Recommendations were published as books, which were named after the color of their covers. The Recommendations, approved by the last Plenary Assembly in 1988, were issued in "Blue Books". The approval time of 4 years and the publication time of 2-4 years for Recommendations was much too long. The procedures were changed so that the Study Group that was responsible for drafting a Recommendation could also approve it. When Recommendations were approved they were published as separate documents. At the end of 2000 there were 2625 Recommendations in force.

In October 2000, the World Telecommunication Standardization Assembly made some additional improvements to the approval procedure by adopting an Alternative Approval Process (AAP). The AAP can only be used for technical Recommendations. Recommendations that have policy or regulatory implications (numbering/addressing or tariff/charging/accounting) will be approved using the Traditional Approval Process (TAP), where a formal adoption by Member States is required. Currently ITU-T has one of the fastest Recommendation approval processes among standardization organizations.

The Alternative Approval Process allows, if a draft Recommendation is sufficiently mature, the Study Group Chairman to request the Director of TSB to initiate the last call. The last call encompasses a four-week time period and procedures beginning with the Director's announcement of the intention to apply the Alternative Approval Process. If no comments are received by the end of the last call period, the Recommendation is considered approved. The approved Recommendation will be available as a Pre-published version on the ITU-T website in a couple of weeks.

ITU-T Recommendations, agreed by consensus in Study Groups, are non-binding standards. Regardless of this, Recommendations are generally complied with because they guarantee the interconnectivity of networks and enable services to be provided on a worldwide scale.

During the reorganization of ITU-T Study Groups at WTSA-2000 a Special Study Group on "IMT-2000 and Beyond" (SSG) was established. The SSG is responsible for studies relating to the network aspects of International Mobile Telecommunications 2000 (IMT-2000) and beyond, including wireless Internet, convergence of mobile and fixed networks, mobility management, mobile multimedia functions, internetworking, interoperability and enhancements to existing ITU-T Recommendations on IMT-2000.

As ITU-R Study Group 8 had earlier established Working Party 8F, which has (on the ITU-R side) the responsibility of developing IMT-2000 and beyond, it was decided that coordination between ITU-T and ITU-R is needed. The SSG collaborates with ITU-R Working Party 8F on the radio aspects of terrestrial elements and with ITU-R Working Party 8D for satellite elements. It was also decided that the SSG shall maintain strong cooperative relations with external Standards Developing Organizations (SDO) and especially with 3GPP in order to avoid duplication of effort or overlaping projects.

The SSG and ITU-R WP 8F are currently collaboratively developing a common ITU vision of systems beyond IMT-2000. Cooperation has also been made with ITU-T SG 11, 3GPP, 3GPP2, ETSI, T1, TTA, TTC and ARIB.

The first Recommendation drafted by the SSG is expected to be approved in the first half of 2002.

In addition to the cooperation of the SSG with other Standards Developing Organizations also other ITU-T Study Groups have valuable cooperation with organizations like ISO, IEC, IETF, ETSI TIPHON, ATM Forum and MPLS Forum. Procedures for the cooperation are given in ITU-T Recommendations A.4, A.5, A.6 and in WTSA-2000 Resolution 7.

Telecommunication Development Sector (ITU-D)

The Constitution of the ITU determines the specific functions of the Telecommunication Development Sector. According to the Constitution the ITU-D shall promote the development, expansion and operation of telecommunication networks and services, particularly in developing countries by reinforcing capabilities for human resources development, planning, management, resource mobilization and research and development. It shall also mobilize resources to provide assistance to developing countries in the field of telecommunications and promote and coordinate programmes to accelerate the transfer of appropriate technologies to developing countries.

Every four years ITU organizes the World Telecommunication Development Conference (WTDC), which serves as a forum for free discussion by all concerned with the Development Sector. The WTDC establishes work programmes and guidelines for defining telecommunication development questions and priorities and provide direction and guidance for the work programme of the Telecommunication Development Sector. The next World Telecommunication Development Conference (WTDC-02) will be held 18-27 March 2002, Istanbul, Turkey.

The Telecommunication Development Advisory Group (TDAG) reviews priorities, programmes, operations, financial matters and strategies for activities in the Telecommunication Development Sector. It also provides guidelines for the work of Study Groups.

The Study Groups in ITU-D study specific telecommunication Questions of general interest to developing countries and prepare draft Recommendations. Important tasks also include the preparation of guidelines, handbooks, manuals and reports within the areas of competence of each Study Group and pay particular attention to the needs and concerns of the least developed countries in furthering the work. There are currently two Study Groups in ITU-D. Study Group 1 deals with telecommunication development, strategies and policies and Study Group 2 deals with development, harmonization and management of telecommunication networks and services. In the final report to the Council the Working Group on ITU Reform recommends that Study Groups in ITU-D should be replaced by project management groups.

ITU Reform

The Plenipotentiary Conference, held in 1998 in Minneapolis, USA, mandated the Council to establish the Working Group on ITU Reform (WGR), which was an open workgroup of Member States and Sector Members. WGR was established in 1999. Its terms of reference were to review the management, functioning and structure of the Union as well as the rights and obligations of the Member States and Sector Members. In addition, WGR should review the contribution of the Sector Members towards defraying the expenses of the Union.

The Strategic policies and financial matters of the ITU are discussed and decided in the Plenipotentiary Conference and in the Council. In these bodies only Member States are represented. This is one reason why the Sector Members are complaining that their role is not sufficiently recognized in the ITU and especially in ITU-T.

The decision-making process and possibilities to act are totally different in the ever increasing number of forums and consortiums from those in the ITU. Forums and consortiums are usually driven by private industry and governments having almost no role at all. In addition, forums and consortiums have very flexible framework for the development of new standards.

The success of forums and consortiums puts more pressure on changing the ITU. The Working Group on ITU Reform submitted its final report to the Council at the June 2001 meeting. WGR recommended to the Council that it should request the Director of the Telecommunication Standardization Bureau (TSB) to provide to the Plenipotentiary Conference 2002 (PP-02) a report of the activities of the Telecommunication Standardization Advisory Group (TSAG) in relation to a Pilot Forum. WGR also recommended that the Council should ask PP-02 to consider whether a Forum should be established and, if so, how. WGR recommended a two-stage approval process for ITU-T technical Recommendations in which both Sector Members and Member States would participate.

Reforming the ITU was the main discussion item in the annual session of the Council in June 2001. No big decisions were made concerning the Reform of the ITU and for instance the much-debated proposal to establish a Pilot Forum for global standardization under the umbrella of the ITU was forwarded to the Plenipotentiary Conference 2002 for further discussion. In order to get some progress the Council agreed that work on the Pilot Forum should continue within TSAG.

ETSI

The European Telecommunications Standards Institute, ETSI is a non-profit organization whose mission is to produce telecommunications standards. It is an open forum that unites almost 900 Members from 54 countries. In its membership it brings together manufacturers, network operators and service providers, administrations, research bodies and users.

ETSI membership is open to interested organizations world-wide. A variety of working modes is available from typical committee to global partnership projects, like 3GPP and including Forum Hosting and interoperability test event (Plugtest) facilities. ETSI was also amongst the first ones to make its deliverables freely available on the Web.

According to its Mission ETSI shall, through international collaboration, including Partnership Projects, pursue the objective of developing globally applicable deliverables, meeting the needs of the telecommunication and electronic communication community while still fulfilling its duty to support EU and EFTA regulation and initiatives.

The objective of ETSI is to produce and maintain widely implemented technical standards and other deliverables as required by its members. As a recognized European standards organization, the objective should be reached in a way so as to support and enhance competition in a unified European market for telecommunications and related areas. On the international level, ETSI contributes to world-wide standardization.

ETSI was founded in 1988. The creation of ETSI was based on changes in European telecom structures when ministry, regulatory authority and network and service provider roles were split in many cases. Also efficiency, openness, quality and global competiveness were considered. An important goal was the harmonization of the European Telecoms market based on common technical standards. The European Commission published (1987) its Green Paper on Telecommunications, which presented the idea of ETSI. In 1987 CEPT (representing at that time adminstrations and operators) also had, in principle, taken the steps and made the decision to establish ETSI in partnership with the European manufacturing industry and users. The original independence, openness and one table concepts were laying firm ground for the future evolution of the ETSI organization.

Today, ETSI is a market-driven organization supporting global needs. Its members directly represent all aspects of the industries, decide on the work programme and allocate resources accordingly. Since its creation in 1988, ETSI´s approach has been to provide comprehensive project and technical support to its Technical Bodies. Also characteristic to ETSI has been its willingness to evolve and develop as an organization in order to meet the requirements of its members.

ETSI Membership as of May 2001:

- 660 Full Members from 35 European countries
  - 7.6 % Administrations
  - 50.3 % Manufacturers
  - 16.1 % Network Operators
  - 22.7 % Service Providers and Others
  - 3.3 % Users
- 174 Associate Members from 19 countries
- 55 Observers from 18 countries

ETSI Structure

As the highest decision making authority in ETSI, the General Assembly (GA) is responsible for determining policy, agreeing budgets, dealing with membership issues, appointing the Board members, appointing the Director-General and Deputy D-G, endorsing external agreements, and approving the Statutes & Rules of Procedure for the Institute. The GA normally meets twice per year (Spring & Autumn).

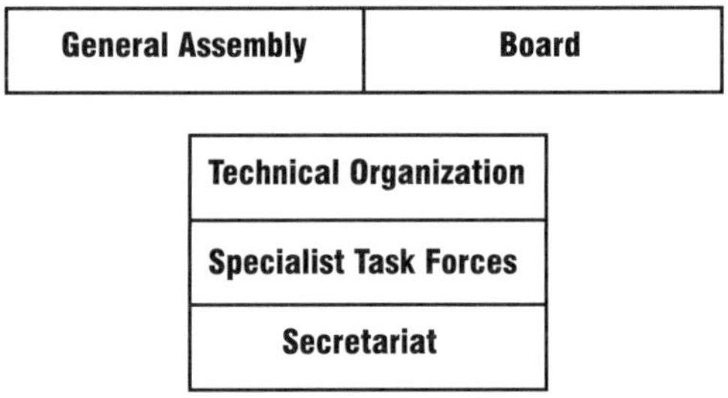

The Board is a body that acts on behalf of the General Assembly between GA meetings by exercising those powers and functions that are delegated to it by the General Assembly.

The Technical organization consists of Committees, Projects, Partnership Projects and Special Committees. One of the Special Committees, the Operational Coordination Group (OCG) will act as a focal point and a forum for co-ordination of the Technical Bodies (TCs/EPs), and between the Technical Bodies and the Secretariat. It shall resolve, as far as possible, any duplication of effort or conflict of technical views between the Technical Bodies to reinforce co-operation within the Technical Organization. Technical Body Chairmen ensure that their Technical Body is properly represented in OCG meetings.

Expert teams (STFs) can be set up to support the Technical Organization to speed up production of urgent deliverables and deliverables requiring specific concentrated expertise, like test specifications do.

STFs experts work together on the ETSI premises of Sophia Antipolis. Funding is provided to compensate companies for providing experts.

The ETSI Secretariat supports the activities of the Institute and is also located in Sophia Antipolis, the headquarters of the Institute. It is headed by the Director-General. From the beginning ETSI has had strong expertise within the Secretariat and is prepared to provide a variety of support functions for the members participating in the standards and specifications development process.

Technical Organization

The following list of ETSI Technical Bodies describes the situation as of September 2001. Some bodies, having activities related with the Mobile Internet, will be briefly described later.

Technical Committees	
AT	Access and Terminals
ECMA TC32	Communication, Networks & Systems Interconnection
EE	Environmental Engineering
ERM	EMC and Radio Spectrum Matters
HF	Human Factors
JTC Broadcast	EBU/CENELEC/ETSI Joint Technical Committee
MSG	Mobile Standards Group
MTS	Methods for Testing & Specification
Safety	Telecommunications Equipment Safety
SEC	Security
SES	Satellite Earth Stations & Systems
SPAN	Services and Protocols for Advanced Networks
STQ	Speech processing, Transmission & Quality aspects
TM	Transmission and Multiplexing
TMN	Telecommunications Management Networks

ETSI Projects	
BRAN	Broadband Radio Access Networks
DECT	Digital Enhanced Cordless Telecommunications
M-COMM	Mobile Commerce
PLT	PowerLine Telecommunications
RT	Railway Telecommunications
SCP	Smart Card Platform
TETRA	TErrestrial Trunked RAdio
TIPHON	Telecommunications and Internet Protocol Harmonization Over Networks

ETSI Partnership Projects	
3GPP	Third Generation Partnership Project
MESA	Public Safety Partnership Project

Special Committees	
ETSAG	European Telecommunications Standards Awareness Group
FC	Finance Committee
JEEC	Joint ETSI/ECMA Committee
NGN	Next Generation Networks
OCG	Operational Coordination Group
SAGE	Security Algorithms Group of Experts
USER	User Group

Highlights of ETSI Technical Activities

MTS (Methods for Testing & Specification)

MTS is responsible for the identification and definition of advanced specification and conformance testing methods, which take advantage of formal approaches and innovative techniques to improve the efficiency and economics of both the standard description and associated conformance testing processes. A recent major achievement is the finalization of version 3 of Tree and TabularCombined Notation (TTCN-3), a flexible and powerful language, which is capable of specifying many types of system tests over a variety of communication interfaces. TTCN-3 was approved by ETSI and was forwarded to the ITU.

TTCN-3 can be used for protocol testing (formobile and Internet protocols), supplementary service testing, module testing, the testing of platforms based on the Common Object Request Broker Architecture (CORBA), the testing of Application Programming Interfaces (APIs) and many other applications. The language is not restricted to conformance testing, and can be used for interoperability, robustness, regression, system and integration testing. Work on the graphical presentation format continues.

A significant achievement was the development of the draft ASN.1 Encoding Control Notation (ECN) for the Third Generation Partnership Project (3GPP) in close collaboration with mobile telecommunications experts. ECN is much more flexible than previous ASN.1 encoding processes and is therefore particularly suitable for UMTS.

BRAN (Broadband Radio Access Networks)

ETSI Project BRAN is responsible for the standardization of Broadband Radio Access Networks. EP BRAN produces specifications for two major standard areas. High Performance Radio Local Area

Networks type 2 (HIPERLAN2) is a mobile broadband short-range access network technology. High Performance Radio Access (HIPERACCESS) is a fixed wireless broadband access network technology.

Among the achievements of the EP BRAN are the completion of the main parts of HIPERLAN2 standard, including the Data Link Control layer extension for home applications, business and home profiles, network management and the generic part of the IEEE1394 Service Specific Convergence Sublayer. In order to ensure the interoperability of devices, EP BRAN has also completed a set of conformance test specifications for the core HIPERLAN2 standards, covering both radio (as a Harmonized Standard) and protocol. Work has started on the specification of HIPERLAN2 interworking with 3G systems.

The importance of the 5 GHz band for wireless local area networking has been recognized and a wish has been expressed for the alignment of current standards in this band, as a first step towards a single global standard. EP BRAN is cooperating in this with the Institute of Electrical and Electronic Engineers (IEEE) in the USA and the Multimedia Mobile Access Communications Promotion Council (MMAC) in Japan.

HIPERACCESS will offer standards for broadband multimedia fixed wireless access, allowing a flexible and competitive alternative to wired access networks. It will be an interoperable standard, promoting a mass market and thereby low cost products. HIPERACCESS is a point-to-multipoint TDMA system. It is targeting high frequency bands, in particular the 31.8-33.4 GHz and 40.5-43.5 GHz bands.

EP BRAN has also agreed to work on the development of standards for Fixed Wireless Access Systems below 11 GHz, called HIPERMAN (High PErformance Radio Metropolitan Area Network).

SCP (Smart Card Platform)

SCP is responsible for, the development and maintenance of a common Integrated Circuit (IC) Card platform for all mobile telecommunication systems, the application independent specifications for the interface with terminal equipment and of IC Card standards for general telecommunications and high security applications. Smart card activity relevant to GSM and 3GPP first passed to the 3GPP Partnership Project, but later ETSI created EP SCP to continue generic work on smart card specifications.

ETSI Project SCP concentrates on the task of creating a smart card platform for 2G and 3G mobile telecommunications on which other organizations can base their system specific applications. This will, in particular, allow users access to global roaming by means of their smart card, irrespective of the radio access technology used. The platform specification will allow the introduction of true multi-application smart cards, like using both a 3G and a banking application in parallel on the same card.

Work in the area of the Public Key Infrastructure is part of the scope, including work on enhancing the administration standard to include generic download of data independent of the transport mechanism. EP SCP is also looking into deriving common requirements from the existing GSM smart card specifications resulting in a generic Card Application Toolkit (CAT) as a new part of the platform.

TIPHON (Telecommunications and Internet Protocol Harmonization Over Networks)

TIPHON is responsible for the standardization of Internet Protocol (IP) telephony including communication between IP-based networks and PSTN, ISDN and GSM. ETSI Project TIPHON has completed the specifications defining the requirements, architecture and protocol profiles for interoperable IP telephony services. Phase 2 covers communication from the Switched Circuit Network (SCN) to IP telephony devices and SCN to SCN, using the IP network as a trunk connection. Release 3 provides a set of standards for operating basic telephony services and covers architecture, services, protocols, protocol profiles, numbering and naming, security profiles and design guidelines for Quality of Service. Release 4 will enhance the details of Release 3.

EP TIPHON has adopted an architecture and protocol framework – TIPHON meta-protocol that is independent of specific call signalling protocols and supports a range of real-time services and capabilities, including the ones currently provided over SCNs across multiple administrative domains. The TIPHON meta-protocol is used to generate profiles for the protocols of given communications network technology, including H.323, SIP, H.248 and BICC.

ETSI and the International Multimedia Teleconferencing Consortium (IMTC) have a co-operation agreement, which is designed to facilitate the technological developments needed to enable VoIP networks to interwork with SCNs. TIPHON is also an active user of the ETSI Plugtests interoperability events service.

JTC Broadcast (EBU/CENELEC/ETSI Joint Technical Committee)

JTC Broadcast is responsible for the standardization of broadcast systems for television, radio, data and other new services via satellite, cable, Satellite Master Antenna Television (SMATV) and terrestrial transmitters, and for the transmission of programmes.

JTC Broadcast is mostly associated with DVB and Digital Audio Broadcasting (DAB). DVB Satellite solutions are adopted world-wide. The basic transmission systems for both DVB and DAB were standardized by ETSI some years ago, and attention has now concentrated on additional features, such as data broadcasting, interactivity or transmission of Internet Protocol packets over broadcast channels. The over 1000 page DVB Multimedia Home Platform (DVB-MHP) specification should be mentioned in particular. It describes a software Applications Programming Interface (API) for interactive broadcast services.

There was considerable conserning around the publishing of both the DVB Cable modem and a European variant of the DOCSIS system developed by Cable Labs in the USA. After some delay, the revised version of the DVB Cable Modem specification was published as ES 200 800 in April 2000, whilst the EuroDOCSIS specification was published as ES 201 488 in November 2000.

3GPP (Third Generation Partnership Project)

3GPP was established by ETSI and regional standards development organizations from Asia and the USA, with participation from industry groups and over 400 individual companies, to develop a complete set of globally applicable Technical Specifications for a 3rd Generation (3G) mobile system. The Specifications are based on the evolved GSM core network and an innovative radio interface known as UTRA, the Universal Terrestrial Radio Access, using Wideband Code Division Multiple Access (W-CDMA) technology. GSM radio access was later included in the Partnership Project.

As of September 2001, the Organizational Partners of 3GPP are ARIB (Japan), CWTS (China), ETSI, TTA (Korea), TTC (Japan) and T1 (USA). The Market Representation Partners are GSA, GSM Association, UMTS Forum, UWCC, IPv6 Forum, 3G.IP, MWIF and WMF. 3GPP also has three observer organizations: TIA (USA), TSACC (Canada) and ACIF (Australia).

At the beginning of 2000, 3GPP completed the first full set of circa 300 specifications, known as Release 99. The accomplishment of Release 99 in just one year represents an unprecedented achievement in standardization. The next release, Release 4, was semented in March 2001.

From the ETSI partnership point of view, 3GPP is treated like one of the ETSI Technical Bodies and approval for specifications in 3GPP meetings makes them ETSI Technical Specifications. ETSI hosts the support functions for 3GPP, called the Mobile Competence Centre, with 27 international team members. ETSI also hosts the 3GPP Web Site. In the beginning, 3GPP inherited lot of content and many of the successful working practices from the former ETSI SMG.

3GPP is described in more detail separetely in this book.

Global Co-operation from the ETSI Perspective

To ensure the widest possible application of its deliverables, ETSI works in partnerships with other organizations around the world. Currently ETSI has over 50 co-operation agreements and MoUs.

As one of the three officially recognized European Standards Organizations (ESO), ETSI is participating and supporting initiatives within collaborative frameworks such as the Information and Communications Technologies Standards Board (ICTSB), Global Standards Collaboration (GSC), and Global Radio Standardization (RAST).

ETSI highlights the need for bridging the gap between different dimensions of traditional telecommunications, Internet Protocol (IP) and broadcasting environments by following the market trends and creating key synergies.

ETSI has a number of co-operation agreements and Memoranda of Understanding (MoUs) with external bodies ranging from the European Commission, European Standards Organizations, European Free Trade Association (EFTA) and other European regulatory bodies, to National/Regional and International Standards Development Organizations (SDOs), to Fora, R&D bodies and Specification Providers. The key to all these co-operation frameworks is that the organizations concerned have a direct interest in telecommunications standardization or in research & development activities that might eventually lead to an input to the standards making process.

Support Functions

The ETSI Secretariat provides a wide range of support functions. Many of these have been essential for rapidly achieving good quality standards. These support functions have also been a requirement for the success of 3GPP.

The Secretariat supports the Standards Making Process and the elaboration of timely high quality ETSI deliverables by providing Technical Officers who facilitate the operations of various Technical Bodies. In addition, editing and approval support services are provided. 1 717 deliverables, representing 173 601 pages, were published during 2000, which corresponds to one published standard per working hour, or one page published every 35 seconds.

With the help of its Members, the ETSI Secreatariat has made efforts to increase the market acceptance of ETSI standards and deliverables throughout the world by promotion, membership recruitment and distribution of deliverables.

In specific cases the set-up of Specialist Task Forces (STFs) is decided by the members and managed by the Secretariat. This possibility to temporarity hire experts to do certain tasks or prepare parts of specifications, has enhanced the quality of the results and made them available faster.

The Secretariat provides support to the GA, Board and OCG, provides legal advice and interpretation and manages the external relations agreements and MoUs with other bodies. The Secreatariat also hosts numerous Technical Body meetings at the headquarters.

FORAwatch is an activity which provides an overview guide with links to telecommunications related fora and consortia.

Operating and developing the IT infrastructure, including the extensive web pages and applications, is also one of the major support functions.

Competence and Service Centres in ETSI

ETSI is applying and developing the model of competence and service centres, which address the needs of the organization in a horizontal manner by supporting several technical bodies within ETSI. This complements the (partnership) project model for vertical technological issues.

The Mobile Competence Centre was set up to support 3GPP and other mobile standardization work. Hosted by ETSI, it is now an integrated team of 27 persons, made up of ETSI Secretariat Officers, assistants and contracted experts, drawn from more than a dozen countries across four continents and supported from both voluntary and funded resources. The MCC has been prolific: during 2000, the Centre was managing 1 327 active specifications (461 on 3G and 1 156 on GSM) and processed a total of 5 716 change requests for GSM and 3G specifications.

The Protocol and Testing Competence Centre provides support and services to ETSI Technical Bodies on the application of modern techniques for specifying protocols and test suites. The PTCC is also responsible for managing the Specialist Task Forces (STFs) which develop test specifications for ETSI standards. In 2000, the PTCC managed/supported 16 STFs, three tasks for the MCC and assisted with the development of test specifications for 3G terminals. The Centre played a major role in the specification of the new version of TTCN-3, the test language developed by ETSI Technical Committee MTS. An informal global TTCN-3 community now exists, centred around ETSI and the PTCC.

The ETSI PLUGTESTS Service is a professional service that specializes in the organization of interoperability testing events for any Telecommunication, Internet or Information Technology standard. Its customers are ETSI Technical Bodies, Internet Engineering Task Force working groups, ITU Study Groups, or any forum or interest group, which is developing a standard. An interoperability testing event is an opportunity for developers to get together to validate a draft standard and its implementations. The benefits are numerous and overall these events save companies and the industry considerable time and money. The introduction of this service is proving popular with engineers and developers.

ETSI's Fora Hosting Service is prepared to provide day-to-day administrative services to interested Fora. Currently the hosting service is provided to the Location Interoperability Forum (LIF). ETSI is now looking to increase the number of hosted fora.

Information Tecnology in Specifications Creation

ETSI has an information technology strategy where the use of electronic and telecom tools in specifications creation are maximized, including document availability.

The definition, development and maintenance of electronic working tools continues, both for use internally within the Secretariat for tasks such as the automation of editing and document management, and for external use, in the interfaces between the Technical Organization and the Secretariat. The introduction of 'paperless meetings' has brought a considerable saving in costs, and the use of Internet servers and services is steadily increasing. Practical solutions include use of LANs and WLANs, distribution of meeting documents via a web server and email and the web casting of some meetings.

Electronic working is now employed throughout the whole standardization process, and provides a speedier, streamlined process where members from all around the world can participate. The ETSI Collaborative Portal has been opened, which makes it easier to access information related to Technical Body activities. One of the goals is to allow organizations and experts to participate and be aware without physically attending meetings.

In the approvals, electronic web based voting is used in ETSI. Approved deliverables are available on the web and on CD-ROM and DVD.

World Wide Web Consortium

The World Wide Web Consortium (W3C) is an industry organization founded to further the development of the World Wide Web by promoting interoperability and encouraging an open forum for discussion. In this role, W3C commits to leading the technical evolution of the Web.

W3C's long term goals for the Web are:

1. Universal Access: To make the Web accessible to all by promoting technologies that take into account the vast differences in culture, education, ability, material resources, and physical limitations of users on all continents.
2. Semantic Web: To develop a software environment that permits each user to make the best use of the resources available on the Web.
3. Web of Trust: To guide the Web's development with careful consideration for the novel legal, commercial, and social issues raised by this technology.

W3C is not an official standards body, but as with many other information technologies, in particular those that owe their success to the rise of the Internet, the Web must evolve at a pace unrivaled in other industries. The consortium has, since its inception in 1994, created a large number of important technical specifications. These specifications have been adopted as de facto standards for the Web. These specifications include HTML, HTTP, CSS, XML, and RDF.

Operation and Organization

W3C is hosted by the MIT Laboratory for Computer Science (MIT LCS) in the USA, the National Institute for Research in Computer Science and Control (Institut national de recherche en informatique et en automatique - INRIA) in France and Keio University in Japan. Work at the consortium is organized around activities, each of which belongs to one of five domains:

- Architecture Domain: develops the underlying technologies of the Web.
- Document Formats Domain: works on formats and languages that will present information to users with accuracy, beauty, and a higher level of control.
- Interaction Domain: seeks to improve user interaction with the Web, and to facilitate single Web authoring to benefit users and content providers alike.
- Technology and Society Domain: seeks to develop Web infrastructure to address social, legal, and public policy concerns.
- Web Accessibility Initiative (WAI): pursues accessibility of the Web through five primary areas of work: technology, guidelines, tools, education and outreach, and R&D.

Each activity consists of one or more working groups, each typically with the goal of producing a new specification. Each specification goes through the W3C process, and the ones the consortium membership approves eventually become official recommendations of the organization.

The day-to-day activities of the consortium are run by the W3C management team, assisted by an advisory board (at the time of writing, the 9 elected members of the board are from Adobe, Boeing, IBM, IPR Systems, Microsoft, Nokia, Opera Software, SAIC, and Sun Microsystems). Ultimately, decisions are sanctioned by the advisory committee, where each member organization has one representative. The current W3C membership consists of over 500 companies and other organizations.

Current Activities of Interest

Several W3C activities, working groups or specifications are related to - and interesting from the standpoint of - mobile Internet technologies.

HTML

HTML is the dominant content markup language on the World Wide Web. The purpose of the HTML Activity is to develop the next versions of HTML. Recent specifications include

- XHTML 1.0 is a recast of the HTML 4.01 specification as an XML application.
- XHTML Modularization is a decomposition of XHTML into a collection of abstract modules that provide specific types of functionality. These modules can be combined and extended by document authors as well as application and product designers to make it feasible for content developers to deliver content on a broader spectrum of diverse platforms.
- XHTML Basic specifies the minimal set of modules required from an XHTML host language document type, plus some other features such as images, forms, basic tables, and object support. It is designed for Web clients that do not support the full set of XHTML features (e.g., mobile phones, PDAs, pagers, and set-top boxes).

Cascading Style Sheets

Style Sheets describe how documents are rendered (e.g., presented on screen, in print, through audio); they allow the separation of document structure and content from how the document is to be presented to users. W3C has promoted the use of style sheets on the Web since the consortium's inception. The Style Activity has produced two W3C recommendations - CSS1 and CSS2 - which are widely (albeit not consistently) implemented in browsers.

XML Protocol

The goal of the XML Protocol Activity is to develop technologies which allow communication in a distributed environment, using XML as the encapsulation language. Solutions developed allow a layered architecture on top of an extensible and simple messaging format, providing robustness, simplicity, reusability and interoperability. Current specifications under development include the next version of the SOAP protocol.

Semantic Web and RDF

The Semantic Web Activity is a continuation of the Consortium's earlier Metadata Activity, which created the Resource Description Framework (RDF) representation standard. The new activity will promote the development of technologies that transition the Web towards richer semantic content. Current work in this activity includes the RDF Core Working Group (with the task of clarifying and refining the RDF specifications) and the emerging Web Ontology Working Group (with plans to turn the DARPA Agent Markup Language into a standard language for ontology creation on the Web).

Graphics and Multimedia

Scalable Vector Graphics (SVG) is a language for describing two-dimensional graphics in XML. It allows for three types of graphic objects: vector graphic shapes, bitmap images and text. Graphical objects can be grouped, styled, transformed and composed into previously rendered objects.

The Synchronized Multimedia Integration Language (SMIL) enables simple authoring of interactive audiovisual presentations. It is typically used for "rich media" or multimedia presentations, which integrate streaming audio and video with images, text or any other media type.

Device Independence

There is significant work underway to integrate Web technologies into various "new" kinds of devices (e.g., mobile phones and TV sets). In this process, we are faced with the possibility that services for those devices may not interoperate with each other or with the existing Web. Not only would that cause fragmentation of the Web space, but it would also make device independent authoring impossible. Through its origins and heritage, W3C has a particular interest in device independent Web access and authoring. The Device Independence Activity was created by merging the Consortium's earlier "Mobile Access" and "TV and the Web" activities.

Earlier work in this area led to the creation of the Composite Capability/Preference Profile (CC/PP), a representation format, which allows web clients (typically "non-standard" web clients such as cell phones) to describe their functional characteristics as well as user preferences.

Location Inter-operability Forum (LIF)

Nokia, Motorola and Ericsson established the Location Interoperability Forum (LIF) as a global forum to address the complexity and multiplicity of solutions and the market situation. The forum should define, develop, and promote an inter-operable location services solution that is open, simple, and secure. This solution allows user appliances and internet-based applications to obtain location information from wireless networks independent of their air interfaces and positioning methods.

LIF's purpose is to define, develop, and promote -through the global standard bodies and specification organisations - a common and ubiquitous location services solution.

Such a solution will:

1. Define a simple and secure access method that allows user appliances and Internet applications to access location information from the wireless networks irrespective of their underlying air interface technologies and positioning methods.
2. Promote a family of standards-based location determination methods and their supporting architectures, that are based on Cell-ID and Timing Advance, E-OTD (GSM), AFLT (IS-95), and MS Based Assisted GPS.
3. Establish a framework for influencing the global standard bodies and specification organisations to define common methods and procedures for testing and certifying the LIF-recommended access methods and positioning technologies.

LIF has representation from a mix of network operators, equipment manufacturers, and service providers responsible for deploying equipment utilizing this solution. Its members will define this solution and submit it to the working standards groups. The members will then support the solution developed in LIF in the appropriate existing global standard bodies and specification organisations and in the deployment of their systems and services.

LIF has currently over 140 member companies, among them key system and handset manufacturers, operators and application developers. More information on LIF can be found in: www.locationforum.org.

IEEE

Most probably every engineer knows what the IEEE ("eye-triple-E", the Institute of Electrical and Electronics Engineers) stands for. If not, one certainly has run across it somewhere when reading a magazine or attending a conference. Most of the main activities of the IEEE concern technical publishing and organizing conferences. With more than 350 000 individual members in 150 countries it ensures a very wide audience and visibility for its products. It is no wonder that circa 30 percent of the books related to IEEE expertise are published by it, or that thousands of conferences are annually organized under the name of IEEE. Nevertheless, IEEE is a non-profit, technical professional association. Through the extensive member group, the IEEE is a leading authority in technical areas ranging from computer engineering, biomedical technology and telecommunications, to electric power, aerospace and consumer electronics.

Technical publishing and organizing conferences is not, however, the whole picture of IEEE. When talking about the main activities of IEEE, one must not forget the third key area in its operation, standardization. In Europe, it might not be as widely known as e.g. ETSI but especially in U.S. it has an established role through its very extensive standardization activities. And once again when talking about engineers, there are quite a few of them who have not heard about Ethernet. Perhaps not all know or have ever thought that it is also a standard created under the IEEE standardization process.

Working Groups

All the standardization efforts within IEEE have been gathered under the IEEE Standards Association (IEEE-SA). IEEE-SA develops and publishes standards that include, but are not limited to, definitions and terminology, methods of measurement and test, systems, products, technology, ratings structures, temperature limits and application guides, recommended practices, and safety.

There are two types of members in SA: individuals and corporate. Individuals, including IEEE members of any grade, or non-IEEE members who are actively involved in standards development are eligible for individual membership. Corporate membership can be acquired by e.g. corporates, universities and user groups willing to actively participate in standards development. The most visible task of the members in the standardization process is voting in the sponsors ballot. That is the final actual phase of the lengthy and well-defined standardization process. For this standard development process there are working groups each of which is responsible for a certain standard.

Traditionally, IEEE working groups have been divided into the following ten main categories:

- Aerospace Electronics
- Broadcast Technology
- Communications
- Electromagnetics
- Information Technology
- Instrumentation & Measurement
- Marine Industry
- Medical Device Communications
- National Electrical Safety Code
- Power Electronics

- Power & Energy
- Quantities, Units, and Letter Symbols
- Reliability
- Transportation Technology

From the mobile Internet perspective, undoubtedly the most important and relevant working groups are within the LAN/MAN standards committee, IEEE 802 that belongs to the Information Technology category.

IEEE 802

The IEEE 802 LAN/MAN Standards Committee (LMSC) develops Local Area Network standards and Metropolitan Area Network standards. The most widely used standards are for the Ethernet family, Token Ring, Wireless LAN, Bridging and Virtual Bridged LANs. An individual Working Group provides the focus for each area.

The first meeting of the IEEE Computer Society "Local Network Standards Committee", Project 802, was held in February 1980. There was going to be one LAN standard, with speeds from 1 to 20 MHz. It was divided into the media or Physical layer (PHY), Media Access Control (MAC), and Higher Level Interface (HILI). The access method was similar to that for Ethernet, as was the bus topology. By the end of 1980, a token access method was added, and a year later there were three MACs: CSMA/CD, Token Bus, and Token Ring. Since that, other MAC and PHY groups have been created, and one for LAN security as well. The unifying theme has been a common upper interface to the Logical Link Control (LLC) sublayer, common data framing elements, and some commonality in the media interface. The scope of work has grown to include MANs. Higher data rates have later been added. Nowadays there are tens of active standardization task groups creating either entirely new standards or amending and enhancing the existing ones in the IEEE 802 project. In the plenary meetings held regularly three times a year (March, July, November), more than 800 individuals are in attendance.

Organization

LMSC is organized in a number of Working Groups (WGs) and Technical Advisory Groups (TAGs) as well as a Sponsor Executive Committee (SEC). They are listed below:

802.0 SEC
802.1 High Level Interface (HILI) Working Group
802.2 Logical Link Control (LLC) Working Group (not active)
802.3 CSMA/CD Working Group
802.4 Token Bus Working Group (not active)
802.5 Token Ring Working Group (not active)
802.6 Metropolitan Area Network (MAN) Working Group (not active)
802.7 BroadBand Technical Adv. Group (BBTAG) (not active)
802.8 Fiber Optics Technical Advisory Group (FOTAG)
802.9 Integrated Services LAN (ISLAN) Working Group (not active)
802.10 Standard for Interoperable LAN Security (SILS) Working Group (not active)
802.11 Wireless LAN (WLAN) Working Group
802.12 Demand Priority Working Group (not active)
802.14 Cable-TV Based Broadband Communication Network Working Group (disbanded)

802.15 Wireless Personal Area Network (WPAN) Working Group
802.16 Broadband Wireless Access (BWA) Working Group
RPRSG Resilient Packet Ring Study Group (RPRSG)

The actual standardization runs in projects within working groups. Each project approved within an existing group is assigned a letter, for example 802.1D for MAC Bridges in the High Level Interface WG. A Study Group (SG) is formed when a new area is first investigated for standardization. The SG can be within an existing WG or TAG, or it can be independent of the WGs. A Task Force develops a new project in an existing group, while a new independent project creates a new WG.

No organization does exist without active members. Membership in LMSC is by WG/TAG, with voting rights after attending two of the last four meetings. Attendance in the interim meetings that are traditionally organized between the plenary meetings can also be counted to some extent. Attendance means going to at least 75% of the 1/2 day meetings of the WG/TAG during the week. Credit is given for attendance at a single group per plenary meeting.

From an Idea to a Standard

Each standard (or recommended practice, or guide) starts as a group of people with an interest in developing the standard. In the first phase, a Project Authorization Request (PAR) specifying e.g. the scope and the purpose of the project is created. Traditionally, a Study Group living from a plenary meeting to the next one is the one writing a PAR. The draft PAR is voted on by the SEC, and it then goes to the IEEE Standards Board New Standards Committee (NesCom) that recommends it for approval as an official IEEE Standards project. Part of the PAR identifies, which outside standards groups there will be liaisons with, for instance ITU for some international standards.

The real standardization work can start after the PAR has been approved. Primarily the process relies on proposals submitted by individual working group members. Proposals are evaluated by the WG, and a draft standard is written and voted on the by the WG. The work progresses from technical to editorial / procedural as the draft matures. When the WG reaches the required consensus on the draft standard, a WG Letter Ballot is done to release it from the WG. It is then approved by the SEC and goes for a Sponsor Letter Ballot (see above, IEEE-SA). After the Sponsor Letter Ballot has passed and the 'No' votes have been answered, the draft Standard is sent to the IEEE Standards Board Standards Review Committee (RevCom). Once recommended by RevCom and approved by the Standards Board, it can be published as an IEEE standard. The process from start to finish can take several years for new standards, and less for revisions or addenda.

Wireless 802 Working Groups

The primary interest for Nokia has been on the three wireless 802 WGs: 802.11, 802.15 and 802.16. In the 802.11 WLAN working group, Nokia has been following the activities since 1997 and for the last two years we have been actively contributing primarily in 5 GHz related task groups. In both 802.15 and 802.16 Nokia has been actively contributing from the very beginning of the working groups since the late 1990's.

802.11 WLAN

802.11 started its work almost ten years ago in early 1990. The goal was to create a standard (PHY and MAC) for wireless LANs. It took several years before the group was ready to agree on all the details and the first 802 WLAN standard was published in 1997. Since then, it has become the widely accepted and used WLAN standard deployed internationally over basically all the continents. The original IEEE 802.11 standard consists of a single MAC but three different PHY layers. The MAC is based on a listen-before-talk access scheme titled CSMA/CA (Carrier Sense Multiple Access / Collision Avoidance). Two of the three PHYs in the original standard were specified for the 2.4 GHz ISM band:

- Direct Sequence Spread Spectrum (DSSS)
- Frequency Hopping (FH)

Both of them provided 1 and 2 Mbps bit-rates. Majority of the products were based on the DSSS. The third PHY was based on infra-red (IR) communication.

Relatively soon after the original 802.11 standard was published, the WG started new projects to enhance data rates on the 2.4 GHz ISM band and to create a new PHY for the 5 GHz U-NII bands. The former resulted in supplement IEEE 802.11b that was published in 1999. It enhanced the data-rates from the original 1 and 2 Mbps to 5.5 and 11 Mbps. Nowadays practically all 802.11 WLAN products use 802.11b.

For a 5 GHz PHY standardization, the IEEE 802.11 WG had a new task group titled TGa. In parallel with the TGb formulating the 802.11b amendment, the TGa created an OFDM based PHY layer for the 5 GHz license-exempt U-NII bands. The effort was equally successful for the TGb and the IEEE 802.11a supplement was published in 1999. With the new PHY the 802.11 was now capable of providing bit-rates up to 54 Mbps.

Since then, the 802.11 WG has been actively defining new enhancements in various ongoing task groups. Recently, 802.11d was approved and published. It defines extensions to operate in additional regulatory domains and thus further enhances the international status of the 802.11 standard family. Today there are five task groups within the WG:

- TGe specifying MAC enhancements for quality of service (QoS)
- TGf creating a recommended practice for the Inter Access Point Protocol (IAPP)
- TGg defining enhanced bit-rates (20+ Mbps) for the 2.4 GHz ISM band
- TGh specifying spectrum management functionalities (DFS and TPC) for 802.11a for the European markets
- TGi enhancing the security capabilities of 802.11 WLANs

For more detailed information and the latest news relating to IEEE 802.11 WG, visit their web-page at http://ieee802.org/11/.

802.16 BWA

The IEEE 802.16 Working Group on Broadband Wireless Access Standards develops standards and recommended practices to support the development and deployment of fixed broadband wireless access systems. All four task groups are currently active and working according to their PARs.

Task Group 1 is developing an air interface for fixed broadband wireless access (FBWA) systems for 10-66 GHz. This TG has been working on the basics of the 802.16 and the work will result in the MAC specification for the entire 802.16 and a PHY specification for the related licensed bands. Task Group 2 concentrates on the coexistence of FBWA systems. Currently it is amending the recommended practice it created and that was published in summer 2001. Even though the basic coexistence standard deals with generic coexistence matters, it primarily addresses the licensed bands 10-66 GHz. The amendment under work will take into account the characteristics of the lower frequencies.

Task Groups 3 and 4 are developing amendments to the basic 802.16 standard created by the TG 1. The focus of TG3 is on licensed bands 2-11 GHz, while TG4 is developing an amendment covering MAC modifications and an additional PHY layer for license-exempt bands. Both of the amendments are anticipated to be published in late 2002.

For more detailed information and the latest news relating to IEEE 802.16 WG visit their web-page at http://ieee802.org/16/.

802.15 WPAN

The IEEE 802.15 Working Group develops Personal Area Network (PAN) standards for short distance wireless networks. These WPANs address the wireless networking of portable and mobile computing devices such as PCs, Personal Digital Assistants (PDAs), peripherals, cell phones, pagers, and consumer electronics; allowing these devices to communicate and interoperate with one another.

802.15 has four active task groups. Task Group 1 is deriving a standard based on Bluetooth™ v1.x Foundation Specification's. Task Group 2 has been working on coexistence issues developing recommended practices to facilitate coexistence of WPAN™ and WLAN devices. Task groups 3 and 4 are actively developing high rate and low-rate standards for WPAN™.

For more detailed information and the latest news relating to IEEE 802.15 WG visit their web-page at http://ieee802.org/15/.

WECA

A few years ago, several WLAN standards and proprietary implementations were competing on the market. WLAN networking components from one vendor would only work with WLAN terminal equipment from the same manufacturer. In other words, customers could not go and purchase a wireless access point from one vendor and use a wireless network adapter from another vendor on the same installation. This was widely recognized as being the main limiting factor in preventing wide scale market acceptance of WLAN technology. The situation effectively limited WLAN usage to certain niche market segments, where multi-vendor capability was not of primary concern.

In 1999, the Wireless Ethernet Compatibility Alliance (WECA) was formed to tackle the interoperability problem by supporting the recently approved IEEE 802.11b standard and by testing and certifying WLAN multi-vendor interoperability. The founding members of the group include 3Com, Aironet, Intersil (formerly Harris Semiconductor), Lucent Technologies, Nokia, and Symbol Technology. It is notable that today WECA has some 120 member companies.

In addition to certifying the interoperability of Wi-Fi™ (IEEE 802.11) products, WECA's mission includes promoting Wi-Fi™ as the global wireless LAN standard across all market segments.

Testing Process

WECA's interoperability certification process can be divided into two main parts:

1) Definition of the test specifications and
2) Testing products from various manufacturers against test specifications to make sure that they conform to the Wi-Fi™ standard

The Agilent Technologies Interoperability Certification Lab (Agilent ICL) (former Silicon Valley Networking Labs) performs WECA testing. Agilent's ICL operates as an independent test facility. However, only WECA members can submit products to the lab for Wi-Fi™ interoperability testing.

When a product meets the interoperability requirements as described in the WECA test matrix, Agilent's ICL notifies WECA. WECA then grants a certification of interoperability, which allows the vendor to use the Wi-Fi™ logo on advertising and packaging for the certified product. The Wi-Fi™ logo is shown in the next figure.

Wi-Fi™ logo.

The idea is that the Wi-Fi™ seal of approval assures the end customer of interoperability with other network cards and access points which also bear the Wi-Fi™ logo.

Organisation

WECA activities are steered by a Board of Directors, which consists of a few representatives from the member companies. The WECA organization is presented in the next figure.

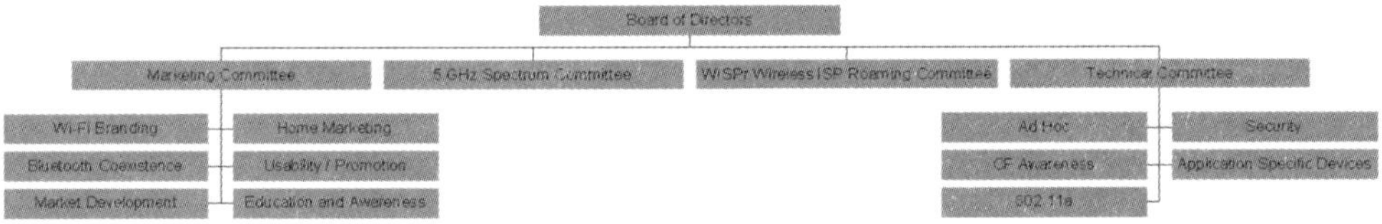

WECA's organization.

For more information on WECA and on the Wi-Fi™ standard, check the Wi-Fi™ web site at http://www.wi-fi.com/. The web site also contains a list of Wi-Fi™ certified products.

Object Management Group

The Object Management Group™ (OMG™) was founded in April 1989. In October 1989, it began independent operations as a non-profit-making corporation. Through the OMG's commitment to developing technically excellent, commercially viable and vendor independent specifications for the software industry, the consortium now includes about 800 members. The OMG is moving forward in establishing CORBA®, the Common Object Request Broker Architecture, as the "Middleware that's Everywhere" through its worldwide standard specifications: CORBA/IIOP, Object Services, Internet Facilities and Domain Interface specifications, UML and other specifications supporting analysis and design.

The OMG was formed to create a component-based software marketplace by hastening the introduction of standardized object software. The organization's charter includes the establishment of industry guidelines and detailed object management specifications to provide a common framework for application development. Conformance to these specifications will make it possible to develop a heterogeneous computing environment across all major hardware platforms and operating systems. Implementations of OMG specifications can be found on many operating systems across the world today. OMG's series of specifications detail the necessary standard interfaces for Distributed Object Computing. Its widely popular Internet protocol IIOP (Internet Inter-ORB Protocol) is being used as the infrastructure for technology companies like Netscape, Oracle, Sun, IBM and hundreds of others. These specifications are used worldwide to develop and deploy distributed applications for vertical markets, including Manufacturing, Finance, Telecoms, Electronic Commerce, Real-time systems and Health Care.

OMG defines object management as software development that models the real world through the representation of objects.

The OMG is structured into three major bodies, the Platform Technology Committee (PTC), the Domain Technology Committee (DTC) and the Architecture Board. The consistency and technical integrity of work produced in the PTC and DTC is managed by an overarching Architectural Board. Within the Technology Committees and Architectural Board rest all the Task Forces, SIGs, and Working Groups that drive the technology adoption process of OMG.

OMG Modeling Specifications

UML™, Unified Modeling Language™, is a graphical language that expresses program design in a standard way, allowing design tools to interchange models. It's an object-oriented language that standardizes an impressive number of diagram types including Class and Object diagrams, Structure diagrams, Use Case diagrams, and more.

The MOF™, Meta-Object Facility, is a standardized repository for meta-data - that is, the descriptions and definitions of the fundamental concepts that applications work with.

XMI™, XML Metadata Interchange, is a stream format for the interchange of metadata including the UML models that you create during your analysis and design activities.

The CWM™, Common Warehouse Metamodel™, completes the OMG data modeling standards. It provides standard models for table-based, file-based, and object-based datastores.

Recent and Forthcoming Enhancements

The Minimum CORBA specification defines a static subset of CORBA. The target is primarily an embedded device; ultimately an ORB printed on silicon. The forthcoming Smart Transducers proposals target even smaller devices—actuators and sensors. The focus is on management interfaces. In telecommunication end-user devices minimum CORBA may be appropriate but a "micro CORBA", that is a GIOP engine and a light-weight activation mechanism, is worth of examination.

The Real-Time CORBA specification addresses timing constraints on ORB-level. The focus is on the allocation of resources and on the predictability of system execution. The interfaces and mechanisms provided by Real-Time CORBA facilitate a predictable combination of the ORB and the application. The application manages the resources by using the Real-Time CORBA interfaces and the ORB's mechanisms coordinate the activities that comprise the application. The Real-Time ORB relies upon the underlying real-time operating system to schedule threads that represent activities being processed and to provide mutexes to handle resource contention. Distributed real-time applications also need real-time transport that is not addressed by the Real-Time CORBA specification but by the forthcoming Extensible Transport Framework for Real-Time CORBA. In addition, the forthcoming Real-Time Notification proposals will be an important infrastructure service for real-time applications.

The Fault-Tolerant CORBA specification aims to provide robust support for applications that require a high level of reliability. The specification supports a range of fault tolerance strategies, including request retry, redirection to an alternative server, passive (primary/backup) replication and active replication which provides more rapid recovery from faults. It allows the users to define fault tolerance properties for each replicated object (object group).

The Wireless Access and Terminal Mobility in the CORBA specification introduces relocatable object reference and terminal mobility transparency by using the bridging concept of the CORBA specification. The Wireless CORBA also supports vertical handoffs between different transport domains through introducing a generic GIOP tunnel.

The forthcoming proposals for GIOP SCTP/IPv6 protocol mapping is an important step towards the possibility to use CORBA as the control plane platform of UMTS. Two RFPs are out targeting middleware interoperability: CORBA/SOAP interworking RFP and CORBA/WAP interworking RFP that are important enablers of middleware interoperability.

To summarize, the OMG has taken several important steps towards middleware for future mobile networks. However, there are still some important issues that have not been tackled. They include 1) CORBA management, that is how CORBA software entities can be managed in a non-vendor specific way, 2) how the evolving Internet Quality-of-Service mechanisms can be incorporated to the Real-Time and Fault-Tolerant CORBA specifications, and 3) how the Real-Time and Fault-Tolerant CORBA specifications can be combined.

Challence of Interoperability

The OMG has been the leading forum in specifying interoperability bridges between CORBA and other middleware platforms.

The OMG, however, has started a comprehensive development of a new architecture denoted as Model Driven Architecture™ (MDA™). The objective is to interrelate IDL specifications, UML modeling, Meta-Object Facility, and XML Metadata Interchange (XMI). The forthcoming MDA might provide a useful starting point for tools supporting interoperability between parts of an application running on different middleware platforms.

Further Readings

All OMG specifications and other material are available from http://www.omg.org/

Wireless Village

The Wireless Village initiative is about building a community around new and innovative mobile Instant Messaging and Presence Services (IMPS). Instant Messaging and Presence is moving from the desktop to the mobile domain. Ericsson, Motorola, and Nokia recognize the need for an industry standard for mobile IMPS. These companies formed the Wireless Village initiative to ensure interoperability of wireless messaging services and IM in particular.

The specifications will be used for exchanging messages and presence information between mobile devices, mobile services and Internet-based instant messaging services, all fully interoperable and leveraging existing web technologies. Through its supporters, the Wireless Village initiative aims to build a vibrant community of end users and global business partners where Internet and wireless domains converge. The IEEE Industry Standards and Technology Organization (IEEE-ISTO) provides day-to-day administrative support to the Wireless Village initiative.

The Wireless Village Initiative IMPS Solution

The Wireless Village Instant Messaging and Presence Service (IMPS) includes four primary features:

- Presence
- Instant Messaging
- Groups
- Shared Content

Presence is the key enabling technology for the Wireless Village initiative. In the desktop world users have been able to announce their status to authorized recipients, facilitating instant messaging. Presence can be seen as a service of its own, or an enabler for other applications, such as Instant Messaging. Instant Messaging is seen to be the first application to utilize presence capabilities.

In the Wireless Village model, Presence takes on a richer meaning. It includes client device availability (my phone is on/off, in a call), user status (available, unavailable, in a meeting), location, client device capabilities (voice, text, GPRS, multimedia) and searchable personal statuses such as mood (happy, angry) and hobbies (football, fishing, computing, dancing).

Since presence information is personal, it is only made available according to the user's wishes - access control features put the control of the user presence information in the users' hands. Instant Messaging is a familiar concept in both the mobile and desktop worlds. Desktop IM clients, two-way SMS and two-way paging are all forms of Instant Messaging. Wireless Village will enable interoperable mobile IM in concert with other innovative features to provide an enhanced user experience.

Discussion groups and chat communities are a fun and familiar concept on the Internet. The Wireless Village initiative enables both operators and end users to create and manage groups. Users can invite their friends and family to chat in group discussions. Operators can build common interest groups where end users can meet each other online.

Shared Content allows users and operators to setup their own storage area where they can post pictures, music and other multimedia content, while enabling the sharing with other individuals and groups in an IM or chat session.

These features, taken in part or as a whole, provide the basis for innovative new services that build upon a common interoperable framework. The Wireless Village initiative will use its community of supporters as a forum in which to test that framework.

Conclusions

The Wireless Village initiative is a community-building effort, both in the sense of building a community of technology companies around a common standard and in the sense of enabling service providers to build their own end user communities.

The Wireless Village initiative is an industry leading coalition and comprehensive solution that leverages a standards-based approach to wireless instant messaging and presence.

Ericsson, Motorola, and Nokia are leaders in wireless communications solutions.

The Wireless Village initiative is open to participation from all industry leaders that desire to support these specifications and help build this community.

Index

#

A

B

C

D

E

F

G

H

I

O

P

Q

R

S

T

U

V

W

X